The Oracle of Night

The Oracle of Night

The History and Science of Dreams

Sidarta Ribeiro

TRANSLATED FROM
THE PORTUGUESE BY

Daniel Hahn

PANTHEON BOOKS, NEW YORK

Grateful acknowledgment is made to the following for permission
to reprint previously published material:

Dell Publishing: Excerpt from "The Real Life of Dreams" from *The Promise of Sleep: A Pioneer in Sleep Medicine Explains the Vital Connection Between Health, Happiness, and a Good Night's Sleep* by William C. Dement. Copyright © 1999 by William C. Dement. Reprinted by permission of Dell Publishing, an imprint of Random House, a division of Penguin Random House LLC. All rights reserved.

Grove/Atlantic, Inc.: Excerpt from *Waiting for Godot: A Tragicomedy in Two Acts* by Samuel Beckett. Copyright © 1954 by Grove Press, Inc., copyright renewed 1982 by Samuel Beckett. Reprinted by permission of Grove/Atlantic, Inc. Any third-party use of this material, outside of this publication, is prohibited.

Harvard University Press: Excerpts from *The Falling Sky: Words of a Yanomami Shaman* by Davi Kopenawa and Bruce Albert, translated by Nicholas Elliott and Alison Dundy, Cambridge, Mass.: The Belknap Press of Harvard University Press. Copyright © 2013 by the President and Fellows of Harvard College. Reprinted by permission of Harvard University Press.

Oxford University Press: Excerpt from "The Dream of Dumuzid" from The Electronic Text Corpus of Sumerian Literature (ETCSL) v.1.4.3. Reprinted by permission of Oxford University Press.

Penguin Classics: Excerpt from "The Forging" by Jorge Luis Borges, translated by Christopher Maurer, from *Poems of the Night: A Dual-Language Edition with Parallel Text* by Jorge Luis Borges. Copyright © 2010 by Maria Kodama, translation copyright © 2010 by Penguin Random House LLC. Reprinted by permission of Penguin Classics, an imprint of Penguin Publishing Group, a division of Penguin Random House LLC. All rights reserved.

Library of Congress Cataloging-in-Publication Data
Names: Ribeiro, Sidarta, author. Hahn, Daniel, translator.
Title: The oracle of night : the history and science of dreams /
Sidarta Ribeiro ; translated from the Portuguese by Daniel Hahn.
Other titles: Oráculo da noite. English.
Description: First American edition. New York : Pantheon Books, 2021.
Includes bibliographical references and index.
Identifiers: LCCN 2020044675 (print). LCCN 2020044676 (ebook).
ISBN 9781524746902 (hardcover). ISBN 9781524746919 (ebook).
Subjects: LCSH: Dreams. Dreams—Physiological aspects. Dreams—History.
Classification: LCC QP426 .R5313 2021 (print) |
LCC QP426 (ebook) | DDC 612.8/21—dc23
LC record available at lccn.loc.gov/2020044675
LC ebook record available at lccn.loc.gov/2020044676

www.pantheonbooks.com

Jacket design by Na Kim

Printed in the United States of America
First American Edition

2 4 6 8 9 7 5 3 1

Because of Vera

For Natália, Ernesto, and Sergio

For Luiza and Kima

In the name of our ancestors,
and of the seventh generation after us:

Dream, Memory, and Destiny

As soon as we were up on our feet
We started to migrate across the savanna
Following the herd of bison,
Beyond the horizon,
To new, distant lands.
Children on our backs, expectant,
Eyes alert, all ears,
Sniffing that unsettling landscape, new and unknown.

We are a traveling species,
We have no belongings, only luggage.
We go with the pollen in the wind,
We are alive because we are in motion.
We are never still, we are nomadic.
We are parents, children, grandchildren, great-grandchildren
    of immigrants.
What I dream is more mine than what I can touch.
I am not from here, but nor are you . . .

              —JORGE DREXLER, *MOVIMIENTO* (MOVEMENT)

But the dreamers move forward, releasing their parrots, dying in
their fires, like children and lunatics. And singing those anthems
that talk about wings, about blazing rays of light—the language of
their forefathers, a strange human language, on these scaffoldings
of the builders of Babel.

              —CECÍLIA MEIRELES, *LIBERDADE* (FREEDOM)

To read is to dream by another's hand.

              —FERNANDO PESSOA, *THE BOOK OF DISQUIET*

Contents

The Oracle of Night

Why Do We Dream?

When he was five years old, the little boy went through a disturbing phase in which he had the same nightmare every night. In this dream, he was living without any relatives near him, alone in a sad city beneath a rainy sky. A good part of the dream took place in a maze of muddy alleys that circled gloomy buildings. The city, which was surrounded by barbed wire and illuminated by insistent flashes of lightning, looked more like a concentration camp. The boy and the city's other children would invariably end up at a scary house where cannibal witches lived. One of the children—never the boy—would go into the three-story building and everybody would watch the many dark windows, waiting for one of them to be suddenly lit up, revealing the silhouette of the child and the witches. There would be a horrifying scream, and that was how the dream ended, only to be repeated, in detail, every night.

The boy developed a terror of sleeping, and informed his mother that he had decided never to fall asleep again, so as to avoid the nightmare. He would lie still in bed, alone in his room, fighting desperately against sleep, determined to remain alert. But ultimately he would always succumb, and a few hours later everything would start over again. The fear of being the child chosen to go into the house was so great that he was unable to prevent the repeating of the narrative, unable to avoid falling into the same oneiric trap. His earnest mother taught him to think about flower-filled gardens as he was drifting off, so as to calm the beginning of his sleep. But after the dark curtain of midnight, the nightmare would return, relentlessly, as if the dawn would never be allowed to return.

Soon afterward, the boy started to have sessions of psychotherapy with an excellent specialist. The memories he retains of this period are of board games kept in an appealing wooden box in the consulting room. At a certain point the psychologist suggested, cleverly, that the dream might somehow be under control. And then the nightmare of the witches was replaced by another dream.

This one also had a disagreeable narrative, though it wasn't a horror story so much as a piece of Hitchcockian suspense with surprising image editing. The gray thriller was experienced in the third person: the boy didn't see the dream through his own eyes, but from outside, as if watching a movie about himself. The dream, which took place in an airport and always ended the same way, was again repeated every night. There was an adult companion with dark hair who was helping the boy to look for a deranged criminal. The boy couldn't find the criminal and ended up leaving the place with his friend. But then, to his great anxiety, the "camera" moved to reveal his quarry, upside-down, hanging from the ceiling of the terminal hall like a huge spider in the gap between the walls . . . The most disturbing thing was not having spotted the criminal earlier, despite his having been there the whole time.

After some more play psychotherapy and more conversations about controlling one's dreams, the boy developed a third dream narrative, this one no longer a nightmare but an adventure dream—still filled with peril but accompanied by much less fear and anxiety. It was about a tiger hunt in the Indian jungle, and the boy featured clearly as the hero, a Mowgli in British colonial clothing, watching from the outside, in the third person. The same dark-haired adult friend was with him at the beginning of the dream as they passed through the thick forest, until they sighted cliffs and a rough sea. On the right-hand side of the field of vision there was a tall island, small and surrounded by sheer cliffs, and in the background the sun was setting, brightly colored against a gray sky. Evening was closing in, and it was barely possible to make out his friend's face. The boy spotted a causeway connecting the mainland to the island, and assuming that the tiger was hiding there, he suggested cornering it. The friend agreed, but explained that from that point onward, the boy would have to make his way alone. The boy advanced, rifle in hand, and started his crossing of the causeway, keeping his balance several yards above a tempestuous, foamy, lead-green sea. The clouds parted, the setting sun appeared, and the horizon was tinged in orange, red, and purple. The boy stepped onto the ground of the island and found himself face-to-face with the deep green bushland, his rifle raised, imagining he was pointing it at the tiger behind the leaves. And then, all of a sudden, he realized that the tiger was behind him, on the causeway. He was the one who was cornered.

Even before the fear came, the boy made a split-second decision to throw himself into the sea. Down he fell, and when he struck the water the dream suddenly switched to the first person, with a vividness that was heightened by the abrupt contact of his warm body with the cold water. He understood that he was dreaming, and with his own eyes he saw the dark sea surrounding him. For a moment, everything was like lead, and then he started to swim around the island. He was afraid, and the fear made him notice a huge shark swimming alongside him. The shock and the tension made time slow down—and then everything was calm. Between the sea and the sky, which were getting ever darker, the boy went on swimming calmly alongside the gigantic shark, and he swam and swam through the night, and nothing bad happened till the following day . . . Not long after he'd started to have the dream about the tiger and the shark, these oneiric narratives left the boy for good, never to return. The nightmares disappeared, the fear of sleeping passed, and nighttime peace was restored to the house.

A CLEAR ENIGMA

How is one to make sense of so many symbols, such a wealth of detail? How might one explain such a reliable repetition of a narrative? What can one say about the very sudden appearance and disappearance of this dream series? How might one try to handle recurring nightmares that are so bad they cause a fear of sleep? If we are to answer these and other questions, we must first understand the origins and function of dreaming.

During our waking hours—whether day or night, but with our eyes wide open—we experience a succession of images, sounds, tastes, smells, and touches. When awake, we live mostly looking outward, since our actions and perceptions are connected to the world that is beyond our inner world. And then, more or less often—whether night or day, but with our eyes firmly shut—we enter that state of unconsciousness in which the screen of reality is switched off. Of this sleep, which is so familiar and restorative, we remember little, which is why it is common to think of it as an absence of thoughts. Sleep is often considered a sort of non-living, a small everyday death, but this is not so. Hypnos, the Greek god of sleep, is the twin brother of Thanatos, the god of death, both of them children of the goddess Nyx, who is the Night. Transitory and generally pleasurable, the sound sleep given by Hypnos is profoundly necessary for every person's mental and physical health.

But something altogether different happens during the curious state of internal life that we call dreaming. There, it is Morpheus who reigns, giving shape to our dreams. Hypnos's brother, according to the Greek poet Hesiod, or his son, according to the Roman poet Ovid, Morpheus brings the gods' messages to the kings and leads a crowd of brothers, the Oneiroi. These dark-winged spirits emerge nightly through two gates, one made of horn and the other of ivory, like bats in flight. When they pass through the gate of horn—which, when polished, is transparent like the veil that covers truth—they produce dreams that are prophetic, divine in origin. When they come through the ivory gate—which is always opaque, even at its thinnest—they prompt dreams that are deceitful or without meaning.

If the ancients allowed themselves to be guided by their dreams, our contemporaries have far less intimacy with these elusive visions. Almost everybody knows what dreams are, but few remember them when they wake in the morning. In general, we see a dream as a movie of variable duration, often undefined in its beginning but almost always brought to a conclusive ending. Roughly speaking, a dream is a simulacrum of reality constructed out of fragments of memories. We participate in our dreams as protagonists, though that does not mean to say we have control over the sequence of events that make up their narrative. By performing in the narrative with no knowledge of its screenplay or its direction, we may experience surprise and even elation as the dream takes its course. However, it also common for a dream to stage situations of great frustration or disappointment.

Despite reflecting the dreamer's preoccupations, the course taken by the dream is almost always unpredictable. The logic of events is fluid and erratic when compared to reality. The succession of images is marked by a lack of continuity and sharp cuts that we don't experience in our waking lives. In a dream, one person or place can be transformed into another with incredible fluidity, revealing the transmutational power of mental representations. The intermittent linking of the symbols establishes a sense of time that is characterized by lapses, fragmentations, condensations, and dislocations, generating layers of multiple—and even contradictory—meanings. A dream's arc of possibilities is absolutely vast, often reaching the unusual, the unrealistic, and the chaotic.

Interpreting a dream presupposes a deep understanding of the cognitive and emotional context of the dreamer, and this interpretation can be transformational. Why did that boy have recurrent dreams about witches, criminals, tigers, and sharks? Would it be enough to say that those dreams were evoking the terrifying encounter with the wicked old witch in Walt Disney's *Snow White,* or the shark in Steven Spielberg's *Jaws,* both of which made frequent appearances on the screens of the day? What do they mean, the elements and plots of these nightmares that are so clear and so filled with emotion? Do they actually mean anything at all? Is there some logic behind the dream? Is the dream an inexplicable fact of human existence or an unfathomable arcane mystery? Is dreaming chance or necessity?

Months before the appearance of the first nightmare, one Sunday at sunset, the boy's father died, struck down by a heart attack. At first, his mother reacted serenely, but a few months later, now a widow with two children to bring up, working every day and taking university courses in her spare time, she fell into a violent depression. It would be months before the boy's younger brother asked where his father was.

It was in this context of family suffering that the terrible recurring nightmare of the witches appeared. It provided a richly detailed illustration of the feeling of orphanhood, as well as the loneliness of the fear of death, which the boy had suddenly discovered to be a real thing. It was an irreversible, chronic situation, and he could see no light at the end of the tunnel. The recurrent dream was an expression of that dead end, which at that moment seemed concrete and inescapable.

The professional intervention was positive. Not long after the therapy began, the dream of the witches gave way to the dream of the detective and the criminal. Horror gave way to suspense, the inexorability of the sacrifice to the witches gave way to a mission, and the boy gained an adult friend with dark hair—like his father and the therapist himself. The setting for the dream was no longer the concentration camp of an orphanage, but an airport, a place from which you set off on a journey to far away.

Then came a third dream, the hunt for the tiger and the swim with the shark: suspense was replaced by adventure, the separation of the father figure was accepted as necessary, and the clarity at the end of the dream left an assurance that the shark was not going to eat the boy. The understanding that our journey is a solitary one was recorded in the memory in orange, red, and purple. The twilight in the dream was painted in the same colors as the moment, on that Sunday as ancient as it is unforgettable, when my father fell.

NOISE, NARRATIVE, AND DESIRE

Even though it can be explained by a single relevant event in waking life, the series of dreams experienced by the boy I used to be includes a fantastical, metaphorical dimension that places it beyond traumatic memory. If the reactivation of memories is at the root of the cognitive functions of sleep and dreaming, what can explain the symbolic complexity that characterizes the oneiric narrative? It is rare to dream an exact repetition of a waking experience. On the contrary, most dreams are characterized by the intrusion of illogical elements and unforeseen associations. Dreams are subjective narratives, often fragmented and composed of elements—beings, things, and places—that may or not be familiar, interacting in a self-representation of the dreamer, who generally only watches the unfolding of a story. Dreams vary in intensity, from faint, confused impressions to intricate epics of vivid images and surprising twists and turns. They can sometimes be entirely pleasant or unpleasant, but on the whole they are characterized by a mixture of emotions. They can even anticipate events from the immediate future, especially when the dreamer is experiencing extreme anxiety and expectation, as in the dreams of students on the eve of a difficult exam, which are often filled with details of context and content.

While it may be impossible to map every dream narrative, there is no doubt that dreams do possess certain elements that are typical. Among the classic plots, we find those dreams that are characterized by incompleteness: a moderately unpleasant dream in which we discover that we are naked, unprepared for a test, irremediably late for a meeting, losing our teeth, separated from somebody important in the middle of a journey, looking for them but unable to find them. As for the characters, we often dream about our relatives, our close friends, and people we deal with on a regular basis, though dreaming about strangers is also possible and can even be a frequent occurrence at certain points of our lives.

Any remotely introspective dreamer will certainly remember three basic types of dream: the nightmare, the pleasurable dream, and the dream of the (usually fruitless) pursuit of some goal. The first corresponds to unpleasant situations we don't have the power to control or avoid. Fear sets the tone for bad dreams, and the nightmare is sustained by the postponing of the feared outcome. Almost nobody experiences their own death in a dream, because we generally wake up before it happens, perhaps because of the great difficulty we have in activating—even in dreams—cerebral representations that are incompatible with a belief in our own lives.

The pleasurable dream is the opposite of the nightmare, presenting gratifying situations stripped of any hint of conflict. This kind of dream often feeds desires that would be impossible in our waking lives, giving the dreamer a complete if unreal satisfaction. But the two extremes of pleasure and terror do not describe most of the dreams we have. Dreaming about such strong emotions requires living through them when awake. It is our memories that give substance to dreams, and nobody dreams without having lived. In the words of Jonathan Winson, one of the pioneers of the neurobiological study of dreaming, "the dream expresses what is happening to you right now."

RELEARNING HOW TO DREAM

Describing one's dreams immediately upon waking is a simple practice that can enrich one's dream-life enormously: in just a few days, somebody who never used to remember their dreams can start to fill pages and pages in their dream diary, a device recommended since antiquity as a way of stimulating oneiric recall. The learned Macrobius suggested in the fifth century that oneiric research depends essentially on recording a reliable dream account. In the twentieth century, the psychiatrists Sigmund Freud and Carl Jung made the interpreting of these records a new science of the mind: depth psychology.

But one need not spend a lot of time on the psychoanalytic couch to recount and interpret dreams. A little bit of auto-suggestion before sleeping is enough, along with the discipline of remaining stationary in bed upon waking, waiting for Pandora's prolific box to be opened. The auto-suggestion can consist of repeating, just one minute before sleeping: "I will dream, remember it, and tell it." On waking, with paper and pencil at hand, the dreamer will first make an effort to remember what they have dreamed. The task will seem impossible at first, but quickly an image or a scene, even if it's faded, will come to light. The dreamer must cling to this, mobilizing their attention to increase the reverberation of the memory of the dream. It is this first memory, albeit perhaps fragile and fragmentary, that will serve as the initial piece of the jigsaw, or the end of the ball of yarn to be unrolled. Through its reactivation, the associated memories will begin to be revealed.

On the first day this exercise may produce no more than a few scattered phrases, but after a week it is common for whole pages of the dream diary to get filled up, with a number of independent dreams collected upon a single awakening. The truth is, we dream for most of the night, and indeed even when we are awake—though we call that imagination.

Dreaming is essential because it allows us to dive into the subterranean depths of consciousness. As we go through this state we experience a patchwork of emotions. Small challenges and anxieties, modest daily defeats and victories, generate an oneiric panorama that reverberates with the most important things in life, but that tends not to make sense in its entirety. When existence flows smoothly, interpreting the symbolic gobbledygook of the nighttime is no easy task.

Not even the superrich can be denied the right or the fate of being tormented by recurring nightmares, which carry intimate existential meaning. But to somebody who survives just barely on the edge of well-being, to somebody who truly fears for their life, day and night, to the billions who don't know whether they will have food to eat tomorrow, or clothes, or a place to live, dreaming can be excruciating on a daily basis. In the life of a survivor of war, or a convict, or a beggar, a dream is a roller coaster of affects in garish shades of life and death, pleasure and pain, at the opposite poles of desire.[1]

The Italian chemist and writer Primo Levi, a survivor of the Nazi extermination camp at Auschwitz, told the story of a recurring nightmare he had after his painful return to Turin:

It is a dream within a dream, varied in detail, one in substance. I am sitting at a table with my family, or with friends, or at work, or in the green countryside; in short, in a peaceful relaxed environment, apparently without tension or affliction; yet I feel a deep and subtle anguish, the definite sensation, of an impending threat. And in fact, as the dream proceeds, slowly or brutally, each time in a different way, everything collapses and disintegrates around me, the scenery, the walls, the people, while the anguish becomes more intense and more precise. Now everything has changed to chaos; I am alone in the centre of a grey and turbid nothing, and now, I *know* what this thing means, and I also know that I have always known it; I am in the Lager once more, and nothing is true outside the Lager. All the rest was a brief pause, a deception of the senses, a dream; my family, nature in flower, my home. Now this inner dream, this dream of peace, is over, and in the outer dream, which continues, gelid, a well-known voice resounds: a single word, not imperious, but brief and subdued. It is the dawn command of Auschwitz, a foreign word, feared and expected: get up, *"Wstawàch."*[2]

With the number 174517 tattooed on his wrist, Primo Levi died in 1987 after falling down the stairwell of the building where he lived. The police treated it as a suicide.

RESISTING THE WORLD'S INSOMNIA

The Portuguese word for dream, *sonho*—from the Latin *somnium*—can, like its English equivalent, mean many other things, all of which are experienced in waking lives and not during sleep. To have fulfilled "the dream of a lifetime," "the American dream"—these are phrases people use every day to mean that they are aspiring to something or have achieved it. Everyone has a dream, in the sense of a plan for the future. Everyone desires something they do not have. Why should it be that *dream*, a nocturnal phenomenon that can evoke both pleasure and fear, is exactly the same word we use to refer to everything we wish for?

Today's advertising repertoire is in no doubt that dreams are the driving force behind our behaviors, the private motivation for our external acts. *Desire* is a more precise synonym for that word *dream*. On one Brazilian radio station, the commercial for the Universal Church of the Kingdom of God makes this quite clear: "This is the place where faith makes dreams a reality." The strength of the link between dreams and happiness is remarkable. In an advertisement for a credit card in Santiago de Chile, the miraculous promise: "We make all your dreams come true." In the departure lounge of an airport in the United States, there is a huge photograph of a happy, smiling couple, sailing across a Caribbean sea on a sunny day. Above it is the enigmatic phrase: "Where will your dreams take you?" Underneath, the logo of the credit card company. One can deduce from the advertisement that dreams are like sailing boats, capable of carrying us to idyllic destinations, places that are perfect, that are supremely . . . desirable. The equation "dream equals desire equals money" has a hidden variable, which is the freedom to go, to be, and most of all to have, a freedom that even the most wretched can experience in the world of lax rules that is nighttime dreaming, but which in the dreams of daytime are the exclusive privilege of those who are the bearers of a magical plastic card.

The routine of daily work and the lack of time for sleeping and dreaming, which affects the majority of workers, is a critical part of the malaise of today's civilization. Although working from home in response to the COVID-19 pandemic may have restored some opportunities for sleep and dreaming, the contrast between the motivational relevance of dreams and their trivialization in the globalized industrial world remains glaring, while its citizens pursue sleep as an elusive prey. In the twenty-first century, the search for lost sleep involves sleep trackers, high-tech mattresses, auditory stimulation devices, pajamas with biosensors, robots to help regulate rhythmic relaxed breathing, and a cornucopia of medical remedies. The sleep health industry, a sector that was growing faster and faster even before the pandemic, recently had an estimated value of between thirty and forty billion dollars.[3] Yet still insomnia reigns. In an era when time is always short, when we wake up every day to the insistent ringing of the alarm clock, still sleepy and already late to fulfill commitments that are renewed infinitely, when so few people remember their dreams because of the simple lack of opportunity to contemplate their inner lives, when insomnia rages and yawning prevails, we have come to a point where the survival of dreaming is called into question.

And yet, we dream. We dream a lot and in bulk, we dream greedily in spite of the city's lights and noises, in spite of the incessant toil of life and the sadness of our horizon. The skeptical ant would say that anybody who dreams as freely as this is a lazy artist, like Aesop's fabled grasshopper. At the start of the seventeenth century, William Shakespeare wrote, "We are such stuff / As dreams are made on."[4] A generation later, in his play *Life Is a Dream,* the Spanish playwright Pedro Calderón de la Barca dramatized the freedom to construct our own destinies.[5] Dreaming is imagination with no brakes, and no control, set free to fear, to create, to lose and find.

In his "I Have a Dream" speech, the Reverend Martin Luther King placed the need for racial integration and justice at the center of U.S. political debate. In a country largely constructed by African slaves, their descendants were forced to build the "American dream" but forbidden from enjoying it. The leader of a peaceful but insistent fight for civil rights in the United States, Dr. King was awarded the Nobel Peace Prize in 1964, and was shot dead four years later. King died, but his dream did not, flourishing and progressively opening up a space for the reduction of racial inequality in the country. In the time of President Donald Trump, almost seven hundred thousand people who had arrived in the United States before their sixteenth birthdays, and who had been approved under the Obama-era program for the legalizing of immigrants, became locked in a desperate struggle to remain in the country where they had spent their childhoods and adolescences. Most of these people were born in Mexico, in El Salvador, in Guatemala, or in Honduras. At the time of writing, they are living in limbo, and they are called Dreamers.

A force this powerful needs explaining. What actually is a dream? What use is it? Answering these questions demands some understanding first of how dreaming began and evolved. To our hominid ancestors, the realization that the dream world is an illusion must have been a mystery that was refreshed anew every morning. But the advent of language, religion, and art certainly gave new meaning to dreams' enigmatic symbols. Curiously, these meanings were very similar in different ancestral cultures. That is an important clue in our attempt to decipher dreams.

The oldest pieces of historic evidence for the occurrence of dreams go back to the very beginnings of civilization. All the great cultures of antiquity reveal references to the phenomenon, marked on turtle shells, on clay tablets, temple walls, or papyri. Dreams were often seen as

oracles capable of revealing the future, offering premonitions, telling
fortunes, or divining the intentions of the gods. Dreaming was taken
very seriously in ancient Greece, placed at the heart of medicine and
politics. The same things happened in more ancient civilizations, such
as those in Egypt and Mesopotamia.

The *Tukulti-Ninurta Epic*,[6] written more than three thousand years
ago, recounts the conquests of an Assyrian king—who has been possibly identified as Nimrod, great-grandson of the biblical Noah—in
his fight against the Babylonian king Kashtiliash IV. The cuneiform
text describes how the gods of various cities under Babylonian control,
overcome by rage at Kashtiliash IV's transgressions, decided to punish
him by quitting his temples. Even the patron god of Babylon, Marduk,
justified the Assyrian attack when he abandoned his sanctuary in the
vast ziggurat that inspired the myth of the Tower of Babel. Surrounded
by the invading army, Kashtiliash IV sought positive omens but found
none. Finally, he despaired: "Whatever my dreams are, they are terrible." This meant that Babylon would fall.

Tukulti-Ninurta and Kashtiliash IV were historical figures, and that
war did in fact happen. In 1225 B.C.E., Babylon was defeated and sacked,
its walls destroyed, its king captured and humiliated. To complete the
raid, Tukulti-Ninurta ordered the removal of the main cult statue from
the temple of Marduk, kidnapping the god himself and taking him on
a long journey that would last many years. This sort of kidnapping was
relatively common, since there was a belief in the concrete existence
of the divinity embodied in the statue. A good example of Assyrian
propaganda, the *Tukulti-Ninurta Epic* illustrates how dreams were used
to give rulers credibility. As a result, it also clearly shows the problem of
secondary elaboration, that is, the fact that we never have access to the
actual dream itself, the primary experience that took place in the mind
of the dreamer, but only ever a subjective account of what the experience might have been according to the person who claimed to have
dreamed it. In the conflict between Tukulti-Ninurta and Kashtiliash IV,
the dream attributed to the loser conveniently legitimized the winner's
conquest.

Accounts of dreams, whether real or not, also occupied a central
place in the administration of the Egyptian state. One concrete example is the Dream Stele, a rectangular block of granite more than twelve
feet tall, positioned between the front legs of the Great Sphinx of Giza.

This stele, engraved with hieroglyphs and dated to approximately 1400 B.C.E., recounts how one day the young prince Thutmose fell asleep in the shade of the imposing statue, which at the time was partially submerged under the desert sands. Thutmose dreamed that the Sphinx promised him the throne if he managed to protect it. According to the inscriptions, the young man ordered a wall to be built around the Sphinx, and acceded to the throne as Pharaoh Thutmose IV. In 2010, some traces of the wall described on the Dream Stele were discovered.

THE ORACLE OF NIGHT

The practice of obtaining divine authorization from dreams to justify actions in reality runs throughout our historical past. The divinatory nature of dreams can be found in the main surviving texts of the Bronze Age (between five thousand and three thousand years ago), such as the Egyptian *Book of the Dead* and the Sumerian *Epic of Gilgamesh*.[7] In addition, it is amply present in *The Iliad, The Odyssey*, the Bible, and the Quran. According to tradition, Maya, mother of the best known of all the Buddhas, became pregnant with him after dreaming that a white elephant with six tusks came down from heaven and entered her side.[8] The white elephant, a sign of the ultimate favor of the gods, foretold the child's special nature. Similarly, as the legend would have it, the conception of the Chinese philosopher Confucius occurred after his mother dreamed of being impregnated by a warrior god.[9] In late antiquity, Artemidorus[10] (second century C.E.) and Macrobius[11] (fifth century C.E.) spread the idea that dreams belong to different categories depending on their content, cause, and function.

Artemidorus was born in in the Greek colony of Ephesus, in what is now Turkey, but he was living in Rome when he became well known as a scholar, a doctor, and an interpreter of dreams. He wrote a classical treatise on dreams entitled *Oneirokritika*, based on his extensive reading and on the oral inquiries that were possible thanks to his travels through Asia Minor, Greece, and Italy, which allowed him to draw on the knowledge of the peoples scattered around the Aegean islands and in the craggy towns on Mount Parnassus. In this five-volume work, which has survived to this day,[12] Artemidorus compiled exemplary dreams and theorized extensively about their causes. He stated that an interpreter must know the dreamer's history, such as their occupation, health, social position, habits, and age, as well as ascertain how the dreamer feels about each of the components of their dream. The plausibility of what the dream contains should be considered, which can only be done with reference to the dreamer themselves.

Artemidorus further claimed that dreams can describe situations that are current (*enhypnia*) or future (*oneiroi*), and argued that they must be categorized correctly for their meaning to become clear:

> The difference between two types of dream, the *enhypnion* and the *oneiros,* is an important distinction . . . An *oneiros* differs from an *enhypnion* in that the significance of the former relates to future events, and of the latter to present events . . . [In] the category of predictive dreams (*oneiroi*), some are theorematic and others allegorical. Theorematic dreams are those where the outcome corresponds literally to the vision. For example, a man out at sea dreamt that he was shipwrecked, and he did find himself in that situation: when sleep left him, the ship sank and was lost, and he barely managed to survive along with a few others . . . Allegorical dreams, on the other hand, are those which signify something by means of something else: here the mind is characteristically speaking in riddles.[13]

Almost two thousand years before Freud, Artemidorus pointed out the importance of the multiplicity of a dream's meanings:

> A man who had a stomach disorder and wanted a prescription from Asclepius dreamt that he entered the god's temple and the god held out the fingers of his right hand and invited him to eat them. The man ate five dates and was cured. The top-quality fruits of the date-palm are called "fingers."[14]

Ambrosius Theodosius Macrobius was a philosopher and grammarian in the period marked by the fall of the Roman Empire and the holding out of the Byzantine Empire. The facts of his birth and career are nebulous, but his work had an impact that was lasting. Not simply a compiler of dreams and dream theories like Artemidorus, Macrobius was a scholar. His study of dreams took as its subject a work of fiction, the *Dream of Scipio,* written three centuries earlier by the Roman consul Cicero. In his *Commentary on the Dream of Scipio,* Macrobius put forward a classification for dreams that was widely accepted in medieval theological thought.[15] Macrobius used the term *visum* (*phantasma* in Greek) to refer to apparitions, considered to have "no prophetic significance" in the transition between waking and sleeping, when the dreamer imagines "specters" around him. *Insomnium* (*enhypnion* in Greek) referred to nightmares, also considered to have no prophetic significance, but rather to reflect emotional or physical problems. *Visio* (*horama* in Greek) was the prophetic dream that becomes reality, *oraculum* (*chrematismos* in Greek) the oracular dream in which a revered person reveals the future and offers advice, and *somnium* (*oneiros* in Greek) the enigmatic dream containing strange symbols, which require the intervention of an interpreter to be understood.

The first two categories listed by Macrobius comprise dreams that are influenced only by the present or the past, with no relevance to the future. The three latter ones all focus on what is to come, whether through the clairvoyance of future events (*visio*), prophecies (*oraculum*), or the symbolic dream (*somnium*) that requires interpretation. Curiously, the attributing of a predictive character to dreams is a recurrent thread in countless so-called primitive cultures in the Americas, Asia, and Africa today.[16] While these societies are very different from one another, they seem to preserve a common ancestral belief in the premonitory possibilities of dreams, considering them a key to destiny for anybody capable of interpreting them, a source of predictions, a tool for divination, a portal for accessing what has not been but is yet to be— but also a warning of spiritual danger. A number of indigenous cultures in North America still make a dream catcher, called an *asabikeshiinh* (a spider in the Ojibwe language), which consists of a net tied to a willow hoop, decorated with feathers, seeds, and other magical objects. Oftentimes the artifact is hung above a sleeping child as a protection, being able—just like a spider's web—to capture any malign force that could cause nightmares.

Amerindian cultures retain some of the best-documented examples of prophetic dreams that are able to guide entire peoples. One exemplary case was the premonitory vision that a Comanche chief had in 1840.[17] Up to that point, Buffalo Hump had been an energetic but minor chief of the Penateka band of the Comanche, the fierce indigenous nation that had held off the Spanish advance in the eighteenth century. For centuries, their people had dominated Comanchería, a territory equivalent to a large part of the prairies in the American southwest, encompassing parts of Texas, New Mexico, Oklahoma, Colorado, and Kansas. Owing to their geographical position in the extreme south of this territory, the Penateka were among the groups of Comanche who were exposed to the closest contact with the whites, leading directly to the disappearance of the buffalo from the southern prairies and to the great epidemics of smallpox and cholera. It is not surprising that Buffalo Hump, like some other indigenous people of his time, avoided contact not only with whites but with anything that had to do with them, including clothes and domestic tools.[18]

Tensions increased in March 1840 with the slaughter of a number of Penateka chiefs on a peace mission in the city of San Antonio. Not long after the massacre, Buffalo Hump had a bloody nighttime revelation in which the Indians attacked the Texans and drove them into the sea, a dream he experienced with great mystical power. In the weeks that followed, Buffalo Hump's vision spread through Comanchería like fire across the prairie. Over the summer, the chief recruited supporters until he had gathered four hundred warriors, along with six hundred women and children to provide logistical support for the attack. At the start of August this army came down from the prairie toward the south, and three days later they invaded the territory of the newly created republic of Texas, which was peopled by white colonists. On August 6, the Comanche launched a surprise attack on the city of Victoria, more than a hundred miles from San Antonio and just twenty-five miles from the sea. They plundered storehouses, burned homes, stole thousands of horses, and killed a dozen people.

In spite of this victory, the dream prophecy had not yet been fulfilled. In order to realize his vision, Buffalo Hump guided his braves on a march toward the coast, and on August 8, the Comanche surrounded the town of Linnville, at the time Texas's second-largest port. When they saw the hundreds of armed horsemen fully adorned for war approaching in an impressive half-moon formation, the inhabitants of this prosperous town grew desperate. After some skirmishes and the death of three of their citizens, the people of Linnville launched themselves out to sea using the crafts moored at the port. Barely able to believe their eyes, the terrified fugitives watched the total destruction of their city, just as had happened in Buffalo Hump's dream. It was the greatest indigenous attack on the white population in the territory of the United States. Linnville never recovered and remains a ghost town to this day.

FROM MYSTICISM TO PSYCHOBIOLOGY

Why did so many different peoples see an oracular function in dreams, and why do so many see it still? Where does it come from, this apparently absurd idea that defies reason itself? Could there be some logical explanation for it, or is it no more than a misconception built on a vast pile of meaningless coincidences and superstitions? Might it be possible to find a scientific explanation for the idea that dream activity anticipates future events? The replies to these questions are not insignificant and can only be attained by considering a large number of interconnected factors. At the origin of this attempted synthesis we find the work of Sigmund Freud, the founder of psychoanalysis.

Freud was born in Moravia in 1856, in what today is the Czech Republic. He was a brilliant child, and by the time he was twenty-five he was a newly trained doctor, insecure but determined. In the late nineteenth century, neuroanatomy was dominated by the thick mustaches of the Austro-German neuropathologist Theodor Meynert and the Italian pathologist Camillo Golgi, two conservative forces who held a great deal of authority. In tune with the vanguard of his day, Freud initially traveled down a similar path as the Spaniard Santiago Ramón y Cajal, who would receive the Nobel Prize in Physiology or Medicine in 1906 for his great contributions to the understanding of the nervous system, such as the discovery of the neuron (fig. 1).

Theorizing in his unfinished work *Project for a Scientific Psychology,* written in 1895,[20] Freud pictured the brain tissue as a network of individual cells allowing the movement of "activity," which today we refer to by a number of synonyms: electrical impulses, the action potential of the neuron, or neuronal firings, the latter being a piece of scientific jargon for the sudden and transient depolarizations of the cell membrane (fig. 1). Freud proposed that the frequent repeated movement of activity down the same paths would lead to the strengthening of that path and the production of memories. This mechanism for long-term potentiation, rather like the reduction in resistance to the passage of water down a stream after a spate, was only shown empirically in 1970, as we will see below.[21]

Despite this considerable insight into the nervous system, Freud did

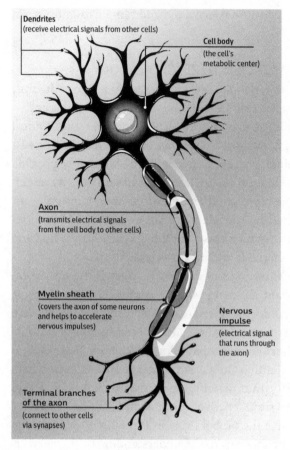

Figure 1. Main parts of the neuronal cell: dendrites, cell body, and axon. Electrical signals coming from other neurons enter the cell through the dendrites; they are integrated into the cell body, transmitted by the axons, and finally moved on to other neurons via the axon terminals. The human brain has approximately eighty-six billion neurons, each with an average of ten thousand contacts with other neurons (synapses).[19]

not become known as one of the founders of neuroscience, but as the creator of a new psychology. Ten years before writing his *Project*, as an apprentice to the neurologist Jean-Martin Charcot at the Salpêtrière hospital in Paris, Freud had witnessed the transitory cure of hysteria by hypnosis. He went deeper into the study of speech production disorders known as aphasias, gave up hypnosis, and finally developed a therapeutic method based on the recounting of dreams and free association of ideas. He came to the concept of the unconscious when, following the death of his father, he began to have dreams that were highly vivid and symbolic, revealing memories and ideas to him that he had not been aware of before. The development of these ideas led to a real revolution.

According to the North American cognitive scientist Marvin Minsky, a twentieth-century pioneer in the re-creation of mental processes in computers, Freud was the first good theoretician in the field of artificial intelligence, conceiving of the mental apparatus as a machine made up of different components, rather than one monolithic system capable of generating all psychic phenomena.[22] When Minsky proposed that artificial intelligence should be a collection of independent parallel systems, he acknowledged the deep influence of psychoanalysis on his work. To Freud, the human mind comprised three distinct agents—id, ego, and superego—whose operations were intimately related but very often antagonistic to one another.[23] The id ("it" in Latin) was unconscious and would produce primitive impulses related to the satisfying of visceral needs, which depend, we now know, on the subcortical parts of the brain that govern the pleasure principle, such as the hypothalamus and the amygdala. This concept has correspondences in the neural circuits that allow us to desire, and especially to seek satisfaction of our desires.[24] To Freud, the id is irrational, it is present from birth, it inhabits the present moment and challenges reality with the compelling strength of need: one doesn't stop being thirsty just because the water has run out.

The ego ("I" in Latin) corresponds to the conscious process that organizes the interface between the id and reality, through perceptional, cognitive, and executive functions governed by the reality principle, which is to say, limited by actual facts. Faced with limitations, the ego attempts to transform them through planned action, striving to shape the future according to past experience. To the extent that the ego consists of corporeal limits, images of the self, and a bank of autobiographical memories, its location in the brain would include the hippocampus, the temporoparietal cortex, and the medial prefrontal cortex.[25]

The prefrontal cortex also participates directly in the third psychic agent of Freudian theory: the superego. Besides governing the body in accordance with the reality principle, the ego must also negotiate the clash between the impulses of the id and the morality exercised by the superego, which reflects the introjection of cultural norms transmitted to children by their parents or immediate caregivers. The superego is the source of censorship, restraint, embarrassment, criticism, and the struggle with the impulses of the id. These functions have their correspondences in the activity of a number of different areas of the prefrontal cortex, which are needed for decision-making, appraising options, and inhibiting unwanted behaviors.[26]

In order to relieve the conflicts between the superego and the id, the ego uses a number of defensive processes that reduce psychic suffering, by means of the repression, suppression, denial, compensation, displacement, rationalization, and even the sublimation of desires. If the id is infantile, the superego is an internal father that manifests itself in habits, in the memory of exemplary episodes, and in explicit rules that can be stated in words.

Because it is cumulative and combinatorial, the bank of autobiographical memories becomes absolutely vast over time, but only a minuscule fraction of those memories occupies the conscious mind at any given moment. Memories are constantly being requisitioned by the ego and the superego, and they make up psychic conglomerates that are transiently animated, at each moment, by the activity of a select group of neurons. Only in the quiescence of the vast majority of the neuronal population, however, can the totality of possible thoughts reside, latent and enduring, the product not only of all the memories acquired during a lifetime, but also of all their possible re-combinations. It is this ocean of mental representations that Freud named the unconscious, identifying dreams as the royal road for accessing them.

The psychoanalytic method was founded upon the receptive attitude of listening attentively to a patient who is invited to talk freely about themselves, recalling dreams and associating ideas. The proposition was to map the patient's latent memories in order to discover clues about the origins of their traumas, which were associated with neurotic symptoms of all kinds. Freud argued that these traumas typically have a sexual tenor and pertain to aversive memories acquired in childhood, whether from situations that were actually abusive or from the reverberation of contradictory feelings between parents and children. In the conflict between the id and the superego, pathological symptoms are generated. In the process of analysis, the ego would become aware of the trauma, which would open up the possibility of overcoming it, pacifying it, taming it.

Lying on the comfortable couch in the famous consulting-room at Berggasse 19, in Vienna, Dr. Freud's patients might not have realized that they were ushering in a new way of treating mental problems in

Europe. The practice of turning inward and simply speaking freely, narrating one's own life out loud, practiced naturally in so many cultures but violently repressed in the patriarchal Austro-Hungarian Empire of the nineteenth century, would take the world by storm in the twentieth century. The reopening of this window to the soul was a great scientific and social event.

Nor could Freud's patients have guessed that their beloved analyst, in their eyes so obviously anointed by the authority of science, would soon be reviled by it and ostracized. From the very beginning, Freud faced opposition from the medical world, which was dominated at that time by neurology. The idea that mental and corporeal symptoms might result from mere thoughts and not necessarily from brain lesions was not remotely palatable to neurologists, though this was not as shocking as Freud's observation that sexuality is present even in small children. Criticized—rightly or wrongly—for personal and professional flaws, attacked by journalists, academics, and moralists of all kinds, and finally pursued by Nazism as a dangerous Jewish intellectual,[27] Freud went into exile in London in 1938 and died a few days after the start of the Second World War.

In the postwar period, psychoanalysis spread across the countries of the Americas and came to have considerable importance in medical schools in the United States, but eventually lost ground almost entirely to psychopharmacology. The difficulty of applying the psychoanalytic method to any and every patient, the advent of drugs capable of suspending a psychotic break apparently without any need to listen to the patient, the tendency of the followers of psychoanalysis toward isolation and fragmentation, along with a good dose of ideological persecution and intolerance, ended up getting Freud's contributions expunged from the scientific establishment. The influential Austrian philosopher Karl Popper, who considered only potentially refutable propositions to be scientific, gave his ruthless verdict in the 1960s: psychoanalytic theories were "simply non-testable, irrefutable."[28] To Popper, psychoanalysis was a metaphysical proposition, without empirical content and as a result totally arbitrary. To twentieth-century science, Freud was at best a poet—and at worst a fraud. He was considered as relevant to neuropsychology[29] as Marx is to the stock market, or Darwin to the neo-creationists.

Notwithstanding their scientific defeat, Freud's ideas won an overwhelming cultural victory. Through psychoanalytic practice, the human sciences, and the arts, his theory of the human mind would deeply infuse Western culture, which would come to incorporate terms like *unconscious, ego, repression,* and *Oedipus complex* into colloquial language. Jungian jargon did not meet with as much success, though nobody would be surprised to hear the expression *collective unconscious* come up in a conversation. A research project carried out on the nature of dreams among students from the United States, India, and South Korea showed that the majority identified with the proposition that dreams reveal hidden truths and allow repressed emotions to emerge. The psychoanalytic conception of dreams was chosen in all three countries in every academic area tested, well ahead of other theories that were in principle more closely aligned with neuroscience.[30]

This does not mean that the people who took part in the research were all psychology students or avid readers of the work of Freud. The exceptional spreading of these ideas happened in spite of—and perhaps even because of—ignorance about them. The trivialization and adaptation of Freud's ideas for a mass audience transformed them into a staple of pop culture, easily penetrating lay circles, but with a legacy that to this day is tenaciously quarreled over by opposing camps of experts. Discredited by twentieth-century science, Freudian ideas were almost entirely accepted outside their own context. If everyone is Freudian, then no one is Freudian.

However, from the end of the twentieth century, and going against the current of the medical establishment, Freudian propositions began to be tested scientifically. One of the most striking examples was the demonstration that the conscious suppression of unwanted memories, described by Freud in pioneering fashion, is a quantifiable cerebral fact. Experiments in functional magnetic resonance imaging published in the prestigious journal *Science* by two independent research groups, headed up by the U.S. neuroscientists John D. Gabrieli and Marie Therese Banich, showed that the deliberate suppression of unwanted memories corresponds to the deactivating of two areas of the brain dedicated to the processing of memories and emotions—the hippocampus and the amygdala, respectively.[31] Curiously, this deactivation is in proportion to the activating of areas of the prefrontal cortex involved with intentionality. This reveals a neurobiological mechanism capable of explaining how a once conscious memory can disappear reversibly into the great scope of the unconscious—not in a process of forgetting exactly, but of burying.

While ideas resembling the unconscious can be identified in a number of their predecessors, it was in the work of Freud and his disciple and rival Carl Jung that the notion of the unconscious came to occupy a central place in psychology. As early as 1948, the Austrian zoologist Konrad Lorenz, a founder of ethology and Nobel laureate in Physiology or Medicine, was warning of the need to take psychoanalysis seriously:

> Another, far more significant branch of psychological research that originated from medical psychiatry remains remarkably isolated and disconnected, although it deserves more than any other field of psychology to be labeled as scientific . . . However much we may reject the theoretical edifice constructed by Sigmund Freud and Carl Jung . . . there can be no disagreement that both of these depth psychologists are *observers,* indeed gifted observers, who saw *for the first time* certain facts that are irrevocable, inalienable components of collective human knowledge.[32]

THE ROYAL ROAD TO THE UNCONSCIOUS

The contribution of psychoanalysis to the understanding of the human mind rests fundamentally on dreams, and marks a vital turning point in their interpretation. In proposing that dream interpretation should be based upon an investigation of the dreamer's subjective experience, Freud identified the memory of waking events as the skeleton around which a dream is constructed. These memories, called day residues in Freudian theory, are the axis around which the dreamer's emotions clump together to generate mental images of great symbolic power. The closely detailed analysis of the accounts of the dreams, when set against the waking context, allowed Freud to develop a new treatment that was based on the patient's becoming conscious of their most intimate motivations. Freud identified dreams as the privileged channel for the investigation of the human psyche, because of their being less subject to the moral censorship that regulates waking thoughts. Conflicts from childhood and from the present day appear in dreams, sometimes resolved by the simple fulfillment of the desire in the fantastical realm of the mind, where it does not need to be compatible with the actually existing world that is accessed while awake.

At the extreme end of this dissonance between dreams and reality, Freud proposed the existence of an intimate relationship between dreaming and psychosis, an opinion shared by the psychiatrists Eugen Bleuler and Emil Kraepelin, pioneers in the study of schizophrenia. Following an intensive and extensive analysis of the accounts of several patients' dreams and especially of his own, Freud proposed that dream activity reflects the desires and fears of the dreamer. He created a therapy that was based on subjective self-reporting, on free association, and on the interpretation of dreams and fantasies, as well as on the conscious identification of memories, desires, and symbolic associations that had been repressed.

Ignored by neuroscience for almost the whole of the last century, Freudian theory did not begin to return to scientific debate on the brain and the mind until 1989, when the electrophysiological correlates of the day residue were first identified. Long before Freud, it was believed that dreams had something to say about the future. After him, they began to be seen as an imprecise but meaningful reflection of the past. More than eighty years after his death, evidence has now been accumulating that both conceptions are correct. Advancing one step at a time, along a winding trajectory, a general theory of sleep and dreaming has been taking shape that reconciles past and future to explain the dream function as a crucial tool for surviving in the present.

That theory forms the spine of this book. In order to present it, we will need to reflect on the pioneering experiments that identified the main phases of sleep, called slow-wave sleep and REM (rapid eye movement) sleep. We will need to unlock the cerebral machinery that turns mental functions on and off without our having the slightest awareness that this is happening. During slow-wave sleep, which dominates the first half of the night, only a small and quite intermittent amount of electrical energy is generated inside the brain itself, which as a result shuffles through memories without vividness. This is a state in which normal thoughts coexist with a lack of sensory images. In contrast with this sleep deprived of light and shapes, REM sleep is marked by a great deal of sustained cerebral activation, reverberating memories with great intensity. This reverberation is the stuff that dreams are made of.

But is there any benefit to dreaming? Is the extravagance of dreaming merely an evolutionary accident, or are there rather profound reasons for it? Freud identified the existence in dream narratives of hidden meanings connected to the dreamer's subjective experience. Swimming against this tide, the English biologist Francis Crick, the Nobel Prize–winner who shared the discovery of the DNA double-helix, claimed in 1983 with the Scottish mathematician Graeme Mitchison that dreams are bizarre, hyper-associative, and apparently devoid of meaning because they derive from the random activation of neurons in the cerebral cortex. The moat that for a century had separated the neural mechanisms for sleep from a comprehensive account of dream subjectivity instigated this anti-Freudian explanatory model, which was disconnected from even the most basic observations available to any dreamer capable of the slightest introspection. To Crick, dreams were just fragments of memory assembled at random. Dreaming resulted from the simple erasing of irrelevant memories, which freed up space for storing new memories. In other words, dreams existed not for remembering but for forgetting, since the random activation of the cortex promoted the relentless erosion of recently acquired memories, generating a learning in reverse (or an unlearning), which was essential for the system not to fill up its capacity for forming memories.[33] One corollary of the theory was that the content of dreams was intrinsically meaningless, an idea that absolved the dreamer of any meaningful connection with their own dreams. This conclusion denied the importance of dreams for the understanding of the human consciousness.

Though ingenious, Crick's idea could not stand up when faced with the fact that it is possible to have recurring dreams over several nights. Repeated nightmares are one of the most common symptoms in people who develop traumas after having lived through aversive situations.[34] Given the vast number of neurons and synaptic connections in the cerebral cortex, it is impossible to explain the recurrence of dreams—and therefore almost identical patterns of neural activation—by means of cortical activation that is purely random. In other words, it would be impossible for dreams to repeat if their genesis were entirely accidental. The need to forget is one important part of sleep, but it doesn't come close to explaining the dream phenomenon in its totality.

VALUABLE SCARS

It is curious that the German word for a dream—*Traum*—so closely resembles *trauma*, which in Greek, with an altogether different etymology, means wound. Memories are scars, and their activation during sleep, in the form of dreams, has cause and meaning. In order to illuminate thoroughly the functions of dreams and the reasons for them, we will have to travel down the long path that runs from molecular biology, neurophysiology, and medicine to psychology, anthropology, and literature, without losing sight of the fact that the evolution of the species, in its most recent phase, covers our entire recorded history.

Any satisfactory theory of sleep and dreams should first consider all the relevant phenomena and not only a part of them. Second, it should distinguish between the various functions of the different stages of sleep and dreaming. Third, it should produce a plausible narrative for how these states offered a genetic and cultural advantage over time, establishing a set of functions that are cumulative and layered upon one another, and that can only be understood in the chronological order in which they appeared. The articulating of all these conceptual tools makes it possible to decipher dreams with clarity. The port of arrival, at the end of the journey, reveals a special state of human consciousness, the lucid dreaming in which the dreamer is not just a main or secondary character, a semi-voluntary actor in every night's internal movie, but also the screenwriter, producer, and director of a spectacular if totally private blockbuster.

Before tackling such special dreams as those, however, it is necessary to recover those dreams that are within everybody's reach, the ones we have every night but to which we pay little attention: the dreams our ancestors cultivated as oracles and which most people today ignore. Jung believed that the prospective function of a dream

> is an anticipation in the unconscious of future conscious achievements, something like a preliminary exercise or sketch, or a plan roughed out in advance . . . The occurrence of prospective dreams cannot be denied. It would be wrong to call them prophetic, because at bottom they are no more prophetic than a medical diagnosis or a weather forecast. They are merely an anticipatory combination of probabilities which may coincide

with the actual behavior of things but need not necessarily agree in every detail.[35]

Our goal then must be to understand, in depth and in detail, in terms of its fundamental mechanisms, how it is that a dream "prepares the dreamer for the events of the following day."[36] How that happens is the subject of this book, a brief history of the human mind with dreams as the connecting thread. In order to make this journey, we will need to consider narratives from all over the world, even while we know that it is impossible to encompass the whole world's narratives. Incompleteness, displacements, condensations, multiplicity of characters, unexpected returns, details that have no apparent explanation, and even a lack of relevant details will be our traveling companions. To weave the weft of stories and conjectures without losing our way, we will need to combine a provisional suspension of disbelief with a commitment ultimately to doubting everything. Above all, it is essential that we do not try to understand too soon, but allow ourselves to be carried by the current until we can get a good perspective on all the evidence assembled, which is necessarily incomplete but nonetheless illuminating.

One final precaution before we set off: the repeated, enthusiastic, necessary invitation to introspection. I hope this book will encourage readers to spend a few more minutes in bed upon waking, in order to remember and make a detailed record of their travels into the deep interior of the mind. The dive into the multiple dimensions of a dream, an almost completely forgotten art in today's world, can and should reactivate the ancestral habit of dreaming and telling.

2

The Ancestral Dream

In contrast with the great majority of other animals, we have a considerable capacity for simulating possible futures based on our memories of the past. We can carry out relatively complex and precise motor functions even while our mind is wandering unconstrained to all kinds of images and situations, on any timescale and on any spatial scale—just as in dreams, but with much less intensity. Might our capacity for daydreaming have originated with the intrusion of dreams into our waking lives?

Answering this question requires that we ask what sort of dreams our ancestors would have had during the immensely long Stone Age. It also requires that we understand how those dreams were transformed as civilizations developed, and how their relationship to waking life would come to be progressively reimagined. It requires, in short, that we piece together how it was that we made a transition from a strict awareness of the present time to a broader awareness of past and future.

Dreaming must have been deeply disturbing on most of the 1.168 billion nights separating us from our oldest great-great-grandparents—those like little Lucy, the fossil of an *Australopithecus afarensis* who lived 3.2 million years ago in what is now Ethiopia. How mysterious and magical must a Stone Age night have been? An incredibly long night starred with oneiric ecstasy and terror through glaciations and thaws, and the timeless morning rebirth of the question: was that really real?

To make some rational speculation about the dreams of our ancestors, we must assume a good amount of continuity between their minds and ours. After all, *Homo sapiens* has been anatomically the same for at least 315,000 years.[1] Besides, there are indications of cultural overlap[2] with the main human subspecies with which they genetically mingled, *Homo neanderthalensis* in Europe and later western Asia, and *Homo sapiens denisova* in Siberia.[3] We can therefore assume that our most distant hominid ancestors, just like us, dreamed when they slept.

DREAMS OF STONE AND BONE

Try to imagine what prehistoric dreams were like. To judge by our ancestors' obsession with stones, it's likely that they dreamed exhaustively about the production of sharp blades from lithic cores and cutting edges, repetitive motor dreams about an activity that was typically carried out on the actual lands where they lived, close to the entrance to the caves. Ever more refined objects of stone and bone attest to the appearance of a cultural ratchet, a concept suggested by the U.S. psychologist Michael Tomasello to describe the almost continuous advance of new technologies and concepts, with no great regressions from a certain given moment in the evolution of the human species onward. To draw an analogy between the body and a computer, we can say that in the last three hundred thousand years, humanity's biological hardware has changed very little, but the cultural software has evolved apace. It's as if the accumulation of adaptive ideas was a ratchet, a gear that can only turn one way. The thing that got us out of the caves was culture. At specific times and in specific places, innovations appeared, were abandoned and rediscovered, but from a certain point the rapid dissemination of adaptive ideas meant that the production of tools expanded to new techniques, materials, and uses.

The prehistoric dream was mostly made of stone, but the picture wouldn't be complete if we didn't go into the most hidden depths of the caves in search of the incredible mural art of the Upper Paleolithic, between approximately fifty thousand and ten thousand years ago. Since there is no certain record of a dream that predates the advent of writing, it is reasonable to speculate that the cave images created by our ancestors represent beings so present in their waking lives that they certainly also appeared in their dream lives. Just like the walls of the caves, the people's minds must have been populated by the enormous variety of fauna that constituted their world: bison, auroch, mammoth, horse, lion, bear, deer, rhinoceros, ibex, and various sorts of birds.

It is no coincidence that there are records of totemic beasts from

Canada to Tanzania, from New Guinea to India, the Pyrenees to Mongolia, in cultures as varied as the Ojibwe, the Masai, the Birhor, the Celts, or the Dukha. Some of the most ancient representations of our species are zoomorphic—that is, they are mixtures of human beings with other animals, often with deer horns or a bison head, as in the

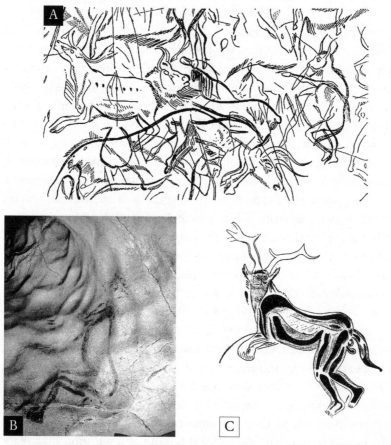

Figure 2. Zoomorphic representations found in the Les Trois Frères cave in the French Pyrenees, dated to fourteen thousand years ago. (A) Drawing of the zoomorphic figure known as "the small sorcerer," which stands out on the right of the picture with a bison head and human legs, possibly playing a flute.[4] (B) Photo of the zoomorphic figure known as the Sorcerer.[5] (C) Drawing of the Sorcerer done in the early twentieth century by Abbé Henri Breuil. The composition of the marks scratched on the rock reveal deer horns, owl eyes, bear paws, the tail of a horse or a wolf, human legs, and an erect penis.[6]

famous figures discovered in the Les Trois Frères cave in the French Pyrenees, dated to fourteen thousand years ago (figs. 2A, 2B, and 2C). These images have been interpreted by scholars as possible evidence of shamanism in the Upper Paleolithic, with the use of masks, furs, and antlers, or of a belief in the transformation into other animals, which remains common today in various hunter-gathering societies. Alternatively they have been seen as a clue to the cult of the Lord of Beasts or Horned God, an ancient being conceived as the protector of good hunting, possibly one of the human species' oldest divinities, a precursor of a number of similar myths (e.g., the Lord of Wild Things, the god Cernunnos, in Celtic myth, and the god Pan in Greek mythology), some of which persist today in hunting communities around the Arctic.

Such closeness to wild animals shouldn't be surprising. Seventeen thousand years ago, when the Lascaux caves in France and the Altamira caves in Spain were finely decorated with the cave paintings that would make them famous, the challenges faced by humans still closely resembled those of any other animal species, and could be summarized in three basic imperatives: eating, not being eaten, and procreating. Animals were essential for obtaining nutrition, bones, teeth, and furs, but they also represented the constant threat of death. Over the millennia, in addition to the dreams of stone, dreams of prey and predator must have been prevalent too, composed of hunger, pursuit, rage, panic, and blood.

Many archaeological sites that are spread across western Europe and east Asia reveal the surprising symbolic and cultural continuity between different Paleolithic populations in Eurasia. The discovery inside the caves of niches filled with bears' bones, in apparently intentional groupings of long bones and skulls, has been interpreted by some scholars as offerings of brains and marrow rich in nutrients, leading to the suggestion that the Lord of Beasts might have received wild animal sacrifices for thousands of years. Ritual deposits of reindeer bones were also found in Siberia and Germany,[7] while mammoth bones were used both for building homes and for ritual purposes in Ukraine and central Russia.[8] Bears' bones painted with ocher, found in Belgium and dated to twenty-six thousand years ago, reinforced the idea of an animal proto-religiosity, which went beyond the simple utilitarian value of the carcasses.[9]

The interring of the animal bones and antlers in order that they might be reincarnated is an ancient hunting ritual, which persists among Arctic peoples[10] and has echoes in cultures as distant as the Nordic (Thor's goats, devoured at night to be reincarnated in the morning) and the Semitic (the valley of dry bones in Ezekiel 37:1–14 in the Bible). Even if one must recognize the opacity of almost all archaeological finds, which typically provide few clues as to the wealth of behaviors that produced them, the magical-religious intention of the hunters of the last interglacial period is hard to deny. It makes sense to presume that this intentionality was connected to the hunters' enthusiasm and therefore to favoring the hunt. A great deal of courage was needed to confront and progressively wipe out the megafauna of the Pleistocene. A good number of the scenes depicted on cave walls by our ancestors represent the extensive bestiary of animals preyed upon by ever more numerous and organized groups of humans, armed with sharpened spears and a mystical certainty that they were fulfilling, in their waking days, a destiny perhaps foretold to them in their dreams.

FIRE, SYMBOL, AND ARCHETYPE

Certain extremely general cosmogonic myths probably date from the Paleolithic period, such as the Creator who goes down into the depth of the primordial waters to bring back matter and create the world, the magic flight up into the heavens, the origin of humans and animals, and the rainbow over the center of the world. In the Upper Paleolithic, the first fertility symbols appear, such as phalluses, vulvas, and many "Venuses." Fire, present in everyday human lives since at least 350,000 years ago,[11] was certainly another important element in the oneiric narratives of the Stone Age.[12] Used as it was in the cooking of food and the warming of bodies, fire became the center of group gatherings, leading to what might have been the first conversation circle. Fire was also the thing that kept predators at bay and protected sleep, creating more safety and more time for dreaming.

The apparent cultural generality of certain symbols found in dreams, such as the association of fire with transformation, seemed to Jung to be the expression of a universal code of instinctive symbols in our species. No biological evidence of this archetypal inheritance has yet been found, but in the past decade there have been notable advances in the understanding of the molecular mechanisms that are capable of promoting the intergenerational transmission of learned behaviors. On the other hand, symbols shared by different cultures (über-symbols) are frequently connected to events of great significance that almost every human being experiences over the course of their lives. Instead of an innate behavioral program, perhaps many of the dreams that are common to different cultures merely reflect the fundamental similarity of human experience the planet over. The mother, the father, the wise old person, the creation, the flood, these are narratives and characters present everywhere in our history. How we live is what regulates our dreams—and the most important landmarks are the same everywhere: birth, puberty, sexuality, procreation, struggle, sickness, and death. This profound life truth has nothing particularly human about it. It applies with equal validity not only to all primates, but to any animal.

One exclusively human thing, however, is the verbal recounting of dreams, along with the accounts of the waking events. Narratives became ever more complex and interesting as our species gained in lexical diversity, complexity of discourse, and the capacity to memorize, recall, and recount. Dreams almost certainly played a prominent role in the growing capacity for narrating human existence, being a source—renewed nightly—of images, ideas, longings, and fears. If dreams reflect what is happening in the dreamer's life, the cavewomen and cavemen would have dreamed about their routines of gathering fruit and roots, making weapons and tools, planning and carrying out hunts, alliances, and conflicts with other humans within and outside the clan, as well as mating, parental care, and death.

Dreams were our ancestors' cinema, and all the more fascinating for being potentially real. In the enormously long dawning of human consciousness, at countless fuzzy moments in the last few millions of years, our prehistoric ancestors must have woken up in wonder at their dreams' limitless world of simulacra. How many of them wouldn't have been furious to discover that the dangerous mammoth, hunted down so gloriously on their dream adventure, had faded away with the dawn, dissolving into the daylight? Or, on the contrary, how many would have woken hugely motivated to hunt a big mammoth in reality? Our prehistoric forebears must have made peace and love, waged war and wrestled with passion, propelled by experiences that probably seemed as real as those in their waking lives. The discovery that dreams are deceptive must have been made countless times at sunrise, but everything suggests that from early on, this discovery was coupled with the certainty that if a dream is not real, it can at least affect the course of reality.

The commotion caused by dreams must have happened on a daily basis. The important decisions of waking life began to depend, at least in part, on the good or bad omens revealed in the images from nighttime. Dreamers whose dreams frequently corresponded to subsequent events must have started to be valued by the group. As is common in so many cultures, our cave-dwelling ancestors might perhaps already have distinguished between ordinary dreams and "great dreams" of an advisory or premonitory nature, which could have a decisive influence on the course of life of the dreamer and their people. The successful navigation of this new universe ended up becoming a social specialization. It was the embryo of shamanism, the great-grandfather of religion, medicine, and philosophy.

A certain moment in the Upper Paleolithic saw the first appearance of the idea of a double, like a "soul" or "spirit," probably by way of dreaming. This was how the nineteenth-century German philosopher Friedrich Nietzsche accounted for that leap in consciousness:

> In ages of crude, primordial cultures, man thought he could come to know a *second real world* in dreams; this is the origin of all metaphysics. Without dreams man would have found no occasion to divide the world. The separation into body and soul is also connected to the oldest views about dreams, as is the assumption of a spiritual apparition, that is, the origin of all belief in ghosts, and probably also in Gods.[13]

To the founder of sociology, the Frenchman Émile Durkheim, a scholar of the religious lives of Australian aborigines, the idea of the soul was suggested to our ancestors by dreaming:

> If during sleep he sees himself talking to a friend he knows is far away, he concludes that this man, too, is composed of two beings: the one sleeping somewhere else, and the other who has made an appearance in the dream. From these repeated expe-

riences the idea gradually develops that in each of us there is a double, another self, which under certain conditions has the power to leave the body it lives in and go wandering.[14]

Around the fires and inside the caves, the shamans could inflame themselves, discover paths, be lighter than air, see in the dark, decode dreams, and cure illnesses. Of all the dream accounts they interpreted, the most unsettling must have been the dreams about dead relatives. How could you not experience turmoil upon meeting your loved ones again after they have gone?

MISSING YOUR ANCESTORS

Although the preservation of skulls and jawbones of *Homo erectus* at least three hundred thousand years ago was discovered at Chu-ku-tien in China, there is no consensus as to the existence of deliberate graves of *Homo sapiens* earlier than one hundred thousand years ago.[15] In Sungir, 120 miles east of Moscow, an archaeological site attributed to mammoth-hunters contains the extremely sophisticated graves of a mature man and two adolescents. Their bodies were buried along with spears, leather clothing, boots, hats, and fox-tooth necklaces, as well as various objects made from mammoth ivory, including bracelets, statues, and thousands of small beads. The tombs, which are conspicuously covered in red ocher, were dated using various methods to around thirty thousand years ago. The painting of bones using iron oxide to symbolize blood and life has spread across the planet since that time, suggesting a belief in life after death. Food and objects began to be placed inside tombs, such as skulls adorned with artificial eyes as well as animal antlers, shells, ornaments, and other symbols of social or magical authority. Graves began to be pointed toward the rising sun, perhaps suggesting an expectation of rebirth. Among hunters and gatherers, the practice persists today. In Amapá, in the far north of Brazil, funeral urns of the Aristé culture were found within a circle of menhirs arranged according to the direction of the sun on the December solstice[16]—an amazing Amazonian Stonehenge.

Ritual burials marked a definitive break between human culture and the mental functioning of other animals. The same is true for cave paintings, something no other animal does. It is possible that the first species to produce cave paintings was *Homo neanderthalensis,* our cous-

ins who disappeared a mere thirty-seven thousand years ago: the oldest cave paintings found thus far, in caves in Spain, date to more than
sixty-four thousand years ago, some twenty thousand years before the
arrival of *Homo sapiens* on the European side of the Mediterranean, in
migrations that came from Africa.[17]

It is also important to consider that the impressive wall paintings
of the Upper Paleolithic are typically found deep inside caves, far away
from the inhabited entrances. These challenging locations, in places
that are hard to access, indicate the occasional and quite deliberate use
of these spaces, suggesting an important ritual function to prehistoric
art carried out in the belly of the earth. Altogether, these elements of
human culture remained more or less similar between thirty thousand
and nine thousand years ago, probably shaping a "cave religion" in
which the belief in life after death was mixed up with a belief in dreams
as a portal between the dead and the living. In some cultures, shamans
have specialized in techniques for passing through this portal and traveling in time and space, seeing what most people are unable to see. This
path of wisdom normally requires a dream-guided initiation that sym-
bolizes death and rebirth. By means of physical trials and privations,
shamans seek to receive numinous visions and increase their knowledge
in the form of songs, true names, totemic guardians, and genealogical revelations.[18] In different cultures, it can also be other dreamers,
unaided by a shaman, who themselves speak with the spirit world.[19]

THE DREAM OF THE END OF HUNGER

Back in the last ice age, maybe twenty-five thousand years ago, a major
change occurred: the beginning of a progressive artificial selection of
new species.[20] The skillful taming and breeding of wild animals triggered this radical change in the relationship between humans and the
rest of nature, producing cultural innovations that forever changed our
place on the planet—as well as our relationship to the spirit world. First
we took advantage of the sociability of wolves to transform them into
dogs, the results of the selection of those genetic varieties best suited to
the support of hunting and protection of the home.[21] Then we domesticated a variety of herbivores and omnivores: pigs, chickens, sheep,
goats, horses, oxen—to provide us with meat, milk, wool, and labor.[22]
This domestication took place in parallel with the selection of breeds
of dogs for shepherding and drafting. The entry of animals into the

home coincided with the beginning of the end of the last glacial period, which had already lasted more than ninety thousand years. With the thawing of huge surface areas came an accelerated development of flora and fauna, creating a real paradise of edible animals and plants for the human gatherers and hunters. At the end of the Paleolithic period, our ancestors fed themselves on anything that moved, from wolves to gazelles, fish to mollusks. Nuts and fruit complemented their diet.

The discovery of edible grains and grasses in the Fertile Crescent, between twenty-three thousand and eleven thousand years ago, sealed the change in our destinies. We began to artificially select new species of plant, as well as of fungi and bacteria used in fermentation. In the millennia that followed, methods were discovered for actively promoting the growth and harvesting of plants. Together, these practices fostered the transition from the gregarious life of the hunter-gatherers to the more or less sedentary lives of shepherds and farmers. The transition from a society of nomadic hunters to an agrarian one, based in a geographical space that is fixed or semi-fixed, involved a telescopic lengthening of the human perception of time. Through the whole Paleolithic period, our ancestors needed to know how to estimate accurately the phases of the moon and its stations, in order to predict the movements of the herds on their seasonal migrations. In the Neolithic, however, what they needed was the ability to carry out a much more complex series of well-ordered actions, whose revolutionary promise was to make seeds germinate and plants grow and bear fruit until there was no more hunger. However, the processes involved in the production of our own food required a much more precise consciousness of the passing of time, and the work involved in farming was as great as its promise—with uncertain results.

Planting and harvesting efficiently, without wasting all of one's months of effort because of unforeseen problems, demanded a great refining of the ability to predict environmental changes and to accumulate knowledge over generations, accelerating the cultural ratchet. If inspiring dreams of bravery and group tactics were important in the age of great mammoth hunts, with the invention of agriculture they must have given way to epiphanic dreams that allowed for the panoramic contemplation of the recurring patterns of nature, with an ever more precise recognition of periods of rain and drought, floods and ebbs, cold and heat. There must have been dreams, too, about the overlapping social interdependencies that were formed in order to plow, fertilize, sow, water, and harvest. It was also the age of majestic dreams depicting the periodic renewal of alliances, not only with the other farmers who lived in the village, but at harvest time with the fertility gods.

Grain could last decades if stored well. Its cultivation saw the development of silos and fixed habitations in villages that were lived in continuously for generations. The high productivity yield of agriculture led to a population explosion, multiplying clans from a few dozen people into cities of hundreds and even thousands. New, complex agrarian tools were invented, such as the plow, ceramics, and looms. The artificial selection of seeds and breeds accelerated domestication and led to countless new varieties. The natural environment gave way to a space that was ever more artificial, planned and constructed: gardens, orchards, buildings, and roads made by creatures who were now creators of the world around them.

This transition brought important symbolic novelties, with increased complexity and inevitable oneiric repercussions. The sculptures, figurines, and paintings of the Neolithic period show feminine figures in abundance, as well as phallic objects, bulls, other domesticated animals, and many circular structures. In agrarian societies, the cult of the dead was expanded into an intimate relationship with the cult of fertility, as death came to be understood as a circular promise of a new life, just like the seed that is buried in the fertile womb of the ground and after "dying" is reborn to bear fruit and die again for people to eat it.

If the idea of circular time is inspired by agriculture and by cosmic cycles, the idea of space—previously as large as the need to migrate—began to be punctuated by settlements and their cultivated fields as its fixed geographical reference-points, leading to the start of a representation of such a thing as the Center of the World. We see the first archaeological evidence of the main fundamental antagonisms of symbolic life: Us vs. Them, Woman vs. Man, Mother vs. Father, Day vs. Night, Summer vs. Winter, Life vs. Death.

THE RISE OF THE DEAD

In the transition from the Paleolithic period to the Neolithic, the cult of the dead spread geographically, as seen in the graves of the Natufian culture in the Levant, where corpses were covered in ocher and interred

in the fetal position. The burying of skulls, which had already been happening in the Paleolithic period, also became a more frequent occurrence. The first temples date from the start of the Neolithic, such as
the fascinating stone constructions of Göbekli Tepe in Turkey, the main
indication we have that religion predated agriculture. This archaeological site on the vast Anatolian plateau, dated to eleven thousand years
ago, contains impressive megaliths almost twenty feet high and weighing twenty tons, engraved with icons of spiders, scorpions, snakes, and
lions. With the proviso that only a small part of the archaeological site
has been excavated, the absence of signs of habitation and the predominance of icons of predators suggests that Göbekli Tepe served
some religious function that was not connected to everyday life. In the
famous formulation of the German archaeologist Klaus Schmidt, "first
came the temple, then the city."[23]

In the archaeological sites of Hacilar and Catal Huyuk, also in Anatolia, burials from nine thousand years ago contain jewels, weapons,
domestic implements, and fabrics, in addition to clay and stone figurines. On the painted walls, there are representations of women, bull's
heads, breasts, horns, and demonic creatures, half man and half beast.
At the same time, in the city of Jericho, which is in Israel today, bodies
were buried beneath the floors of houses, but their skulls were later
removed, covered in gypsum, decorated with shells in the shape of artificial eyes and painted as if with hair and mustaches, in a clear attempt
to emulate life after death. In various other excavations in Palestine,
dated to approximately 4,500 years ago, female figurines were found
with human bones, suggesting a mixing of the cults of the dead and
of fertility.

At that time, there was a spread in the sacrifices of domestic animals,
as well as in the use of fundamental symbols of the sun, the snake, and
the wavy curves of water. The surplus food made possible by agriculture allowed a deepening of social specializations, with members of
the group ever more dedicated to each function: tilling, shepherding,
hunting, fishing, cooking, taking care of the children, teaching them,
fighting, praying, or governing. This division relied upon two new techniques that were inherently transformational, whose impact reached
every part of life: metallurgy and pottery. In many different cultures
another new and mysterious divinity came into being, a Lord of Fire
who was the patron god of the miners and the blacksmiths but also of

the magicians. From that time to this day, the production of objects in metal has been at the heart of economic, military, and technological activity. In the biblical dream of Nebuchadnezzar II, emperor of Babylon, in the sixth century B.C.E., gold, silver, bronze, and iron appeared as the nomenclature for different historical ages. During antiquity this metallic classification of history spread as far as India and Europe, and survives with some variations to this day.

The new way of living brought with it new symbolic elements. Underground mines, besides being a source of moldable minerals, were also a representation of the world of the dead. Monumental tombs and temples were constructed everywhere, originally in the shape of mounds and then, with ever-increasing enthusiasm, as proper artificial mountains whose purpose was eminently funerary, whether in Egypt or in Mexico or Peru.

PYRAMIDS AND CEMETERIES

In Egypt, care for the dead began and spread through the building of mastabas, "houses of eternity," which predated the pyramids and were constructed for people who were not a part of the royal family. They have preserved a record of the great stratification of luxury after death, since they were used right across a broad social spectrum, from the highest officials to quite subordinate serfs. The wealthiest mastabas contained rooms that were fully equipped for life after death, including tables with a funeral banquet, games, tools, weapons, pots, jugs, decorated trunks filled with clothing and wigs, bathrooms with toilet utensils, cosmetics, and basins for face-washing, in addition to miniatures, dioramas, and murals painted with representations of all these things. In some mastabas, stone heads were found that would have served as substitutes for a mummy, providing a material substrate for the spirit to inhabit in the event that the tomb was looted.

In Ur, close to a huge ziggurat of brick and bitumen as tall as a five-story building, excavations uncovered a cemetery dated to 4,500 years ago and containing about two thousand tombs. Among them, sixteen tombs attributed to royalty stood out for the presence of abundant food, carts with oxen, recreational games, musical instruments, and cosmetics. The treasure included sculptures and rings made from gold, silver, lapis lazuli, shells, and bitumen; they were decorated with bulls, lions, gazelles, and goats carrying out typically human actions, as well as hybrid creatures, half human and half beast, such as the alarming scorpion-men. A big pit preserved evidence of large-scale human sacrifices, with the remains of seventy-three people, including armed men and very richly adorned women. The group burial suggested the need for a retinue of soldiers and servants, apparently necessary for life after death, to accompany those dead who were of high social significance.

During the Neolithic and Bronze ages, the cult of the dead flourished around the world, as contact between geographically distant groups increased. In the south of Egypt, funerary pyramids stretched up the Nile as far as Nubia, in what today is Sudan. To the north, on the island of Malta, about seven thousand skeletons were deposited in an underground complex of interconnected chambers, dug into the rock six thousand years ago. This is the Hypogeum of Hal Saflieni, a necropolis constructed by a Mediterranean Neolithic culture notable for its enormous tomb megaliths, which can be found on Crete and in Troy and which resemble the dolmens and menhirs of northern Europe. They were dwellings for souls that were able to leave at night, wandering randomly. The luxuriousness of a tomb dated to the fifth century B.C.E., found inside a burial mound close to the French town of Lavau, bears witness to the huge importance attached to the dead at the end of the Celtic Bronze Age. In addition to containing a two-wheeled chariot, jewels, princely clothing, and the remains of a banquet washed down with wine, the burial includes objects of Mediterranean origin, such as an Etruscan cauldron and a Greek jug depicting the god Dionysus.

It was no different in the Americas. Between about 8,000 and 1,400 years before Christ, burial mounds spread from the mouth of the Amazon to the source of the Prata River. These are structures up to a hundred feet tall, made of shells (sambaquis) or earth (cerritos).[24] At the Jabuticabeira II archaeological site, in the Brazilian state of Santa Catarina, there is a huge sambaqui nearly one hundred feet high and over 1,300 feet in length and 800 feet in width. It is estimated that more than forty-three thousand bodies were buried there over the course of a millennium of continuous occupation.[25] Burial mounds were prevalent from Canada to Tennessee, while on the Yucatán peninsula, in Mexico, people were sacrificed and thrown into caves called cenotes. These very beautiful and frightening networks of semi-submerged caves, created by water infiltrating the limestone weakened by the impact of the asteroid that wiped out the dinosaurs, represented Xibalbá, the underworld governed by the gods of death. The Mayan book, the *Popol Vuh*, tells the story of the journey taken by twin heroes Xbalanké and Hunahpu to Xibalbá, where they defeated the gods before returning to the upper world triumphant and so bringing about the creation of the sun and the moon. In the version that was translated not by a European but by a Mayan, Xbalanké and Hunahpu are doubles, two aspects of the same hero, or perhaps the hero and his soul.[26]

A PERCH FOR THE SOUL

The prevalence of the idea of the soul transcends cultural barriers. In Jamaica in the seventeenth and eighteenth centuries, the very high mortality rate for both slaves and whites was accompanied by a considerable preoccupation with funeral rites.[27] In western Africa and the parts of the Americas touched by the black diaspora, especially in Brazil, Cuba, and Haiti, concern about the souls of the dead was lively. The occurrence of similar beliefs is striking among the Bahian Candomblé, Cuban Lucumi, and Haitian Voodoo. In Brazilian Umbanda, which is marked by religious syncretism—catalyzed from its very beginnings by a deep unease in Christianity about the fate of souls—it is believed that dreams are the portal for communicating with divine beings and the souls of people who have passed away.[28]

Accounts by Christian missionaries traveling through central Africa in the seventeenth century testified to the Mbundu belief in the transmigration of souls, which are immortal and even able to pass to the wives or children of the dead.[29] Since there was a belief that life was affected, regulated, blessed, and often cursed by the dead, dreaming about people who had died required carrying out rituals of veneration, offerings of food at the tombs or the Houses of the Dead, temples that were generally far away from homes, as well as sacrifices of domestic animals and of people. The birthdays of the deceased were marked by similar rituals, complex funeral rites that lasted several days.[30] Tombs were frequently topped by small pyramids equipped with little windows to allow the soul to see its surroundings.

Despite many regional variations, western Africa saw a general belief in two different kinds of supernatural beings. The first type comprised universal or territorial divinities, distant and powerful beings, connected to a whole culture and not to one family in particular and inhabiting prominent geographical features such as mountains, rivers, and lakes. The second kind of being corresponded to the souls of the relatives of a specific family, inhabiting their tombs and sometimes objects such as altars, reliquaries, and amulets.

As early as the end of the nineteenth century, the British ethnographer Mary Kingsley reported a belief among the Fang of central Africa that each person had four different souls: the one that exists after death, the one that is the body's shadow, the one that inhabits a wild animal, and the one that ventures out of the body every night, traveling through dreams and encountering other spirits.[31] Their return upon waking was considered essential for the person's health, and a great deal of harm could arise from the use of magical objects equipped with hooks, which were capable of catching unwary souls when they were traveling outside their bodies. Such situations demanded the assistance of sorcerers to free the captured dream soul and blow it back into the sick person.

Among the Fang, the cult of wooden Byeri statues was traditionally devoted to the ancestors they represented. In order that they might be able to protect the living, the statues were placed in reliquaries containing the skulls and fingers of ancestors, vessels with medicinal herbs, and offerings of blood and meat. Prayers and animal sacrifices allowed each family to consult their forefathers on the issues of the greatest significance to their communities, such as hunts, wars, and displacements. As a rule, the answers from their ancestors would come via dreams and visions induced by psychedelic plants. The Byeri cult went into decline in the twentieth century and gave way to another ancestral cult, the Bwiti.[32] This religion mixes African beliefs from the south of Gabon with Christianity, in a syncretic blend that includes the receiving of spiritual messages following the consumption of iboga, a potent psychedelic root.

In every continent, with more or less intentionality, the practice existed of preserving the dead by means of extracting their organs, and then desiccating and embalming the bodies.[33] Human culture's striking fixation with the dead has a distant echo in the mourning behavior of chimpanzees, observed in captivity,[34] but also among animals living in the wild in the African jungle.[35] Immediately following a death there is a great commotion through the whole group, and the closest relatives remain sorrowful for hours. There are well-documented cases of the mother of a baby that has died carrying the dry corpse around as though it were alive, behavior that can go on for weeks after the death.

The persistence of the connection to the past is clear. It is impossible not to see a parallel in the mummification of ancestors, carried out by people as different as the priests of the Necropolis of Memphis or the ingenious Ibaloi of the Philippines, who used heat from fire to speed up the drying of the body. There is ample evidence of a divinatory cult of the dead in antiquity, whether in Mesopotamia, in the Nile Valley, or in sub-Saharan Africa. A very similar phenomenon occurred later in Central America, in the Maya and Aztec civilizations. The Incas treated their mummified rulers as if they were socially alive, valuing them as repositories of authority and knowledge of the past. On feast days or in the presence of foreign visitors, the mummies were exhumed, transported, fed, and then listened to.[36]

More than a thousand years before the Incas, in the northern part of the Peruvian Andes, the Moche mummified prominent people and made sacrifices of companions who were placed inside wealthy tombs or near to them. Five thousand years before the Moche and two thousand before the Egyptians, the Chinchorro learned to bury their dead in the Atacama Desert. The attempt to keep the bodies of the dead alive, which is extremely widespread among the human species, is an expression of the concreteness of our primate reasoning.

Along with the spread of the cult of the dead, the move to farming led to a number of important myths, such as the flood, which appears in both the Sumerian and Hebraic traditions and in the *Popol Vuh*.[37] If torrential rain and lightning could frighten the cavemen and inspire the cult of the thunder god, a storm followed by flooding could devastate an entire harvest as well as the irrigation canals and silos, destroying months or even years of work, potentially wiping out whole cities. The subject appears in one of humanity's oldest texts, *The Instructions of Shuruppak*, produced by Sumerian scribes in a city close to the Euphrates River, in what is now Iraq. Written 4,500 years ago in cuneiform characters on clay tablets, the text describes the advice and teachings of the King Shuruppak, final ruler of the city-state that bore his name, to his son Ziusudra. He is the Sumerian Noah, halfway between a biblical myth and a historical character, since the city of Shuruppak did in fact exist and was destroyed by a flood about five thousand years ago. It is no surprise that dreaming appears in *The Instructions of Shuruppak* in connection with a divinity: Ziusudra dreams about Enki, the god of wisdom, who warns of the flood and gives instructions for building an ark to save Ziusudra's family and a couple of each animal species.

The art of eternalizing stories, conversations, and norms using symbols engraved on clay tablets or in blocks of stone seems to have arisen quite suddenly in Sumer and Egypt. The invention of writing accelerated the process of accumulation of knowledge even faster, changing the course of the evolution of human consciousness. From that moment on, the multiplication of new symbols became unstoppable, propelling the cultural ratchet with such force that in less than five thousand years we have arrived at computers and the Internet.

THE ORIGIN OF THE GODS

Technical advances did not prevent dreams, divination, and necromancy from accompanying us on our historic progression until very recently. Many of the texts that come to us from ancient Egypt are instructions not for living but for dying. The *Book of the Dead*, or, in its literal translation, *Book of Emerging Forth into the Light,* is basically a collection of papyri with prayers, magical incantations, and practical guidance on how to move safely through the gap between finite life and eternal existence, which is granted only to the just. A mixture of guidebook and passport, the *Book of the Dead* clearly reveals the idea of guilt, since the deceased must present a negative confession to Osiris, a declaration of "nothing on record" regarding guilt and misdeeds. Osiris is a murdered god reborn in his son Horus. This rebirth has a direct correspondence with the dynastic succession of the pharaohs. When the spirit of the deceased ruler was transported up to heaven, transformed from living god to dead god, he bequeathed his princely heir the supreme command over Egypt.

Judging by the ample historical evidence, the gods commanded human actions for millennia, and their influence persists to this day in the minds and behavior of billions of people. Unless we just put our hands over our ears and eyes, it is quite clear that belief in gods is a striking fact that demands explanation. The U.S. psychologist Julian Jaynes of Princeton University noted that there are plenty of records of visions and verbal commands communicated directly by gods to rulers from the beginning of history until around three thousand years ago, the period that includes the formation, development, and collapse of many city-states, including Homeric Troy. If we are to take this historical evidence seriously, we must explain why our ancestors heard voices and saw hallucinatory images.

As an explanation for the omnipresence of divinities during the first two millennia of history, and undoubtedly prior to this, Jaynes suggested that the first gods had their origins in mental representations of departed ancestors, which continued to reverberate in the minds of their relatives, in their waking hours but especially in their sleep. An Egyptian inscription from four thousand years ago proclaims "Instructions which the Majesty of King Amenemhet I justified, gave when he spoke in a dream-revelation to his son."[38] This idea has a Freudian inspiration: "The primal father of the horde was not yet immortal, as he later became by deification."[39] When the head of the group died—or the father of the primitive horde, in the terminology proposed by Darwin and adopted by Freud—there would inevitably be dreams in which the deceased, still as central to social life as he had been when alive, would appear to dreamers in spite of being dead. After one of these nighttime visions of the head of the group, the other members—amazed—would be inclined to consider him still alive in some parallel world. The belief in life after death received its most eloquent confirmation whenever the dead person gave orders, warnings, or useful advice.

VIKING DELIRIUM

Tens of thousands of years after the Paleolithic and almost four thousand years after the ancient Egyptians, the fatalistic Norse culture developed its own notion of a divine destiny that could be visualized in dreams. Such dreams offered a trustworthy premonition, distinct from the confused information jumbled up in a *draumskrok*, an oneiric illusion that is absurd and devoid of any sense.[40] In the *Poetic Edda*, an important collection of Norse poems compiled before the eleventh century, the poem "The Ballad of Skírnir" makes the Norse concept of a preestablished future explicit: "My destiny was fashioned down to the last half-day, and my whole life determined." The Norse sagas contain hundreds of symbolic dreams, many of them prophetic.[41] One of the most famous is attributed to Ragnhild, a historical figure from the ninth century who was married to King Halfdan the Black, from a kingdom in the south of Norway. She dreamed that she was in her garden when a thorn caught on her cloak. Ragnhild pulled the thorn

Figure 3. Queen Ragnhild's Dream *(1899) by Erik Werenskiold.*

out with her hand, but it grew till it had become a huge tree whose
roots plunged deep into the earth, while the branches rose up so high
that the queen could barely see them through the thick foliage (fig. 3).
The part beneath the tree was red, while the trunk was green and the
branches white. Its majestic upper branches spread boundlessly, covering all Norway and beyond.[42]

Years later, Ragnhild would interpret the tree as a symbolic premonition of the future importance of her descendants in the history of
Scandinavia, since her son Harald Fairhair would in 872 become the
first king of all Norway. The red symbolized the blood spilled in the

conquest of power, the green represented the lushness of the future kingdom, and the white branches represented Ragnhild's descendants who would rule Norway for many generations. This story fits into the tradition of dreams reported by mothers and fathers of great leaders,[43] generally announcements of their future greatness, such as those we can find in the stories of the lives of the Buddha, Confucius, and Jesus.

CONSULTING THE SPIRITS

The preservation of the memories of the dead as representations of beings that are alive, wise, and filled with authority resulted in a great accumulation of knowledge in a short span of time. It formed a memory bank that might not have been directly accessible by an individual but that was active in them via mental processes unprecedented in other animals: inspiring, admonitory, or therapeutic conversations with ancestral spirits received through the dreams or visions produced by an active imagination. The cult of the dead remains as alive today in the Egungun rituals on the island of Itaparica as it was two thousand years ago in Spartacus's Rome. In Tibet, dream practice is rooted in a long history of spiritual work through dreams, which exists in popular pre-Buddhist beliefs as well as in the Bön religion and Buddhism itself. When a Tibetan is facing difficulties that are ascribed to spirits, dreams are often used for contacting protective spirits and consulting oracles.

The cult of the dead and the veneration of gods was the foundation stone of religion, making communication with these beings the main function of dreams. Dreams fulfilled a central role in the mythological narratives of the first great civilizations—Sumer, Egypt, Babylonia, Assyria, Persia, China, and India. The first manuals for the interpretation of dreams appeared in the Assyrian Empire three thousand years ago with the production of collections of premonitory dreams like the *Iskar Zaqiqu,* which drew connections between dream events and their hypothetical implications in reality. Over the ages, augurs whose divinatory practice was based on dreams began to prosper. The belief in the divine or demonic inspiration for good or bad dreams was widespread, as well as the belief in the possibility of incubation, i.e., "planting a seed in the mind" to produce specific desired dreams. From the Chinese dream dictionary of the Duke of Zhou[44] to the traditions of Islam,[45] from the cuneiform texts of Mesopotamia[46] to the Upanishads of Vedic philosophy on the banks of the Ganges,[47] a belief in the power of dreams to predict the future spread across the planet. Lying alongside this capacity was the idea of a predetermined destiny, which fueled oneiromancy over the ages.

In the *Epic of Gilgamesh*, for example, written about four thousand years ago, the mythical king of the Sumerian city of Uruk learns from a dream of the existence of his rival Enkidu, a Mesopotamian version of the ancient Lord of Beasts. They fight, they become friends, and together they carry out great heroic acts. Then, filled with arrogance, they challenge the fertility goddess Inanna, worshipped as Ishtar by the Akkadians, Babylonians, and Assyrians. Soon afterward, Enkidu dreams that he has been condemned by the gods, falls sick and dies. Gilgamesh is desperate, and becomes obsessed with his own fear of dying, until he decides to travel to the kingdom of the dead to win his immortality. When he crosses the waters of death he meets Ziusudra (the Sumerian Noah), who tells him: "Go on, see if you can stay awake for six days and seven nights!" But Gilgamesh keeps falling asleep and he fails in his initiation to immortality.[48]

In the Hellenistic tradition, divinatory dreams are found overlapping in the most ancient of narratives. In Homer's *Iliad*, dreams play a fundamental role in the plot that causes the destruction of Troy by the Greeks.[49] Following the birth of the third child of Priam, the king of Troy, Queen Hecuba dreams that the child is a flaming torch who will destroy the city. This is Paris, who much later will abduct Helen and spark the conflict. Virgil's *Aeneid* describes how, right at the end of the war, when Ulysses's warriors hidden in the wooden horse open the city gates for the Greek army to invade, the dead Hector appears in Aeneas's dream to warn him of the disaster under way. Fleeing from Troy toward Italy and the establishment of the Roman lineage, Aeneas looks at his city in flames and witnesses the fulfillment of Hecuba's dream premonition.

But not all Homeric dreams are predictions of the future. Sometimes apparitions merely end up in disappointment. During the siege of Troy, Zeus sends King Agamemnon, the Greeks' military leader, a deceptive dream that promises a great victory over the Trojans in the event of an immediate attack. Agamemnon carries out the attack and suffers a terrible defeat. Oh, divine, tricky oracle . . .

DREAMED EMPIRES

If dreams fulfill a key role in the journeys of mythological characters, they also play a central role in the histories of rulers of flesh and blood. The story of the rise to power in the third millennium B.C.E. of Sargon of Akkad, unifier of Mesopotamia and the first emperor in human history, pivots on a disturbing dream that Sargon had about Ur-Zababa, king of Kish, whom he served as cupbearer. In the dream, the goddess Inanna drowned Ur-Zababa in a river of blood. When he learned what the dream had been about, a terrified Ur-Zababa gave the order for Sargon to be killed, but Sargon ended up prevailing.⁵⁰

The Akkadians were a Semitic people who succeeded the Sumerians and appropriated their civilization, turning the cultural ratchet of cuneiform characters and Mesopotamian divinities. Sargon's daughter, Enheduanna,* was high priestess at the most important temple in the empire, dedicated to Nanna, god of the moon, in the city of Ur. Enheduanna wrote hymns, prayers, and poems that earned her the first acknowledged authorship in history, that is, recognition of her work as the creation of a specific person. In the poem "Inanna, Lady of Largest Heart," written in the first person, Enheduanna described a magical dream in which she was raised up through a celestial gate, and she praised above all gods Inanna, the planet Venus, goddess of love, for having "a great destiny throughout the entire universe."⁵¹

The close contact that the Babylonians had with the Hebrews and other peoples to the west disseminated Enheduanna's work sufficiently widely that it would influence the psalms of the Bible and the Homeric hymns. This cultural continuity is connected to the great importance assigned to dreams in the narratives of the Torah, the Bible, and the Quran. The cultural flow between east and west involved wars and migration in both directions. In the mythological Mesopotamian past, Abraham the patriarch was born in Ur and then migrated to regions that today are to be found in Turkey and Israel. In the sixth century B.C.E., King Nebuchadnezzar II took Jerusalem and deported thousands of Jews to Babylon. In this ancient metropolis nearly six hundred miles away, the Jews suffered almost sixty years of captivity until Cyrus the Great, founder of the Persian Empire, took the city and freed

---

* *En*: priestess; *hedu*: ornament; *anna*: celestial.

them. When they returned to the Levant, the Jews disseminated the rich Babylonian culture through the words of Enheduanna.

Ever since the beginning of written records, dreams by members of the ruling elite have been preserved for political and religious purposes. The use of dreams for the communication between gods and kings persisted through the ages, and left a tangible cultural legacy. This use is well documented in the largest Sumerian clay cylinders ever discovered, made around 2125 B.C.E. by King Gudea of Sumer, with cuneiform inscriptions representing the longest known Sumerian text as well as one of the oldest written records in all human history.[52] Nearly two feet tall and hollowed out in the center so that they could be turned while they were being read, the cylinders tell of an extraordinary dream of King Gudea, which begins with the appearance of a man as tall as the sky and with the head of a god, the wings of a bird, and a huge wave in the lower part of his body. The giant was flanked by lions and he seemed to want to say something, but Gudea could not understand him. As the dream continued, the king thought he had woken up in the morning to see a woman with a glowing stylus consulting representations of the starry sky on a clay tablet. Then there came a warrior with a tablet of lapis lazuli on which he drew the plans for a building. The warrior handed him a mold for bricks and a new basket, while a purebred donkey pawed at the ground with its hooves.

When he actually did wake up the following day, Gudea was confused about the meaning of the dream. He decided to consult Nanshe, the Sumerian goddess of prophecy and dream interpretation. He performed a series of rituals on his way to the goddess's temple, and when he arrived there, he recounted his dream. He received the explanation that the giant represented the god Ninurta ordering the construction of a temple in honor of the god Eninnu. The woman represented the goddess Nidaba, who recommended that the temple be aligned astronomically according to the sacred stars. The warrior was the architect-god Nindub, with specific instructions for the plan of the building. The donkey was Gudea himself, impatient to raise up the architectural work that had been thus revealed. The details of the foundations and the construction materials were specified in subsequent dreams, which were incubated through propitiatory rituals. The temple was constructed in the city of Girsu, and it was under its ruins, which still exist in Iraq, that the Gudea cylinders were found.[53]

The construction of large buildings was for a long expanse of time a divine matter. More than fifteen centuries after King Gudea, in the sixth century B.C.E., clay cylinders engraved with cuneiform characters recount a dream of King Nabonidus, in which Marduk appeared to guide the king in the reconstruction of the important temple of Sin, the god of the moon. The reconstruction did in fact happen and the ruins are located in the city of Harran, in southern Turkey, which corresponds in the Bible to the city to which the patriarch Abraham traveled after leaving Ur.

Although not all the Hebrew prophets recognize the divinatory potential of dreams, oneiric accounts including the presence of the Hebrew god Yahweh play a key role in the stories of Jacob and Solomon.[54] The sacred books of Judaism, Christianity, and Islam also state that an Israelite called Joseph was made vizier of Egypt for having correctly interpreted two disturbing dreams of the pharaoh's.

In the first dream, the pharaoh was on the bank of the river Nile when seven fat cows appeared, followed by seven thin cows, which then ate the fat cows. In the second dream, the pharaoh saw seven fat ears of wheat springing up, followed by seven small, parched ears that swallowed up the larger ears. Joseph interpreted the dreams as having identical messages: seven years of abundance followed by seven years of scarcity. His advice to the pharaoh was that he construct storehouses for grain. It is believed that this story is related to a devastating drought that occurred in the Nile Valley about four thousand years ago, and to the measures taken by the Egyptian state to mitigate it.

Many centuries later, another pharaoh had a disturbing dream, which was interpreted by his wisest advisers as a grim prophecy: a newborn child would grow up to free the Israelites held in captivity and take the throne. In response to this dream the pharaoh had all recently born Hebrew baby boys drowned in the Nile, but a baby who had been placed in a basket and left in the river was found by the pharaoh's daughter and adopted under the name of Moses. When he became a man he fulfilled a part of the dream prophecy, leading his people's exodus out of Egypt toward Canaan.

Dreams also had a prominent role in the history of Persia, where the Zoroastrian magi were considered expert interpreters of their symbols and meanings. According to the Greek historian Herodotus in the fifth century B.C.E.,[55] Astyages, the king of the Medes, dreamed that his daughter Mandane urinated so much that she flooded all of Asia. The magi interpreted the dream as an ill omen that Mandane's son would supplant Astyages, and because of this, the king married her to a Persian man of middling social status. When Mandane gave birth to a healthy child called Cyrus, Astyages had a second dream in which a huge vine came out of his daughter's womb and spread until it had covered all of Asia. The dream was interpreted by the magi as a prediction that the grandson would rebel against his grandfather. Astyages gave the order for Cyrus to be executed, but the child survived, grew up, toppled the king from his throne, and built what up to that time was the greatest empire the world had ever seen.

Thirty years later, just before he was to die on the steppes of central Asia in a battle against the Massagetae, Cyrus dreamed that Darius, son of a Persian ruler, opened up enormous wings until all of Asia and Europe were in his shade. The prophecy made Cyrus afraid of a revolt, and he had Darius arrested, but soon after, Cyrus was killed in battle. As the dream had foretold, the young man ended up ascending to the throne and bringing the Persian Empire to its peak. In the decades that followed, Darius and his son Xerxes, Cyrus's grandson on his mother's side, took part in the legendary Persian invasions of Greece, with vast consequences for the cultural syncretism between East and West. Just as the arrogant Agamemnon in Homer had received a falsely prophetic dream from the gods, so Herodotus relates that recurrent dreams of world domination tempted Xerxes to embark on his doomed effort to conquer the Hellenic peoples. Xerxes told his dreams to Artabanus, the minister of war, who responded skeptically that he did not believe dreams could be oracles: they were merely mental pictures. Then Xerxes asked Artabanus to sleep in his bed to check whether he received the same dream. The next morning, emerging from the same astonishing dream, Artabanus had become a convert to the emperor's disastrous plans for war. After years of preparation, the Persians invaded Greece and burned Athens, but were ultimately beaten back.

A century and a half after Xerxes's defeat, the Macedonian king Alexander III reversed the direction of the invasion and in short order conquered Syria, Egypt, Assyria, Babylon, and the Persian Empire itself, reaching as far as India. Alexander the Great's frantic trajectory was infused by a range of highly symbolic premonitory dreams. During the bloody siege of Tyre, a strategic Phoenician port in what today

is Lebanon, Alexander had dreams about Hercules that anticipated
the herculean effort needed to take the city. After seven months of
violent clashes, Alexander had a second dream in which he was trying repeatedly to capture a satyr that was escaping from his assaults.
Finally, the satyr, which was dancing on his shield, was captured. Alexander's favorite seer interpreted this as meaning that "satyr"—*satyros* in
Greek—could be broken up into *sa* and *Tyros,* meaning "Tyre is yours."
Alexander redoubled his efforts and took the city.[56]

DREAMS OF HEALING

In antiquity, the social repercussions of dreams were also intimately
linked to their therapeutic use. The restoration of a person to health
was often attributed directly to dreams, as in the *Poem of the Righteous
Sufferer,* an Akkadian narrative of the misfortunes of the protagonist
Tabu-utul-Bel, who was stricken with countless deformities and illnesses. When this Babylonian Job was about to die, a series of dreams
revealed to him that the god Marduk was going to save him. In a trance,
he watched Marduk's battle with the demons, and in this way he was
finally cured.

It is no exaggeration to say that the main civilizations of the Mediterranean during the classical period developed their therapeutics
under the influence of dreams.[57] In Greece and later in Rome, marvelous temples were built to Asclepius, the god of medicine, which would
be visited by pilgrims in search of diagnosis, cure, and divine guidance.
Each sick person was subjected to a ritual of dream incubation (*egkoi-
mesis* in Greek, *incubatio* in Latin), with instructions to go to sleep in
the temple in order to foster the receiving of a divinatory vision.[58] On
waking, the sick person would relate their dream to one of the temple
priests, who would listen carefully for any signs that might suggest the
correct treatment for the illness. Sometimes, in special circumstances,
the treatment might be determined in the dream by Asclepius himself, son of Apollo, the god of truth, healing, and prophecy. The large
number of votive offerings in terra-cotta or clay representing parts of
the human body found in his temples attests to the frequent attribution of a cure to the god.[59] Very similar rituals persisted for centuries
in classical Egypt around the deity Serapis. Similar practices took place
in the Byzantine Empire during the Middle Ages, and—with specific
alterations—also in Islam.

DELUSIONAL ROME

In ancient Rome, the influence of dreams in social life reached hitherto unprecedented levels. Owing to the widespread belief in oneiric communication with the gods, accounts of dreams began to be used freely to legitimize or delegitimize political actors. The Roman biographer Suetonius made countless references to dreams to point to the divine origin of the first Roman emperor, Augustus. His mother Atia, a notable patrician, had gone to the temple of Apollo one night, and there she had fallen asleep in her litter. She was visited in her dreams by Apollo in the form of a serpent and became pregnant by him. During the period of gestation, Atia dreamed that her entrails "reached up to the stars and spread right across the earth and the sea," while her husband dreamed that the sun was born from his wife's belly.[60] Roman senators had also dreamed in the same year about the birth of a king who would save the Republic, and Julius Caesar had a dream that persuaded him to make his adopted son Augustus, then called Octavius, his political heir. Years later, at the Battle of Philippi, which led to the death of Julius Caesar's main assassins and opened the way for Augustus's rise, he managed to escape being ambushed in his own tent upon being warned by a premonition in a friend's dream. It is unsurprising, then, that as emperor Augustus was extremely susceptible to dreams. Inspired by one of these, he began the habit of disguising himself as a beggar to ask for alms once a year, and he even passed a law declaring that any person who had a premonitory dream should share it in the public square.

Perhaps the most dramatic case of dream accounts being used to overlap historical fatality with the deification of rulers relates to the first of the Caesars and his wife Calpurnia. In the days before he was assassinated, Julius Caesar received a dire premonition from a seer, warning him of mortal danger on the religious feast day of March 15. The prophecy spread across Rome by word of mouth until it reached the Senate, where there was growing discontent over the politician's ambitions. What especially alarmed the senators was the growing cult of personality around Julius Caesar, manifest in statues and effigies, commemorated in overly extravagant festivals and deified by a religious sect. Indeed, his own family claimed to be descendants of the Trojan Aeneas, son of Venus, and the rumor was spread that his spectacular military victories were an indication of the gods' favor. Julius Caesar's aggressive political and religious rise led the Senate to conspire to kill him.

On the night of March 14, 44 B.C.E., Caesar dreamed that he was being transported magically through the clouds, raised up into the heavens and received by Jupiter, who warmly clasped his hand. It didn't seem like a bad dream—on the contrary, it was a magnificent one. Beside him in bed, however, Calpurnia had a horrible nightmare. She dreamed that the front of her house was collapsing, that Julius had been stabbed and she was mourning his bloody body.[61] The following morning, she urged her husband not to go to the Senate. He contemplated giving up his plans, but he was talked out of the idea by one of the conspirators and by the favorable auguries of his soothsayers. When he arrived at the Theater of Pompey, Julius was surrounded by dozens of men—many of them senators—and butchered with twenty-three knives.

The funeral caused huge popular upset. There were executions, sacrifices, and the cremation of the body right in the Forum itself, with weapons, amulets, jewels, and clothing thrown into the fire by the crowd. The commotion got out of control, and the flames were so great that they almost destroyed the Forum. Faced with a popular reaction on that scale, Julius Caesar's killers were unable to prevent him from being the first Roman historical figure to be officially deified. He came to be represented as Divus Julius (Julius the Divine), and Augustus took on the title of Divi Filius (Divine Son). What psychological journey could our ancestors have taken that would have caused such a fantastical story to be considered normal from time immemorial until relatively recently? To answer this question, we will need to carry out a more detailed survey of our path from prehistory to history.

3

From Living Gods to Psychoanalysis

In the beginning was the longing: nostalgia. Firmly anchored in dreams, the funerary care of the dead, which began hundreds of thousands of years ago during the Paleolithic period, became more complex over the millennia up to the end of the Neolithic. Having begun with small piles of stones and shells, our ancestors arrived at the Bronze Age on the vast scale of the pyramids and ziggurats. The cult of the dead was one very successful mode of mental processing, which allowed human groups to reach a state where hundreds of thousands of people were living under the direct rule of a living god (in Egypt) or his direct representative (in Mesopotamia). The rulers were nourished and inspired by the knowledge accumulated from the whole dynasty, as far back as memory could go, with the preservation of the foundational myths about an original father or mother. Equipped with these beliefs and all the secular power that existed at that time, freed from any real physical work but encumbered with incredibly heavy spiritual, administrative, and military work, a typical pharaoh might have lived in a permanent trance, floating between sleep and waking, in a sustained delirium of real and fictitious power.

Over the millennia, this new form of consciousness grew not only in the pharaohs but probably in ever larger portions of society. People became more and more capable of realizing protracted flights of the imagination without moving a muscle, or better still, while carrying out movements with no connection to the scenes being imagined. It literally became possible to dream in certain parts of the brain and not in others, creating a mental space that was versatile and almost always available for simulating—while awake—the consequences of one's actions in the real world and of one's ideas in the symbolic one. This new type of consciousness, characterized as a "waking dream," was as useful in the planning of wars as it was in the production of food, making it a driving force for new areas of knowledge, such as the storage and trade of grain and the engineering of buildings and of means of transport, along with astronomy, mathematics, and writing itself. From the primitive horde to the crowning of the pharaohs on the banks of the Nile, the capacity for constructing fictions, disseminating them through a large group, and then implementing them in real life leveraged the expansion of pyramid-structured societies.

The invention of writing extended the limits of central power in time and space, as symbolized by the rapidly spreading custom of erecting enormous stone steles carved with divine commands and laws, communicated by the rulers to their subjects. The use of the steles allowed for the demarcation of much more extensive cultural territories, and this geographical extension encouraged the cult of multiple divine beings. The rise of literature bears dynamic testimony to this process, as gods and spirits of dead people appear with some frequency in the most ancient texts.

Paradoxically, however, writing was also the beginning of the end for the worship of gods and ancestors, and the start of the decline of dreams. It was no longer necessary to go into a trance to hear the hallucinatory voices of the gods, propitiated by sleep, statues, prayers, fasting, sacrifices, and substances. It was now possible to read the words of the gods and their direct representatives—or, as stated in the oldest records, it was now possible to hear divine words—just by looking at these strange handmade marks. Carved into stone so that they might last millennia, the words of authority could be heard with complete accuracy at multiple different locations around the empire. The knowledge that had been accumulated across countless generations and stored up in oral form through divine commands, experienced inside the brain as a succession of sounds, became progressively more obsolete with the advance of writing. When our ancestors invented ways of recording the auditory commands of the gods in stone or clay, they were creating the necessary condition for the increasing irrelevance of these commands, until for most of the population their potency finally disappeared entirely.

In Egyptian and Mesopotamian texts, the story of the death of the gods happens at the very start of the written record, but the complaint that the gods had fallen silent only became prevalent around 1,200 to 800 B.C.E. It was a period of huge social, economic, and environmental crises, with a population explosion, migrations, wars, famines, droughts, plagues, and natural disasters,[1] which led to the collapse of cities and empires such as Knossos (c. 1250 B.C.E.), Mycenae (c. 1200 B.C.E.), Ugarit (c. 1190 B.C.E.), Megiddo (c. 1150 B.C.E.), Egypt (c. 1100 B.C.E.), Assyria (c. 1055 B.C.E.), Babylon (c. 1026 B.C.E.), and Troy (c. 950 B.C.E.). In almost all of these cases, the cities and empires rose back up again or were reorganized with new gods or resuscitated ones. However, the growth in the population of those who were imbued with the belief in the revived gods—that is to say, with similar cultural software—created totally new contradictions.

In Egypt, a new consciousness began to expand among the lower classes, who also wanted their own tombs and especially eternal life. This catalyzed a social conflict, as these lower-class Egyptians angrily expressed their perception of inequality in the treatment of the dead. The literature of the period documents the despair of those people who were terrified at the imminent threat of an ending without a safe-conduct to the other world, while the promise of eternal life was a privilege only for those who could pay for the elaborate confection of the protection and guidance spells contained in the *Book of the Dead*.

The geographical spread of the words of the gods served to trivialize them. It was now no longer necessary to hallucinate divine voices in order to have access to their knowledge, since this was now externalized in persistent, solid objects, which were capable of spreading words from one mind to another, with no dreaming or trance or madness required. Moreover, the occurrence of catastrophic events that were hitherto unprecedented, as well as the fragility of the supply chains that sustained such large societies, made divine wisdom obsolete, malfunctioning, old, incapable of finding solutions to the new problems. This period coincides with the great civilizational collapse that ended the Bronze Age, when a number of central powers dissolved. It was the end of Troy, of Mycenae, and of the Minoan civilization on Crete. A time of droughts, floods, tidal waves, shortages, migrations, and wars. In this context of chaos and unpredictability, the gods no longer had answers, and they fell silent. Men now needed to solve their problems alone.

After a few centuries of rough transition, a marvelous cultural change took place approximately between 800 and 200 B.C.E., in what the twentieth-century German philosopher and psychiatrist Karl Jaspers called the Axial Age. This period bore witness to the flourishing of civilization in a number of places in Afro-Eurasia, including Athens, Rome, Babylon, and the Persian, Macedonian, and Mauryan Empires. Hundreds of ancient literature's key texts date from this period, such as *The Iliad, The Odyssey,* Plato's *Republic,* the book of Genesis, *The Avesta,* and *The Mahabharata.* Multicultural development and integration accelerated thanks to the consolidation of alphabetic writing, new literary traditions, and the first institutions for higher education, such as Plato's Academy in the fourth century B.C.E. and the Library of Alexandria in the third. The world was belonging less and less to gods and more and more to men.

The Iliad and *The Odyssey* exemplified this shift very clearly. Achilles, who has no plans for the future and acts mostly upon orders from the gods, is a typical example of the mind of long ago, while Odysseus with his new mentality attains his objectives by using stratagems, persistently imagined while he is awake. A new introspective mentality that can still hear the voices of the gods but starts to construct a powerful internal dialogue, practical and utilitarian, for imagining the future and thereby allowing him to shape it. A human being like ourselves, able to daydream a great variety of plans that are subtle but highly effective, whether for building a huge wooden horse and voyaging to return to the arms of a long-lost beloved, or leaving work early and surprising a girlfriend with a candlelit dinner. A human being who hardly ever listens to the gods anymore, but is constantly engaged in conversation with himself.

While it might seem fantastical, the theory that human introspection is a relatively new phenomenon earned some corroboration from a semantic analysis of Judeo-Christian and Greco-Roman texts, carried out by a team of Argentine researchers at the Thomas J. Watson Research Center at IBM. If the psychologist Julian Jaynes is right, the transition to the conscious self, the habit of mind in which a human is conditioned to listen to himself and not to the gods, is sufficiently recent that it should appear in the historical record, that is, in those texts produced by humanity since the beginning of writing. The self that is introspective, reflective, that imagines itself, would be only about three thousand years old.

To test this hypothesis, physicists Guillermo Cecchi and Mariano Sigman teamed up with computer scientists Carlos Diuk and Diego Slezak to study ancient texts using a mathematical technique that allowed them to measure the distances between words in an objective fashion, quantitatively and automatically. Their method is based on the observation that, when a very large number of different texts are examined, pairs of words that are semantically close (*cat* and *rat*, *mother* and *daughter*, *love* and *passion*) tend to occur in the same texts, which is not something observed with words that are distant from one another (*cat* and *helicopter*, *rice* and *poetry*, *flower* and *zenith*). Using this method, the semantic distance between any pair of words corresponds to a number, making it possible to calculate the average distance between any particular keyword and all the words contained within a text. In order to test their hypothesis, the researchers chose for their specific keyword the word *introspection*—a term that does not even appear in ancient books and for that very reason was ideal for probing the diffuse, implicit presence of the word in each work. In this way, for each text analyzed, the researchers measured the distance from each word in the text to the word *introspection*, and calculated an average of these distances taking account of every word in the text.

The results[2] showed that the concept of introspection became increasingly prevalent in both literary traditions, showing accelerated growth in the period of each civilization's cultural expansion (fig. 4). While it is not possible to prove that introspective behavior similarly became more prevalent, the results allow us to imagine the people of Homer's day (eighth century B.C.E.) as much less introspective than, for example, those in Julius Caesar's (first century B.C.E.). As we will see later, other recent research on the structure of ancient texts also supports the idea that human mentality has radically changed in the past three thousand years.

THE GLORY AND DECLINE OF DREAMS

The slow but inexorable loss of the importance of dreams is one of the clearest examples of this change. The intermittent decline of the belief in the efficacy of dreams ran through the first millennia before and after Christ. On the one hand, in the fifth century B.C.E., Gautama Buddha gave great existential breadth to the oneiric problem, in asserting that all of life is a dream. The idea that reality itself is a dream has

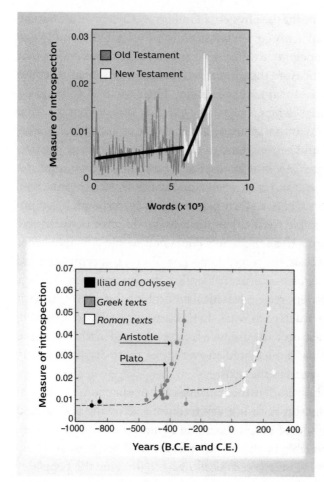

Figure 4. A semantic measurement of introspection increases with the passing of time in Judeo-Christian and Greco-Roman cultural records.

very ancient roots in India. One traditional representation of the Hindu god Vishnu shows him reclining on the snake Shesha while he "dreams the universe into reality."

But the Buddha introduced the symbolic interpretation of dreams into his culture, too. When he was about to abandon the privileged Kshatriya life to follow the strictest asceticism, the young prince Siddhartha Gautama offered totally nonliteral interpretations of a premonitory nightmare experienced by his wife, Gopa.

According to tradition,[3] the prince who would one day be Buddha was avoiding his wife, who was suffering terrible distress. When she finally managed to get to sleep, she dreamed that the mountains were trembling under a wild gale that was tearing trees out of the ground. On the horizon, the stars had poured down from the sky. Gopa saw herself naked, stripped of her garments, adornments, and crown. Her hair had been cut, her marriage bed was broken, and the prince's clothes, covered in precious gems, were scattered all over the floor. Meteorites were falling on a dark city.

The terrified Gopa woke her husband: "My lord, my lord," she shouted, "what will happen? I have had a terrible dream! My eyes are full of tears, and my heart is full of fear." "Tell me your dream," the prince replied. Gopa told him everything that she had seen in her sleep. The prince smiled. "Rejoice, Gopa," he said,

Rejoice. You saw the earth shake? Then one day the Gods themselves shall bow before you. You saw the moon and the sun fall from the sky? Then you shall soon defeat evil, and you shall receive infinite praise. You saw the trees uprooted? Then you shall find a way out of the forest of desire. Your hair was cut short? Then you shall free yourself from the net of passions that holds you captive. My robes and jewels were scattered about? Then I am on the road to deliverance. Meteors were speeding across the sky over a darkened city? Then to the ignorant world, to the world that is blind, I shall bring the light of wisdom, and those who have faith in my words will know joy and felicity. Be happy, O Gopa, drive away your melancholy; you will soon be singularly honored. Sleep, Gopa, sleep; you have dreamed a lovely dream.[4]

Days later, Siddhartha slipped quietly out of the house during the night.

For six years, exposed to rough weather on the steep banks of a river, the young man led a life of meditation, isolation, and fasting among the wild animals. He attracted disciples, but when he decided to renounce the mortification of the flesh they abandoned him. Siddhartha then had the series of dreams that announced his enlightenment:

Night came on. He fell asleep, and he had five dreams.

First, he saw himself lying in a large bed that was the whole earth; under his head, there was a cushion which was the Himalaya; his right hand rested on the western sea, his left hand on the eastern sea, and his feet touched the southern sea.

Then he saw a reed coming out of his navel, and the reed grew so fast that it soon reached the sky.

Then he saw worms crawling up his legs and completely covering them.

Then he saw birds flying toward him from all points of the horizon, and when the birds were near his head, they seemed to be of gold.

Finally, he saw himself at the foot of a mountain of filth and excrement; he climbed the mountain; he reached the summit; he descended, and neither the filth nor the excrement had defiled him.

He awoke, and from these dreams he knew that the day had come when, having attained supreme knowledge, he would become a Buddha.[5]

It's worth pointing out that the interpretation Buddhism gives for his dreams is symbolic, quite unlike the literal interpretations that were typical of ancient Brahmanism. The dream about the mountain of excrement, for example, would demand complicated purification rituals if it were interpreted literally. Seen through the Buddhist lens, the dream represents the clothing, objects, and especially relationships left behind by Siddhartha, the stripping away of the desires and expectations that constrain the spiritual experience.

In China, the cradle of a deep-rooted tradition of oneiric divination,[6] the philosopher known as Master Zhuang presented the dream problem in a new way in the fourth century B.C.E.:

Once upon a time, I, Zhuang Zhou, dreamt I was a butterfly, fluttering hither and thither, a veritable butterfly, enjoying itself to the full of its bent, and not knowing it was Zhuang Zhou. Suddenly I awoke, and came to myself, the veritable Zhuang Zhou. Now I do not know whether it was then I dreamt I was a butterfly, or whether I am now a butterfly dreaming I am a man.[7]

This fine philosophical doubt never occurred to Plato, who concluded that there was no place for dreams and madness in the management of the State.[8] To the great Athenian philosopher, the truth could come only from the logical exercising of thought, with precedence given to the deduction of perfect forms of reality, which can go beyond the illusory veil of appearances. Platonic truth is the product of rigorous thought in the waking hours, not of dream hallucinations induced by sleep, sickness, or intoxication.

Nor did the millennia-long tradition of oneiric magic prevent Aristotle from recognizing the biological nature of dreaming, which is both marvelous and prosaic.[9] Plato's main disciple pointed out the value of observable facts over theory, that is, the supremacy of induction over deduction. Like other philosophers of antiquity, Aristotle attributed the determining factor in the explanation of dream content to waking experiences, identifying what Freud, more than two thousand years later, would call day residue. The dream, then, would be an inexact copy of reality, a memory of past events, a vivid involuntary recollection.

Between the Greek idea of logos and the Enlightenment belief in reason, there was a long transitional period in which the historic influence of dreams varied hugely, with both highs and lows. Dreams played a key role in the genesis and development of Christianity, as a powerful instrument for the revelation of divine will. In his gospel, Matthew states that on several occasions, God used dreams to protect the infant Jesus, warning the three Magi through a dream not to return to Herod, but to go back to their own countries, and sending angels to offer guidance to Joseph in his sleep.[10] It was a dream angel who persuaded Joseph to accept Mary as his wife in spite of her having become pregnant before marriage, arguing that the conception had been performed by the Holy Spirit—zoomorphized into a dove—and instructed him to adopt the son, as the boy would one day redeem the people of their sins. Dreams about the Lord's angels also guided Joseph's decisions to flee to Egypt, then to return to Israel, and finally to head for Galilee in an effort to protect his newborn from the wrath of King Herod, who had given the order for male infants to be killed.

While there are no records of dreams by Jesus himself, the gospels do present the report of a dream that could have changed the course of his life, and, therefore, of history. Accused of being "king of the Jews," Jesus was brought before governor Pontius Pilate. According to Matthew, while Pilate was sitting in judgment over Jesus, his wife sent him a message: "Have thou nothing to do with that just man: for I have suffered many things this day in a dream because of him." But the crowd decided to condemn Jesus to be crucified, and Pilate washed his hands of the matter.[11]

According to the Acts of the Apostles, about twenty years later, when Paul and his companions were traveling around Asia Minor preaching Christianity, his journey was profoundly altered by a dream. While he slept, Paul visualized a man from Macedonia who asked for his help. When he awoke, he concluded that this was a call from God, and he set off for Macedonia, carrying out a successful mission of evangelism that spread the Christian faith well beyond the Jewish population.[12]

In Islam, the interpretation of dreams has always enjoyed a good reputation. The prophet Muhammad himself recognized it as a spiritual exercise capable of allowing real communication with Allah. In one famous story, Muhammad dreamed that he was following first a flock of black sheep, then a flock of white sheep. After a while, the two flocks had become completely mixed together, and it was impossible to separate them. The black sheep were interpreted as symbolizing the Arabs and the white sheep the non-Arabs, leading to the conclusion—clearly political in tenor—that Islam would spread across the world beyond ethnic differences.

Dreams appear in Islamic history in contexts of prophecy and divination; they are frequently used to legitimize rulers, and may be invoked to reach a solution to specific problems (istikhára). The importance of dreams reaches its apex in Sufism, an introspective strand of Islam whose practitioners often seek to attain a mystical trance and pursue dreams of the prophet or other spiritual counselors in order to guide waking behavior. The twelfth-century scholar Najm al-Din Kubra founded a Sufi order based on his visionary experience of dreaming, and wrote important treatises on the subject. Another Sufi thinker, also a founder of one of its orders, Shah Ni'matullah Wali from Kerman, is revered by Sunnis as a saint. He composed poems inspired by dreams that have been accounted prophetic at various historical moments, from the rise of the Turkic conqueror Tamerlane in the fourteenth century to the jihad of Syed Ahmad in the nineteenth, the dissolution of the Ottoman Caliphate in 1924, and the religious conflicts in Pakistan in 2010.[13]

If dreams remain important in Islam to this day, in Christianity their importance was gradually eroded over the course of the Middle Ages, with the development of ecclesiastical Christianity, which began to see traces of paganism in dream divination. These were centuries of transition, which brought with them a recognition of the fundamental opposition between *somnium coeleste* (dreams of the future as revealed by God) and *somnium naturale* or *animale* (dreams with physiological or psychological causes). The theologian and philosopher St. Augustine, born near the Mediterranean in 354 in a region that today corresponds to northeastern Algeria, had a profound influence on the adoption of Neoplatonism by the Church. His vast work touched upon many psychological subjects, such as the origin of memories, dreams, desires, suffering, and guilt. One of the subjects that concerned him was erotic dreams, which he was unable to avoid despite—or more probably because of—his celibacy and the repression of sexual thoughts in his waking life.

Addressing God, St. Augustine explains his amazement at the autonomous nature of dreams:

> During this time of sleep surely it is not my true self, Lord my God? Yet how great a difference between myself at the time when I am asleep and myself when I return to the waking state . . . Surely reason does not shut down as the eyes close. It can hardly fall asleep with the bodily senses. For if that were so, how could it come about that often in sleep we resist and, mindful of our avowed commitment and adhering to it with strict chastity, we give no assent to such seductions? Yet there is a difference so great that, when it happens otherwise than we would wish, when we wake up we return to peace in our conscience. From the wide gulf between the occurrences and our will, we discover that we did not actively do what, to our regret, has somehow been done in us.[14]

St. Augustine's solution to the problem of oneiric eroticism was to consider dreams not as a human action subject to one's willpower, but as an involuntary event for which one has no responsibility or guilt. Dreaming of a sin is not, therefore, a sin.

MONKS AND DEMONS

Up until our great-grandparents' generation, most people would go to bed very soon after sunset. Since time immemorial the night had always been something to be feared, especially in the absence of moonlight, and all the more so during wintertime, when the darkness seemed unending. In antiquity and in the Middle Ages, the night belonged to drunkards, thieves, highwaymen, murderers, and occasional invading troops—in addition, of course, to wild animals. For this reason, at nighttime people would congregate around a fire and behind walls, inside houses, farms, castles, inns, taverns, and brothels. Over the course of the Middle Ages, a belief began to spread that demons, called incubi and succubi, could invade people's dreams in order to have sexual relations with them. Faced with the dangers of nighttime and the fantastical nature of dreaming, it is not surprising that the period of darkness was marked by terrifying fantasies and by the protective use of meditations, prayers, and charms.[15]

It was common for adults to split the night into two parts, called first and second sleep, with a brief waking interval at around midnight, which was used for praying, dining, spinning, talking, or lovemaking. But the nighttime habits of Christian monks were rigorously controlled: the first sleep ended at two a.m. for morning prayers. This meant that Benedictine monks were deprived of REM sleep, a dream-rich phase prevalent in the second half of the night. Curiously, the total deprivation of REM sleep provokes a vigorous compensatory sleep rebound, with a subsequent increase in sleeping time and an intrusion of intense dreams. The Order of St. Benedict, the most ancient Catholic monastic order, forbade the second sleep but tolerated an evening nap. In the eleventh century, the monk Raoul Glaber, a superstitious and sleepy French Benedictine, left a record of being assailed by a demon whose temptation consisted of whispering in his ear that he should ignore the bell and surrender to the "sweet repose" of the second sleep. But if it was common to fear the activities of nighttime demons capable of sexual seduction, dreams were also expected to reveal divine intentions through the apparition of angels and saints.

From the twelfth century onward, France began to see the operation of Catholic institutions dedicated to persecuting heresies. These institutions, which were almost always linked to the Dominican order, came to be known as the Inquisition. In the centuries that followed, the persecutory frenzy spread across Germany, Spain, and Portugal, before proceeding to the colonies in the Americas, Asia, and Africa. Thousands of people were prosecuted by the Inquisition, accused of witchcraft, tortured, and executed in the name of an infinitely kind god. The twelfth century was also the time when individual confession became institutionalized within the Church, making the priest the knower of the private secrets of entire communities.[16] Placed simultaneously in the roles of absolver and accuser, the Catholic priest needed more than ever to face the dilemma of dream interpretation. Should heretical dreams be considered a sin? Could someone be condemned and punished while awake for thoughts they experienced while asleep?

Faced with this terrible question, St. Thomas Aquinas, the great thirteenth-century defender of reason within the church, the person mostly responsible for the revival of Aristotelian induction after nearly a thousand years of Neoplatonism, answered with an emphatic "no." After all, not all dreams are "true." To underline the importance of dreams in his work, the word appears seventy-three times in the *Summa Theologiae,* one of his most influential texts. In it, the Lazio-born theologian states that

> it is unreasonable to deny the common experiences of men. Now it is the experience of all that dreams are significative of the future. Therefore it is useless to deny the efficacy of dreams for the purpose of divination, and it is lawful to listen to them . . . As stated above, divination is superstitious and unlawful when it is based on a false opinion. Wherefore we must consider what is true in the matter of foreknowing the future from dreams. Now dreams are sometimes the cause of future occurrences; for instance, when a person's mind becomes anxious through what it has seen in a dream and is thereby led to do something or avoid something: while sometimes dreams are signs of future happenings, in so far as they are referable to some common cause of both dreams and future occurrences . . . "If there be among you a prophet of the Lord, I will appear to him in a vision, or I will speak to him in a dream." Sometimes, however, it is due to the action of the demons that certain images appear to persons in their sleep, and by this means they, at times, reveal certain future things.[17]

In this passage, Thomas Aquinas added a new dimension to the problem of dream interpretation, when he stated that the predictive accuracy of a dream is not evidence of its divine origin. The Church began to cultivate an increased skepticism toward the possibility of allowing oneself to be guided by dreams, though their divinatory nature did continue to be recognized. *The Mirror of True Penitence,* a collection of edifying sermons on virtue and sin written in the fourteenth century by the Italian Dominican friar Jacopo Passavanti, concludes with a "Treatise on Dreams" stating that "dreams that are made round about dawn . . . are the truest dreams of all, whose meanings can be best interpreted."[18] This view is echoed by Dante Alighieri in the *Divine Comedy* when he declares that the prophetic dreams are the morning ones.[19]

THE PROTEST OF THE EAGLE

The German theologian Martin Luther, the great reformer of Christianity, also had an ambivalent relationship with dreams. At the start of his monastic career, Luther discovered the sermons of Jan Hus, a religious leader from Bohemia burned as a heretic a hundred years earlier for preaching the rejection of the Catholic indulgences. The young monk was struck by the story of the reformer's death: when the executioner was approaching Hus to light up the flames, he said, "Now we're going to cook the goose"—*Hus* means goose in the Bohemian dialect. The condemned man then uttered a mysterious prophecy: "Yes," he replied, "but there will come an eagle in a hundred years that you will not reach."[20]

Hus became a reference point for Luther, who shared his dislike of the clerical system of selling indulgences. When Luther nailed his theses, with their strong criticism of ecclesiastical corruption, to the door of the church of Wittenberg Castle on October 31, 1517, he knew that he was walking a dangerous path. After all, many people had been burned at the stake before Luther, and many would burn after him. Pope Leo X ordered the retraction of the theses, but Luther's reply left no room for doubt as to his rebelliousness: he set fire to the papal bull. The German was then excommunicated by the pope and condemned by the Holy Roman emperor, Charles V. Prince Frederick III, elector of Saxony, who was holding Luther in custody, was supposed to hand him over to the punitive fury of his enemies. Against all likelihood, however, Frederick protected Luther, the theologian's ideas survived, and the Protestant Reformation spread across Europe. The amazing story of how this happened involves an important dream.

According to chroniclers from the period, the night before Luther was to nail his theses to the church door, Prince Frederick had a revelation in a dream. In Frederick's own words:

I again fell asleep, and then dreamed that Almighty God sent me a monk, who was the true son of the Apostle Paul. All the saints accompanied him by order of God, in order to bear testimony before me, and to declare that he did not come to contrive any plot, but that all that he did was according to the will of God. They asked me to have the goodness graciously to permit him to write something on the door of the church of the Castle of Wittemberg. This I granted through my chancellor. Thereupon the monk went to the church, and began to write in such large characters that I could read the writing at Schweinitz. The pen which he used was so large that its end reached as far as Rome, where it pierced the ears of a lion that was crouching there, and caused the triple crown upon the head of the Pope to shake. All the cardinals and princes, running hastily up, tried to prevent it from falling. You and I, brother, wished also to assist, and I stretched out my arm;—but at this moment I awoke, with my arm in the air, quite amazed, and very much enraged at the monk for not managing his pen better. I re-collected myself a little; it was only a dream.

I was still half asleep, and once more closed my eyes. The dream returned. The lion, still annoyed by the pen, began to roar with all his might, so much so that the whole city of Rome, and all the States of the Holy Empire, ran to see what the matter was. The Pope requested them to oppose this monk, and applied particularly to me, on account of his being in my country. I again awoke, repeated the Lord's prayer, entreated God to preserve his Holiness, and once more fell asleep.

Then I dreamed that all the princes of the Empire, and we among them, hastened to Rome, and strove, one after another, to break the pen; but the more we tried the stiffer it became, sounding as if it had been made of iron. We at length desisted. I then asked the monk (for I was sometimes at Rome, and sometimes at Wittemberg) where he got his pen, and why it was so strong. "The pen," replied he, "belonged to an old goose of Bohemia, a hundred years old. I got it from one of my old schoolmasters. As to its strength, it is owing to the impossibility of depriving it of its pith or marrow; and I am quite astonished at it myself." Suddenly I heard a loud noise—a large number of other pens had sprung out of the long pen of the monk. I awoke a third time: it was daylight.[21]

It is possible to believe that this dream had a profound influence on Frederick, who defended Luther bravely against the pope and the emperor. We can also imagine that the story was made to order so as to justify Frederick's support for Luther, for strictly political reasons. Either way, Luther himself remained highly skeptical as to the truthfulness of dreams, reserving his faith for a very limited minority of visions that were truly considered divine.

THE IRRELEVANT DREAM

With the formation of Europe's nation-states and the initial stages of mercantilism, the interpretation of dreams moved out of the public sphere for good. By the sixteenth century, Christianity had already begun to consider oneiric revelation as a source of blasphemy and damnation at worst, or irrelevance at best. As exemplified in the trials that led to the imprisonment of the theologian Giordano Bruno and his execution in 1600, dream visions began to be seen as a sign of heretical influences. The discrediting of dreams deepened in the eighteenth century, with the rationalism that is at the root of both science and capitalism. It was neither rational nor commercially justifiable to have recourse to dreams when making important trade decisions, and any kind of augurs lost their importance in the courts of kings and queens. It is no accident that many currents of Protestantism—Calvinism especially, which was ever so pragmatic in its pursuit of sacred prosperity—moved away from dreams quite significantly. Within a few centuries, a profound transformation took place in the understanding of what dreams are, or what they mean.

From transcendent inspiration to visceral commotion, dreams fell off their pedestal and came to be seen as mere reflections of the residual feelings of a passively sleeping body; they were attributed to a lack of stimulation and a trivial mirroring of the current corporal state, whether it be hunger, thirst, or any other need of the moment. The scatology of the sixteenth-century French writer François Rabelais, who interpreted bad dreams as the inevitable product of indigestion, and the skeptical objectivity of the Enlightenment philosopher and mathematician René Descartes trivialized the predictive powers of dreams to the same degree. In spite of important dream revelations that the latter reported having experienced himself in his youth—powerful dreams on the banks of the Danube, which according to him inspired analytic geometry and his method of systematic doubt—in his more mature years Descartes would define a dream as a mere state of illusion derived from the impressions of the waking day.[22]

On the other hand, the same years saw a great increase in popular treatises of dream explanation, centered on a predetermined interpretation of their constituent parts. The invention of printing provided the conditions for the commercialization of a product that to this day can be found on any newsstand: a manual for dream interpretation based on fixed keys for the decoding of symbols—a far-distant echo of the Assyrian *Zaqiqu*.

MESSAGES FROM THE UNCONSCIOUS

It was in this context of the dream phenomenon being relegated to cheap serials that Sigmund Freud developed his theory, in which dreaming was born as an object of rational study, a biological phenomenon of the greatest significance for the understanding of the human mind. Psychoanalysis marks an open-eyed return to the oneiric practices of antiquity, considering dream interpretation as an essential tool for exploring symbolic networks and their Gordian knots.

Freud's invaluable contribution in repositioning dreams at the center of human life began from the observation that they are very revealing of the structure of the dreamer's mind. A particularly rich source of symbolic relationships, they make it possible to understand mental life by a process of attentive listening aimed at mapping word associations with therapeutic relevance. Published in 1900, *The Interpretation of Dreams* was the foundation of psychoanalysis with its focus on nighttime experience as a way of deciphering memories from waking life.[23]

In this book, Freud stated that dreams are "the royal road to the unconscious." He also posited that dreams contain day residues from the waking hours, which go some way to explaining their content. The deeper motivations, however, are generated by repressed desires—that is, not by things that have happened, but by things that are desired and that have not happened yet or may never happen. In dissecting the day residue that is present in a dream, Freud destroyed any possibility of accepting fixed keys for dream interpretation, instead asserting that dream interpretation is only possible when it is carried out by the dreamer himself, or by somebody very well informed about his most intimate mental context. At the same time, Freud, who was Jewish by birth, rejected the trivialization of the dream, and recognized and restored its profound significance to the dreamer. Like the Talmud, the central text of Judaism, psychoanalysis understands that "a dream that has not been interpreted is like a letter that has not been read."[24] A letter made up of images of the past and guided by the desires of the present, whose careful reading might even be able to change the future.

By taking a step back to gain perspective, this chapter has presented the historical context for establishing dreams as a cornerstone of human consciousness. Now, to take the problem on a step further, the next section requires that we understand how we dream today.

4

Unique Dreams and Typical Dreams

It was not until the invention and spread of the electric light, at the end of the nineteenth century, that it became normal to occupy the first hours of darkness with typically diurnal activities. It is estimated that in the United States, the average length of time that people spend asleep fell from nine hours in 1910 to seven and a half hours just sixty-five years later.[1] Artificial light has produced negative effects that overlap with the effects produced by the light-dark cycle, causing a misalignment of the circadian rhythms, that is, the biological rhythms synchronized to the rotation of the Earth on its axis over a period of twenty-three hours, fifty-six minutes, and four seconds. The ever greedier occupying of the nighttime by waking life has made it harder to separate the night's sleep into two distinct sessions, producing the single period of sleep, six to eight hours long, that is common today in most parts of the planet. It is in this private, reserved, compact mental space that we develop our capacity to dream.

Contemporary dreams, in general, evoke and interweave fragments of experiences, from simple images of things or people right up to quite vivid, specific scenes, experienced as if they were lived situations. They might have a single subject, or be made up of a number of thematic units that are connected to one another more or less surprisingly. Traumatic dreams tend not to be metaphorical, instead reverberating singular memories in a realistic, intrusive fashion—while everyday dreams that don't feature any terrible frights are a hodgepodge of minor events all mixed together.

The first person to measure these properties of dreams in any systematic way was the U.S. psychologist Calvin S. Hall, who over the course of his life collected more than fifty thousand dream reports. Hall completed his doctorate in psychology at the University of California–Berkeley in 1933, studying under Edward Tolman, a visionary scientist who posited some degree of intentionality to explain the complex cognitive skills observed in rats ("purposive behaviorism"). After a brilliant start to his career studying behavioral genetics in rodents, Hall became head of the psychology department at Case Western Reserve University, and decided to steer his research toward the contents of human dreams. In a search for thematic patterns, Hall developed a system of dream codification for recording and quantifying scenarios, characters, objects, interactions, frustrations, and emotions, among many other factors. Hall's work was expanded, and is continued to this day, at the University of California–Santa Cruz by the psychologist William Domhoff, who completed his doctorate under Hall in 1962. Domhoff and his colleague Adam Schneider made an incalculable contribution to the science of dreams when they set up the DreamBank, a publicly accessible database containing more than twenty thousand dream reports (www.dreambank.net).[2]

In recent decades, a number of other researchers joined the effort to collect dream narratives on a large scale, such as the U.S. neuroscientist Patrick McNamara, from Boston University, who oversees the Dreamboard platform, with its more than 250,000 dream narratives (www.dreamboard.com). The main conclusion of the research carried out using large data sets is that people's dreams are more similar to one another than they are different, even if there is considerable cultural variation.[3] There is often a thematic continuity between waking and sleeping, which corroborates the Freudian concept of day residue. But dreams are also a privileged space for the simulating of counterfactual situations,[4] that is, things that did not happen but could have happened—and may still.[5]

When the dreamer's context is a comfortable one, marked not by a big problem but by a myriad of small day-to-day snags, dreams appear to make no sense, and they are hard to interpret. They become patchwork quilts of life, with each square having its own pattern and internal logic but with no overall cohesion. However, when the context is very challenging, such as in a situation of serious illness or violent dispute, dreams may offer a solid and vibrant expression both of the lived experience and of the essential directives for acting against the imminent threat. For this reason, it is crucial that they be interpreted correctly.

As stated above, traumatic dreams tend to be monothematic and not metaphorical, reverberating singular memories in a reliably scary way. On the other hand, dreams of great significance use powerful metaphors to mark the transitional phases of childhood, adolescence, adulthood, and old age, as well as important changes in social status, whether downward or upward. These "big dreams" are characterized by containing an extensive series of representations, threaded together with such poignance that all the symbols seem to slot together perfectly.

I heard one lovely example of the "big dream" in a contemporary context from the mother of my sons, on the first night of her labor with Sergio, our second. After the regular contractions had started, Natália lay down in a hammock, fell asleep, and dreamed of her maternal grandmother, whom she never got to meet. It was a vivid dream, highly charged with emotion in spite of—or because of—one extraordinary detail: her grandmother was embodied in the hammock that gently rocked her, or rather, her grandma *was* the hammock. And while being this hammock she smoothed down her granddaughter's hair and told her affectionately, with an old woman's gentle voice, that she would have liked to have met her, that grandmother and granddaughter had similar temperaments, that she was going to go on being a calm mother, a good mother, and that everything was going to work out, because she had already given birth in this incarnation and in others and everything had always gone all right. Natália woke from her dream sobbing with happiness at the encounter, feeling blessed and filled with courage to face the future optimistically—which came in very handy, as the labor lasted an extraordinary forty-three hours of ever stronger and ever more frequent contractions that nonetheless did not cause any dilation, until we decided to go for a caesarean intervention.

The epic dimension of this particular adventure echoes mythological dreams. It recalls, for example, the dream of the Viking queen Ragnhild, which I described in Chapter 2. The thorn that catches Ragnhild's cloak grows until it has become a vast tree with roots that bury themselves deep in the ground and branches that rise till they have covered all Scandinavia and beyond, symbolizing the dynasty that her son and his descendants would found in the fertile kingdom of Norway.

This interpretation must have suited the interests of Ragnhild's family like a glove, strengthening their claim to power and acting as a lever for the prophecy itself. For this reason, the subjective experience of the dream—if it really happened—must have been highly emotional for Ragnhild herself, overtaken as she was in her everyday garden by powerful, arcane symbols until she had a panoramic vision of the geopolitical consequences of her own destiny.

It is worth comparing the account of Ragnhild's dream with Princess Mandane's dream in the early years of the Persian Empire, in which a vine covered with fruits sprang out of her genitals and spread until it had covered Asia. Almost 1,500 years apart and with half the world between them, Mandane and Ragnhild experienced very similar dream symbologies, relating fertile trees of planetary dimensions with the noble dynasties that they began. As we will see, the political use of dream narratives runs through the whole historical record and raises doubts as to their trustworthiness.

THE DIVERSITY OF TYPICAL DREAMS

Mythology and history are crammed with dream narratives of alliances and conflicts, triumph and impotence, joys and disappointments, successes and failures. Might the plots of our dreams today be comparable to these incredible examples from the past? In order to understand the logic of today's dreams, we must contemplate their great diversity, their cultural specificities, and the intimate connection to the contexts in which they occur. In Africa, for example, there is a recognizable phenomenon of dream triangulation, in which one person receives messages in their dreams that are intended for other people.[6] Beyond any cultural differences, it is particularly necessary to identify the dreamer's own anxieties and expectations, which look ahead to imminent reality and can simulate possible solutions or alternatives for problems in the present. The series of dreams that will be presented below, collected either in the DreamBank or by me from relatives and friends, illustrates a number of these possibilities. Any significant changes to lifestyle tend to set off dreams that are easy to interpret, such as the next example.

A twenty-eight-year-old woman, after spending some months on a training course outside her city in which she had quite a lot of freedom in ideas and behavior, was getting ready to return to her regular job, which involved a lot of work, discipline, and sameness. A few days before resuming her former role, she dreamed that she had gone back to study at middle school, where she had to wear a uniform and sit through tedious classes. She skipped a few classes and got annoyed at not being allowed to wear golden sneakers. The dream presented her with a clear feeling of regressing to the behavioral restrictions imposed upon children, which quell the desire to shine and incite rebellion.

Dreams relating to exams are commonplace, too, whether depicting the rehearsal of specific knowledge or skills, the fear of a bad result, or the celebration of success. People engaged in writing books, articles, theses, and dissertations often experience intense phases of dreaming, visualizing problems that need to be resolved and potential solutions, and these dreams tend only to disappear when the person does indeed manage to produce the material promised. On the eve of a thesis defense or a presentation to a hiring committee, it is very usual to dream that the computer has broken down, the projector has blown, or some other technical problem is jeopardizing the presentation. Dreams of this kind warn against accidents and basic pieces of negligence and resemble those dreams that prepare the dreamer not to repeat on the following day those mistakes they might have made in the past.

HE LOVES ME, HE LOVES ME NOT

If there are some dreams that can be interpreted as real keys to resolving problems, most of the time dreams are merely metaphorical reflections of the emotions governing us. In terms of the capacity for mobilizing dreaming, there are not many experiences that can compete with passion, particularly in adolescents. Dreams collected during this stage provide clear displays of social anxiety, ambiguity of affects, contradictory desires, hesitation between possible suitors, internal conflicts over how to act, and alternating between passive and active roles in relationships, anticipating the frustrations of love and its cyclical game of liking and not liking. As an example, let us look at this narrative from a thirteen-year-old girl:

> I was beautiful and popular and P. asked me out to the prom, so of course I said YES!! Then the next day a really cute guy comes to school named JC, and then he asks me to the prom so I said YES!! Then I realized I said Yes to both and they were both so cute and nice, they both were singing my favorite songs. It was

a major dilemma . . . which one . . . I called my friend and asked her which one and she just hung up on me and then I woke up!! . . . The weirdest thing is at a party a guy named Jeremy I met, I fell in love with him, and I found out he liked me. It was almost like my dream predicted it. Good dream!

The start of a young person's love life involves the discovery of very interesting relationships with people who are almost completely unknown, and creates the necessity of fitting them into preexisting social relations, including a web of family members living and dead. A young woman of nineteen dreamed she was in a university dorm and was saying goodbye to a boyfriend who was heading off to another university.

He was trying to kiss me goodbye, but I hesitated because there was a carload of my friends looking at us, and they didn't approve of our relationship. He left and I went back into the room, and it was filled with things all of a sudden. Then my new roommate was there and a strange guy came out of the shower and took off his towel and I got a shocking eyeful. Then I went to my mom's house, where I found my dog; she is actually dead in reality.

Those emotions that are peculiar to the entry into adult life mix with images of the past in the creation of unsettling dream narratives. The themes include the powerful attraction of sex, its reproductive and professional consequences, the need for approval by the group, with the potential moral judgment from the dream representations of one's parents, social inadequacy, and a fear of rejection. Having to choose, getting dumped, not being loved—these are all universal themes that just appear, connected by sudden transitions, characters who show up without needing to make an entrance, places that change abruptly, and a merging of people known and unknown.

STEALING A HEART AWAY

It is increasingly common for passion to involve threesomes, four-somes, and, in these days of polyamory, polyhedral n-somes. Even so, it is almost always when caught between two mutually exclusive loves that a person in the grip of desire suffers and comes apart with jealousy, regret, and nostalgic yearning. The discovery of a new love that rocks the structure of the old love is a much more arcane plot than any in the Greek tragedies. Dreams have this incredible capacity for capturing the signs of passion right at the very start, detecting the deepest internal revolutions, whose emotional repercussions are very often incubated for days, weeks, or months before exploding into epic conquests, separations, and reconnections. Any reader who has never written a ridiculous love letter should skip this section.

The fact is, dreams are finely tuned sensors of changes in the course of a person's affects, even when these are not visible to the naked eye, even when the actual dreamer is not aware of what they feel. A married man without children fell secretly in love with a younger woman, who was also married without children. At the point when he had the dream described below, he had only seen this young woman a few times, and always with several other people around, in professional situations. There was no reason to believe that the two of them might one day be transformed into a couple—this was clearly, merely, a fantasy, an erotic friendship with no consequences beyond candor and onanism. Even in the few weeks after meeting her, however, he dreamed that a lynch mob was approaching his home with sticks and stones, walking threateningly down a dark, scary, beaten-earth road, wanting to tear his skin off. The head of the mob was this young woman's then husband. A year later, she separated from her husband and went off to have a torrid, unstable romance with the dreamer, till things settled down and they had children.

THE STAGES OF UNLOVING

Dreams related to romantic breakups are a category of their own, as there is a typical pattern for the progression of dream reports over the course of the separation, which includes both nightmares of loss and dreams of the pure satisfaction of desire, whether by restoring the relationship or replacing the partner with another person. The U.S. psychologist Rosalind Cartwright, from the Rush University Medical Center, studied some recently separated people who were subjected to multiple sessions of polysomnography and woken from their REM sleep to collect accounts of their dreams. The data showed that the degree of concern about the ex-spouse is in proportion to the percentage of dreams in which this person appears. Participants in remission from depressive symptoms reported a larger number of dreams that were well put together, rich in congruent affects and associations, than patients who remained depressed, who described dreams that were impoverished. In addition, patients who dreamed of their ex-spouse more frequently also did so in a way that was more distant or incidental, showing a better prognosis than patients who dreamed of the ex-spouse rarely but, when they did, were overwhelmed by negative emotions. The examples below illustrate the metaphors and images used by the dream process to explain and to get around the difficulty of adaptation to loss.

Following a blazing passion, with some adventures and international travel for romantic encounters, a couple set a date to start living together abroad. Weeks before the agreed-upon meeting, the guy started having frightening dreams about venomous snakes coming out of the fridge. Soon afterward, one sad evening, he was at the receiving end of a definitive breakup on the telephone, which the girl attributed to virulent attacks made on his character by her relatives and friends. Hours after being thus summarily discarded, the guy dreamed that he was adrift in the sea at night, in a huge bay where the lights on the shore were visible in the distance. He swam and swam in the sea that was dark with the oil spilled from the enormous ships that passed silently by—though a fear of sharks almost paralyzed him. Finally he reached a ruined quay, stepped out onto the streets, in his swimming trunks, beneath the yellow streetlights, wet and dirty, altogether a walking disaster, to present himself to his beloved. At the end of the dream—by which time he was already half-awake—he twisted its outcome to make the young woman take him back. The enforcing of his desire shaped the dénouement of the dream, but when he awoke it was to a bitter taste. In hindsight, the snakes he had dreamed of seemed to be warnings of the destruction of his reputation, while the dream of the sea and the quay provided a richly detailed illustration of the feeling of abandonment, fear, inadequacy, and downfall that marked the abrupt end of the relationship.

One series of dreams that particularly demonstrates the adaptation to an affective loss was collected during the turbulent separation of a couple who parted and came back together several times; they had many conflicts over the course of many years, as, in spite of the fact that they loved each other, they were also both in love with other people. Immediately after the start of his extramarital affair, the husband dreamed that he was getting together with his new girlfriend to perpetrate a terrorist act: blowing up the former couple's car with a bomb. Then he dreamed of his ex-wife looking very beautiful but gradually turning into his girlfriend. He also dreamed that he was losing the company of many other important people in his life, and he was headed toward a bedroom where his ex-wife was with her new boyfriend. He tried to open the door, but he restrained himself, since what was happening inside was no longer any of his business. In another dream he saw himself hugging his ex-wife, both of them with backpacks on for a journey, convulsed with weeping as they said their goodbyes. In another still, the girlfriend did appear, but in the end he left hand in hand with his ex-wife, thinking about how to resolve things with her. Once he dreamed of being at a crowded social gathering with his ex-wife and then losing sight of her. He looked for her and couldn't find her; he tried calling but realized that the cellphone he was holding was hers and so there was no way of making contact.

A year after the separation, the man discovered that his new wife was pregnant. Right after this, he dreamed about his ex-wife and two other unidentified people. At one given moment they all decided to have a subcutaneous injection of poison. He injected his ex-wife but not himself, and she died calmly beside a swimming pool. Soon afterward, in real life, professional circumstances led to an unintentional meeting, filled with animosity, between the ex-spouses. Then he dreamed that he had died, but later reappeared alive again, which made his ex-wife furious.

This surfeit of variations on the same theme testifies to the husband's difficulty in accommodating the representations of his new wife and his ex-wife's new husband into his symbolic landscape. Torn between loyalty to the past and the decision to live through a new future, vacillating painfully between two opposing destinies, the man suffered through his separation in many different ways. All the comings and goings of the process reflect the fact that symbolic death, unlike real death, is not irreversible.

DREAMING GOODBYE

The dreams that follow the physical disappearance of a person who is dear to you are quite different from the narratives of romantic separation, forming a category of their own. On the first night following the death of a very close relative, a man dreamed about a small car moving along a dark coastal road. It is the night of a new moon, and the scene is pictured from above, in the third person. Inside the car, now in the first person, the dreamer sees the coming of a huge wave that licks the shore. The car resists and then goes on, another two similar waves form, and then the same thing happens again: the car goes on.

Strong emotional reactions, some of them morally reprehensible, appear in the dreams that happen after the death of somebody loved: "relief that it wasn't me who died," "panic that I'm the one dying," "panic at the death of someone I love," as well as the actual feeling of missing the one who has been lost, the guilt and the denial of the absence. These dreams, powerful as they are, can complicate significant emotional problems or solve them.

A man had, in childhood, lost his father to a murder. He learned that the corpse had been embalmed, and it made a great impression on him as he pictured that lifeless body. He grew into a very happy, extroverted adolescent; he got married and lived many years in harmony as if the murder were something he had gotten over. However, when he was about to turn forty, his wife suddenly left him. He went into a depression for the first time in his life, losing his hair and beginning an intense process of psychotherapy. He dreamed, then, that he was standing with his therapist at his father's tomb, not the simple grave that it actually was, but a large stone vault. The therapist encouraged him to use an ax to break his way inside. He started breaking it, chipping away at it, sometimes wanting to stop, but the therapist urged him on, until they got inside and found only a skeleton. Dead, truly dead, at peace. The dream marked the beginning of the end of the depression.

WHEN THE NEW COMES

Dreams about offspring are paradigmatic of pregnancy and of the period around the birth. Researchers at Johns Hopkins University asked 104 pregnant women to try to guess the sex of their babies using any method of their choice—popular sayings, dreams, hunches, the shape of their bellies, etc. On average, the mothers-to-be managed to be correct in 55 percent of cases, which is not appreciably different from what one would expect from pure chance—50 percent. However, when the data were analyzed separately according to each specific method, an intriguing result did appear. Unlike a range of other methods whose effectiveness was random—"abdominal position of the fetus" (52 percent), "just a feeling" (56 percent), and "comparing it to a previous pregnancy" (59 percent)—hunches based on dreams got the babies' sex right 75 percent of the time. Among women with more than twelve years of education, dreams managed a 100 percent accuracy rate.[7] Even with the proviso that the sample was a small one, these are intriguing results.

Expectation, fear, and delight stand out in those dreams typical of pregnancy. One mother, seven months pregnant, was distressed at the fact that her child's name had not yet been chosen. She dreamed then that the baby was born and was lying in her arms, and she was asking her husband what the child was to be called. An expectation of the resolution of this very concrete problem clearly motivated the plot of the dream. The worries that come with the arrival of a new member of the family also affect the dreams of fathers. A couple who were expecting the birth of their child at any moment decided finally to buy a crib. The night after they assembled the new piece of furniture, the father had a dream that caused him to wake up in a state of great agitation, babbling that his wife had to get up and feed the baby who had not yet been born.

Dreams produced by the couple over the course of the gestation period reveal those concerns typical of the situation through a wealth of imagery. One couple decided very enthusiastically to have their first child. At six weeks of pregnancy, the mother dreamed that she was walking into a bedroom and she knew she was going to see her baby. She pulled aside the net curtain and saw a little girl sleeping. She watched as the girl's face took shape, the nose changing, the mouth changing, the eyes which seemed blurry as they were changing a lot, and she was scared her daughter might have conjunctivitis. She picked the child up in her arms and the girl winked at her, wiping away the image of the sick eye. The dream was declaring the mother's perennial anxiety about their child's health, an anxiety that tends to increase as the birth approaches. At eight months pregnant, the same mother dreamed that she was surrounded by friends who were laughing as she was going into labor. At a certain point the baby was born and almost fell on the floor; it looked right at her like an adult and said, "Honestly, Mom!" Even before the birth, the baby's subjective representation in the mother's mind had already begun to take shape.

The dream processes of a first-time father tend to take a little longer to start, but they end up presenting with great clarity. A week after the birth I have just described, the father had his first dream about his son. The boy appeared in his dream aged about three and learning to say particular words, as if the father was simulating a son with whom he could already have a symbolic relationship apart from the mother. But the son's arrival also represents a paternal displacement. Two months after the birth, the baby started to sleep in the couple's bed. On that first night the three of them spent together, the husband dreamed that his own parents were roaring with laughter and that they persuaded him to leave his wife, who was thrown into some quicksand and disappeared. He was then seized with remorse and couldn't stop searching for her, until he suspected that she was staying in an enormous hotel. He knocked on each of the doors in turn, trying to find her, searching in vain down long hallways with an infinite number of mysterious doors. When all seemed lost, the dreamer's mother appeared again and pointed toward the door behind which his wife was to be found. The husband knocked, waited, and finally she appeared through the gap only to say that she couldn't let him in, since she was with another man. The scene is a precise diagnosis of the family from that time onward: another man had appeared in the family, and his wife only had time for this newcomer now.

As fatherhood matures, this change also leaves its marks on the plots of dreams. On the first night after learning that his wife was again pregnant, the same husband dreamed that he was driving very fast on a highway, the moving images defined with perfect clarity. He felt the centrifugal force increase as he went around a bend; realizing that the car might skid, he was afraid, he thought about his family and finally took his foot off the gas. The arrival of children tends drastically to increase the sense of responsibility and fear of accidents, turning even the most intrepid adventurers into parents who are cautious, calculating.

Mothers' dreams about their babies are dense and significant. The labor can be accompanied by altered states of consciousness that are close to dreaming, where past events are mixed up with rage, fear, loneliness, a paradoxical attachment to the pain and a light at the end of the tunnel. There is often contact with very old memories, while there are other instances of "white dreams," in which the mother knows that she has had a dream but cannot remember a single detail.

Many mothers report that, following the birth of their child, all their dreams start to include them. Extreme situations are simulated in these dreams, often just expressions of their fear of not performing well at their parental care. A very successful child occupational therapist gave birth to twins. For several nights following the birth, the same nightmare recurred, with horrible images in which she dropped her babies on the floor. Other dreams are just expressions of the delights of the situation. One young woman described a dream in which she was contentedly licking her youngest child, because they were made of vanilla. In the dream's space of flexible associations, a "delicious" child does indeed taste good.

FEAR AND POWER

The basic dichotomy of life, like that of dreaming, takes place between the fear of perishing and the power to adapt. The unfolding richness of the next dream makes it worth analyzing its lengthy narrative.

In the seventh month of her pregnancy, a mother woke in distress in the middle of the night. She had dreamed that she was pregnant and that she was going into a shopping mall feeling alone, trying to remember who was supposed to be there with her. She couldn't remember her mother, her husband, or her first child. She made an effort to recall who her relatives were, but she couldn't do it. She sat down to have a coffee and bumped into her *comadre*, a very close friend, but there was no intimacy between them. She saw people wandering about, and tried to remember her family. She spotted somebody who looked like a much-loved aunt, but it wasn't her, it was only somebody with her appearance. The way the double behaved was quite different from her aunt, wandering along as though she didn't know her, walking with other people, busy with her own life. Then her *comadre* said, very differently to the way she would say it in reality, that there was a childcare service and that she had left her baby son there since his birth. The mother found this strange because she thought him very young for this, but she had a vague sense that this was normal. Then she began to feel as though she was going into labor. There was a shift and suddenly she was in the hospital, and she had given birth. A medical procedure was carried out, she wanted to breastfeed but the nurse said that the baby was in the newborns' ward: "Don't worry, you'll soon get your superpowers and then you'll be able to hear him."

At that moment, the dreamer almost became aware that she was dreaming because she commented to herself that this didn't make any sense. She even thought it might be a dream, but then forgot all about it. Her memory was weak. She knew it was wrong that she wasn't with her child, and she missed her family, but she couldn't remember who the absent people were. Alone in the recovery room, on a saline drip and with other clinical paraphernalia, she began to hear her child crying, her milk started to gush out all of a sudden and her breasts started hurting. She knew she needed to feed, so she said: "I know how it goes, I'll go there." She started to walk, saying, "I can hear him crying." The nurses didn't want her to get up; they told her she still needed to stay lying down because of the medical procedures. But, she thought, if she could already hear her son, she must have acquired those superpowers they'd told her about and, therefore, she must be well: "I can hear him from here, I need to know where he is, I'll go there, my milk is coming."

She walked, unhurriedly at first, following the signage, but the hospital corridors kept filling up, until it had become difficult to make any progress, especially to get from one floor to another: there were huge lines at the elevator and staircases. People passed by without looking at her, rushing, making no contact. An infinite succession of signs went by, one after another. She kept going down to lower and lower floors, her breasts gushing milk, aching, throbbing, the signs telling her to keep going down, the feelings of strangeness increasing as she thought, hurt: "That's not how the world was meant to treat somebody who's just given birth."

Then the young mother began to despair. The child's crying was expressing great pain now. She went into a panic thinking that maybe they were mistreating it or kidnapping it. She started to run, bumping into people, afraid because of the surgery she'd just been through, but she ran all the same, hanging from the plants, leaping down stairs and over obstacles, like a real parkour tracer. The leaves snapped with her weight, but she fell trusting in her superpowers. She could hear the crying in the distance, and her breasts were still gushing with milk. The frenzy increased, images followed images dizzily, she ran and ran but never arrived.

Until she awoke, with a start.

The mother's distressing dream clearly shows the difficulty in completing the most biologically significant task—breastfeeding the newborn baby—in a process in which the weakness of her memory partly determines how the plot of the dream develops. The interrupting of one dream scene and the start of another, like a piece of video editing, produces a displacement that creates a sense of time passing. The world represented in the dream is a fragmented one, it contains only a part of what it ought to contain, and the dreamer is aware of this throughout the dream. The plotline is dominated by two opposing emotions: fear of failing in her duty of maternal care, and confidence in her own powers to carry it out. The dream also amply illustrates an important property of dreams: the characters are developed to different degrees, and they sometimes seem mere shells, superficial images of people who really exist. As Shakespeare's Hamlet said, "A dream itself is but a shadow."[8] In Jung's depth psychology, this concept was given the name of imago, a thing that sometimes seems like a person or being of vast power and wisdom, and sometimes not.

INESCAPABLE INCOMPLETENESS

Just as literature spans genres from the shortest poem to the longest novel, just as there is a fundamental link between the frozen pictures of photography and the moving images of cinema, so dreams represent a diverse range of experiences, from imagetic haiku to monumental sagas. There are even people who report that certain dreams can be the condensation of a whole life, like a polysemous weave that can transmit all the meanings of the journey in a single narrative. Each episode of the dream is one particular instantiation of the electrical activity in the dreamer's brain, a fragile, unstable web of symbols that can be interrupted at any moment. But it can also persist and develop into a broad range of possibilities, from a narrative made up of illogical shadows, faded and sad, to a torrent that can be sustained, fed, and developed, ever further and ever more deeply, till it has been transformed into a work of art in itself, a complex experience that is infused with beauty and insights vital to the dreamer.

The electrical reverberation of certain memories is increased by the emotion and the desire associated with them, and so those people who are dominated by fear will be assailed by nightmares. What is created is a wound, a scar, a trauma, an attractor of negative emotions, a deep valley of strong synaptic connections linking toxic thoughts, imbricated with one another in sad and ugly ways. The reverberation converges here, fermenting the bile. When the process passes a certain threshold, the dream is no longer able to get out of that symbolic mesh. It is caught, trapped, ruminating, hurting, and deepening the trauma. It is then necessary to interrupt the reverberation and seek a way out, overflowing electrical activity into other neuronal networks that offer solutions for life.

Fortunately the possibility always exists of a breach being opened in the despair, whether through a liberating, transforming dream in which the desire is benignly satisfied through effective work, or a dream that merely presents the satisfying of the desire in an unexplained, magical way, with no work carried out on the part of the dreamer to overcome the difficulties—a dream that sustains hope, though not equipped with any solutions to the problems of waking life.

Dreams frequently deal with the feeling of impossibility in attaining an objective, sometimes with a plot so outlandish that the dream images can take on a frankly comic appearance. One good example is the dream in which a man got off a bus and saw an enormous, hairy white pig walking alongside the vehicle. He concluded that the pig belonged to a friend who had stayed behind in some other city. He decided to drive the pig back home, but halfway there he lost his way and no longer knew which bus to take. All of a sudden, the pig rolled around in a gully, it got dirty, turned red, and the dreamer started to worry about returning it. He made his way along the street, fighting with the pig, pulling and pushing the heavy animal until they arrived at a lagoon into which the pig launched itself with great commotion. He thought: "Pigs like water, not a problem." But the creature started to sink. The man, calling for help, waded into the water and began trying to get the pig out of the lagoon, and finally succeeded in doing it. He thought: "And now I'm going to need to give the pig mouth-to-mouth." Even in this ludicrous dream narrative—or perhaps because of it—what stands out is the incompleteness of the dreamer's subjective experience. The initial mission is never completed: the plot unfolds into unexpected complications, which make the man's chances of attaining his objective increasingly remote.

DEAD-END ALLEY, GATE TO THE FUTURE

At their worst, dreams are reverberations that are deeply disagreeable, sometimes useful for warning against avoidable risks, at other times terrifying. Being afraid of real danger and simulating possible negative consequences is a healthy behavior that is at the root of survival and adaptation. The broad range that spans from cinematographic nightmares to dull dreams of anxiety and frustration make up the palette of dreams of the wretched, the enslaved, the incarcerated, the tortured, and those condemned to death. This is true literally as well as metaphorically, since somebody made wretched by love and somebody tortured by a bad job operate according to the same symbolic keys as people who really have been ill treated or those who are real pariahs—with a lesser intensity, but sometimes with the same qualities of affects and images.

At their best, dreams are the actual source of our future. The unconscious is the sum of all our memories and of all their possible combinations. It comprises, therefore, much more than what we have been—it comprises all that we can be. "The Library of Babel" by the Argentine writer Jorge Luis Borges describes a collection of all the possible books, generated by the totality of possible orthographic combinations, the infinite shuffling of all the letters of the alphabet, recombining to form all that is yet to become.[9] In the same way, a dream is the possibility of imagining potential futures by means of a mechanism capable of exploring past experience and forming new psychic conglomerates, assembling old ideas in a new way. Everybody who had successful ideas and transformed the world, each of those people who have managed to transform themselves into what they yearned for, all of them without exception, and by definition, experienced days and nights when they had not yet realized any of these things. And then they dreamed.

Halfway between the best and the worst, when life is not going through any extremes, a dream is a badly defined collage of seemingly unconnected pictures and echoes of countless uncompleted desires. For the majority of the working masses, in their day-to-day lives of waking up early and going to work without thinking about the dreams they have had or telling them to anybody, who tend to go to sleep without planning anything specific for the following day, there is no such thing as a habit of asking dreams for inspiration, the way the ancient Greeks did and the way hunter-gatherer peoples still do. So it is common for today's manual workers to dream a mixture of mental sketches, which offer descriptions of the present moment more than future possibilities.

But this dynamic changes completely when life really does become complicated. Dreams can warn about an imminent illness, preceding the first clinical symptoms by weeks, months, or even years. The U.S. neurophysiologist William Dement, the first person to provide a rigorous characterization of the increase in dream activity during REM sleep, described his own experience as follows:

> I used to be a very heavy smoker. What started as an occasional indulgence in my army days had, by the early 1960s, become chain smoking . . . One day in 1964 I was coughing into a handkerchief and noticed with a chill that the little flecks of sputum on the white cloth were reddish pink . . . I sought out a radiologist friend and asked him to order a chest X ray. The next day I went back to his office, full of dread. I will never forget the grim expression on his face as he motioned me to the light box behind his desk. Without a word, he turned and clipped my chest film onto it. Immediately I saw that my lungs harbored a dozen white spots—cancer. The wave of anguish and despair that I felt was overpowering. I could barely breathe. My life was over. I wouldn't see my children grow up. All because I hadn't stopped smoking, even though I knew all about smoking and cancer. "You utter fool," I thought. "You've destroyed your own life!"
>
> And then I woke up.
>
> The bloody sputum, the X rays, and the cancer had been a dream—an incredibly vivid and real dream. What a relief. I was reborn. I had been given the chance to experience having inoperable cancer of the lung without having it. I stopped smoking right then and have never lit another cigarette.
>
> . . . To some it may seem amazing that people will take drastic action as a result of something that didn't even happen. But the emotional impact of dreams can be so powerful that they might as well really have occurred.
>
> . . . The logical part of the awake brain knows that the dream was not real, but the emotional part of the brain cannot set it

aside. As far as our brains are concerned, what we dream really happens to us.[10]

BETWEEN LIFE AND DEATH

What gives dreams their power, both perceptually and in terms of the emotions and symbolic associations they involve, is the concentrating of desire. A conglomeration of psychic formations that is more cohesive, with more internal coherence, is much more meaningful and impactful than the juxtaposition of a handful of unconnected memories. This is why dreams were actively sought out in antiquity, and they remain so in hunter-gatherer cultures, through rituals that are able to prepare the mind for dreaming. But even without preparation, the mind will trigger very striking dreams when it is a matter of life and death. The sequence of dreams described below exemplify the parallelism between the dream content and waking reality.

At the age of forty, a university professor was stressed about the poor working conditions in his department, and he was no longer able to stick to the athletic habits he had been keeping up since his teenage years. The horizons of his life had turned gray after the sudden death of his father, one year earlier. An only child, he lived hundreds of miles from his mother. At the start of the year, he underwent a complete medical checkup, and nothing was found. He set off for a scientific congress at a ranch close to Luján, epicenter of the great Catholic pilgrimages in Argentina. On the first day of the congress, not long after lunch, he had a very serious heart attack, with a total occlusion of the lateral branch of his left coronary. He was roaring with pain, trembling, sweating, his whole body writhing. His arrhythmic heart almost stopped, to the despair of all those around him. Half an hour of torment waiting for the arrival of the ambulance, another interminable half hour to reach the nearest hospital. The catheterization was successful, and he went into the I.C.U.

On the first night, he dreamed he was in his living room, which was full of people. Suddenly, he saw his father, sitting on the other side of the room. He was surprised and pleased, since the old man was smiling, but at that same moment he remembered that his father was dead. In desperation he went over to him, and when he touched him, everything vanished. He saw himself in the hallways of the old building of his first doctor, with dark, old elevators. He felt totally lost—and woke up. On the second night, he dreamed that he was screwing the nurse. The horny man saw her smiling face as she told him that she loved him: *"Te quiero."* He thought, "Man, I'm having a heart attack! I'm going to die here, we've got to stop." But she smiled and he didn't stop, even knowing that he might die. On the third night, he dreamed that he was in some arid land, on a desert ground with furrows formed by the lack of irrigation. He was doing Capoeira, the exquisite Afro-Brazilian martial art and dance, with a tall, graceful woman. Three people were playing the berimbau, an Afro-Brazilian instrument: a friend who had helped to save him at the time of the heart attack, and two people who had already died, his father and Mestre Pastinha, venerable master of Capoeira de Angola. The game with the tall woman was progressing energetically inside a ring, until a certain point when she kicked a perfect *meia-lua de compasso.* He dodged it, though only barely, and spun away. Mestre Pastinha shouted one of his best-known phrases: "Capoeira is everything that the mouth eats." The dreamer's friend turned to him and said, "That's life." His father bent down and said, "Calm the game down and go back into the circle." They started singing the traditional song "A Pebble from Aruanda," and he woke up.

The professor made a gradual recovery from the heart attack. On the night of the first dream his condition was still a cause for concern, and he was overwhelmed by the idea of death. On the night of the second dream, his vitals had already recovered considerably: that was when he had the dream that began with full libido and was then conflicted with the fear of death. On the night of the third dream, the patient's health had improved. The plot was complex and illustrated with images close to his heart: the dance with death, the nearly fatal blow, and good advice for accepting the setback and moving forward, rounded off with a musical soundtrack with great poetic impact, which demonstrated the need to be humble before Aruanda, the spiritual plane where the ancestors live.

This sequence of dreams has meanings that are simultaneously biological and psychological. Just as the dreamer's immediate context has the power to elucidate the oneiric elements, dream narratives make it possible to understand what is happening in their life. In the 1920s, among the indigenous Kwakiutl people of Canada's Pacific coast, the German anthropologist Franz Boas collected dreams that were dominated by the recurrent themes of hunting, fishing, and fruit-gathering.[11] The anthropological replication of this finding in different cultures makes it easy to suppose—as we have done in previous chapters—that our Paleolithic ancestors' dreams had matters of very direct expression: being hunted, hunting, engaging in sexual courtship, conquest and the sexual act, pregnancy, childbirth, providing parental care, loving, suffering, dying.

Those dreams most studied by Freud were clearly constructed on these ecological foundations of dreaming, but they presented many other new problems, deriving from the libidinal castrations of Viennese bourgeois life. Those dreams preferred by Carl Jung, meanwhile, were different in nature: in extreme situations, when very important events take place, the dreams that emerge are memorable, moving, and full of details. At first examination, they do not seem to derive from the dreamer's immediate concerns, because they symbolize them broadly, more philosophically or poetically, with ample spatial and temporal scope. Beyond the theorematic dimension in which the outcome corresponds literally to the images of killing, fleeing, and mating—instructive and exemplary in themselves—such dreams express these images' profound symbolic repercussions: arcane and numinous archetypes of life and death.

GREAT DREAMS

It was such "great dreams" that the ancient Greeks and Romans prized most highly, the epic journey through the interior of the self, capable of expanding the very limits of existence and inspiring important changes in the course of events. One classic example of this type of dream was experienced in 1909 by Carl Jung, when he was accompanying Freud on his historic trip to the United States. In this period when the two men are deep in an intense discussion about the structure of the human psyche, Jung dreams that he is in a house, which is unknown to him but which nevertheless "is his." He decides to take a look at the lower floor and is amazed to discover medieval chambers. He goes down another floor and arrives at some Roman constructions that make him intensely curious. Through a trapdoor he goes down into the depths of the house and along a tunnel that takes him to a dusty little cave. Inside he finds primitive remnants: bits of bone, pottery, and two human skulls. On waking, Jung realized that the house was a metaphor for the stratification of human consciousness, with the oldest and deepest levels representing those parts of the mind that go back to our ancient past. This dream was crucial for the development of the idea of the collective unconscious as a source of phylogenetic memories (instincts) and transcultural memories (archetypes).

Even in contemporary urban life, great dreams take place at moments of major change in relations with the environment, such as in the child's conquest of language and of a mentality that is well adapted to the adult world, in the adolescent embodiment of the need to find relationships outside the family group, in the discovery of sex, in the beginning and repetition of motherhood or fatherhood, in the chance collisions with the risk of death, in menopause and in the transition to old age. On these occasions, dreams frequently illustrate astonishment at the irreversibility of time—dreams that do not deal with day-to-day problems, but are surprised at the inexorable change in everything; dreams that are special, mythic, which tend to arise within specific age ranges but can also occur at any moment in life that is directly touched by a recollection of finiteness; dreams that evoke ancient memories from archetypal cycles and, even if they might be clothed with the impressions of the day, that mark those major symbolic transitions on our uncertain path of being, perhaps procreating, and ultimately disappearing.

Up to this point, we have been discussing a varied display of dream accounts in order to delineate the broad frontiers of the dream phenomenon. Our next step is to start to understand the mechanisms that allow dreams to reflect the dreamer's problems and to offer possible keys for solving them.

5

First Images

In order to understand how the human mind—as a rememberer of the past and imaginer of the future—is born and develops, we need to understand how dream narratives vary from babies to old people, shifting and changing through childhood, adolescence, and the various shades of adulthood. Even though an adult, in general, is likely to have dreamed thousands of times in their life, few people remember when they dreamed for the first time. Try to remember your first dream. It is almost certain that it happened after the age of three, when you were on the threshold of using grammar and syntax.[1] If dreams do exist before that, they are rarely remembered. The earliest memory I have of a dream comes from when I was four years old, and it is a classic dream of wish fulfillment. In real life I wanted a particular kind of tricycle, and in the dream that was just what I received as a present from my parents. The dream was delightful, but to this day I recall the huge disappointment I felt when I woke up and realized it was an illusion. An experience common to all, recalled with nostalgia in the lines of the Spanish poet Antonio Machado: "There was a boy who dreamed / About a cardboard horse / But when he opened his eyes / The little horse he couldn't see."[2]

Dream narratives develop over time, along with the maturing of perception, motility, language, and socialization. When exactly do we start to dream?

FIRST SYNAPSES

This apparently simple question proves more complicated when we take into consideration the fact that the brain changes substantially over the course of a lifetime. While the fetus brain is practically formed in the thirtieth week of pregnancy, major changes continue to happen after birth. During this early stage of development, more neurons and many more synaptic connections are formed than are generally found in the adult brain. This happens because the maturing of the brain corresponds to an extensive, intricate process of neuron death and synaptic pruning, with the increase being followed by a reduction of cortical thickness between birth and the end of adolescence.

The overabundance of synapses that characterizes the beginning of life outside the uterus is thinned out very many times as the person grows and learns about the world through the senses, movement, and reason. This capacity develops like a sculpture, starting with a shapeless block of stone—equivalent to a vast quantity of neurons—and ending up in a particular shape, with much less stone and for this very reason much more information: a smaller group of neurons with specific connections shaped by experience. The elimination of synaptic connections continues up to adulthood, in parallel with the forming of a small number of new synapses with each night's sleep, as we will see below.

To carry this reasoning further, it is important that we make a more detailed examination of what synapses are. While certain connections between neurons occur through direct contact of their membranes, allowing the flow of ions between adjacent cells via electrical synapses, a large proportion of the connections takes place across chemical synapses, incomplete approaches between the membranes of two neurons that never quite touch. In the minuscule space that is the gap separating this kind of synapse, the freeing of neurotransmitter molecules creates connections between the neurons, transferring signals from one cell to another by briefly replacing the electrical impulse with chemical substances such as glutamate or dopamine (fig. 5).

By and large, the synapses involved in carrying out successful behaviors are strengthened, while synapses activated by unsuccessful behaviors are weakened. A good part of this process happens during sleep, which is one reason why children sleep much more than old people. Babies tend to spend most of the time asleep, devoting more time to REM sleep than they will at any other stage of their lives.

DREAMING BABIES

Interviews with the mothers and fathers of babies frequently reveal the opinion that they dream even as newborns. Careful observation

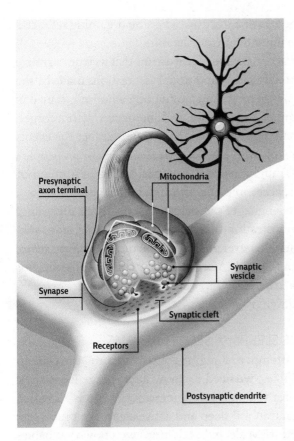

Figure 5. A chemical synapse presents vesicles full of neurotransmitters that are released into the synaptic gap when an electrical impulse takes place.

of the facial expressions and body movements of a sleeping baby makes it possible to tell not only when they are dreaming, but often what kind of affect is dominating that mental activity. Smiles, tutting, and grimaces are frequent occurrences, suggesting the presence of emotions. Pregnant women can sometimes even deduce at what point the fetuses might be dreaming in their bellies, in periods of motor unrest.

If mothers and fathers are certain that babies dream, science is more skeptical. The main impediment to examining the content of childhood dreams comes from the worsening, in the preverbal phase, of the difficulty inherent in any study of dreams: having access only to their secondary elaboration. If the raw material of this kind of study is always an *a posteriori* narrative produced by somebody who is awake, edited more or less consciously in order to fill up lacunae and increase internal consistency, how is one to know what the dreaming self actually experienced in the inner depths of the mind?

It is only during the actual course of the dream that its true narrative can be accessed, still free from any additional associations made by the conscious mind upon awakening. And as anybody who isn't trained in the art of remembering dreams knows, waking brings with it forgetting. If even dreamers themselves find it hard to remember the events experienced in their dreams with any fidelity, researching other people's dreams is even more fraught, since it depends on the examining of secondhand narratives, accessible only through language.

If they do not have language, can babies describe their dreaming? Is any convincing demonstration that sleeping babies experience subjective realities possible? However abstract these questions may seem, they deserve to be confronted. After all, understanding the origins of dreams might be a crucial step in deciphering how our own self-awareness arises and develops.

ACTIVE SLEEPING, SECRET DREAMS

If we are to make any attempt to imagine the contents and the dynamics of the dreams of babies, we need to know at what age REM sleep is established. In the last ten weeks of pregnancy, there is already a differentiation between tranquil sleep and active sleep, the precursors of slow-wave sleep and REM sleep. In addition, we know that babies sleep much more than adults, especially where REM sleep is concerned, which in adults coincides almost perfectly with the occurrence of the dreams experienced by the sleeper. None of this, however, allows us to deduce that adults' dreams resemble those of babies.

The amount of sleeping time that is spent in REM sleep comes to 33 percent of the total in newborn babies, but it is gradually reduced to stabilize at around 10 percent after the age of three, a level not unlike that of a young adult. By this age, there is already a regular alternation between slow-wave sleep and REM sleep, making up a complete sleep/ wake cycle that becomes longer with age. Despite these similarities, the correlation between dreaming and REM sleep in children is not always high.

There are no verbal accounts of dreams before the age of three, but this does not mean that babies don't dream. If it is not possible to ask them directly how they dream, we can at least watch their sleep closely.

Babies move around quite a lot during REM sleep, which can give us some clues that their dreaming consists of a rich subjective experience. In order to imagine the dreams of babies, we need to evoke images that are compatible with the infant experience: hot and cold, dryness and damp, tastes, smells, sounds, colors, movements, textures, and shapes, until such time as they are able finally to make out people and objects.

FIRST ACTS AND OBJECTS

In the first eighteen months of life outside the uterus, babies progress down a crucial cognitive path—actually the most important of their lives. They learn to use their senses and their muscles. They learn to see, hear, touch, taste, move, and communicate. They start learning to learn. Bit by bit, the objects of the world take shape. Childhood is the stage of post-birth life with the greatest neural plasticity, that is, with the utmost malleability of the synapses. Even so, for all this time, babies do depend on maternal care for survival. It is the mother that is the source of the first object of the world to be represented, the first source of material and psychic nourishment, the first embodiment of reward, the original trigger of the libido: a breast full of milk.[3]

The fragility of the human baby is extreme. Never again, until they reach very old age, will a human being experience such weakness and dependency. And yet it is this weakness that allows humans, when they are healthy, loved, and well cared for, to get through the period of greatest neural plasticity—childhood—given up to the carefree and innocent discovery of reality. The first contact with reality beyond the placenta, at the time of birth, launches a whole carnival of new experiences, which are shapeless at first, as the cerebral apparatus of perception is itself immature. Therefore, the dominant feelings in the dreams of newborns can only be primary ones, such as hunger and satiation, the perception of dampness, temperature, high-contrast images and sounds, the locating of a touch to the skin, the feeling of gravity and an awareness of the positions of their limbs and head.

THE DEVELOPMENT OF THE DREAM NARRATIVE

Are there any such things as characteristic dreams for certain ages? Are children's dreams more frightening, delightful, or banal than those of adults? Are there substantial differences between the accounts of dreams experienced by girls and boys? To what extent is the capacity for dreaming related to intellectual and emotional development? The U.S. psychologist David Foulkes carried out pioneering studies with dozens of children on whom he did extensive research over several years, which allowed him to investigate the dream dynamics of individuals between the ages of three and fifteen.[4] Each child was subjected to psychological exams and polysomnographic records of brain activity, muscle activity, and eye activity for nine nights each year. Accounts of the children's dreams were collected systematically after they woke, during both REM sleep and slow-wave sleep. This made it possible to gain reasonably direct access to the content of the dreams, with only a minimal amount of interference from the waking mind.

The children's intellectual and emotional development was monitored via the serial application of psychological tests. The systematic observation of the same children in the laboratory, in school, and at home, with a focus on games and spontaneous stories told by them, made it possible to interpret the results in a way that was both broad and deep. Despite being more than forty years old, Foulkes's research remains the most complete longitudinal study on the development of human dreams.

The observations were revealing. Between the ages of three and five, dreams are infrequent and typically poor, with little imagery, without strong emotions and movement. This is why the accounts of terrifying nightmares, while common at this age, often reflect subjective experiences arising right after waking: a fear originating not in the dream itself, but in the disorienting—and therefore frightening—experience of waking up in a dark room. Dreams in this age band reflect a cognitive system that is immature, with great limits to representation that do not support symbolisms that are complex, peculiar, or fantastical. There is a striking lack of social representations—parents, siblings, aunts and uncles, cousins—which in subsequent ages will come to occupy the center of the dream narratives. In Foulkes's study, associations related to their own bodies among children of three to five do not seem to be dominated either by the maternal breasts, so essential to the physical and psychological health of the newborn, or by the areas posited by Freud as milestones in the development of the libido in one's own body, such as mouth, anus, vagina, and penis. Instead what emerged most clearly in these dreams were the basic physical needs, such as sleeping or eating.

A MINIMALIST BEGINNING

The dream accounts from the children of preschool age researched by Foulkes presented a rather limited organization of thoughts, not unlike that revealed in their waking lives. We do not know whether children's dream accounts are so impoverished because their dreams actually are simple, or whether their apparent poverty resides in their limited capacity to recall their dreams or to express what they have remembered. As long as the limitations of language directly affect the account, any conclusions based entirely on young children's own accounts will continue to be called into question.

In those children researched by Foulkes, being woken from REM sleep often didn't correspond to any dream experience. When the children did recall a dream, it tended to be a description of a scene that was simple and static. The representation of the dreamer's own self seemed diffuse. One dream described by Dean, a four-year-old boy, illustrates this kind of dream content:

> DEAN: I was asleep in the bathtub.
> RESEARCHER: Was this in your bathtub at home?
> DEAN: Yes.
> RESEARCHER: Was there anyone else in the dream besides you?
> DEAN: No.
> RESEARCHER: Could you see yourself there?
> DEAN: Uhhh, no.
> RESEARCHER: I mean, did you have like a picture of the bathtub and you could see your body inside the tub?
> DEAN: No. . . .
> RESEARCHER: How did you feel?
> DEAN: Happy. [5]

As one might expect, Dean was also laconic when it came to talking about daytime events, as Foulkes documented over the years. Compare the dream account above with Dean's description of a drawing of an adult woman scolding a crying child holding a headless doll:

> DEAN: His head came off.
> RESEARCHER: Can you tell me anything else that is happening?
> DEAN: That's all. [6]

NARRATIVES AND SIMULATIONS

The greatest transformations in the content of dreams recorded by Foulkes happened between the ages of five and seven, a period in which the children no longer reported simple images but rather successions of connected scenes, like a movie. While a number of important characteristics of dreaming are only established after this period, this is where the fundamental structure of dreams as *narratives* is formed. It is also at this age that the fantastical nature of dreaming emerges, as the capacity to mentally represent the objects of the world increases. Some of the distortions typical of adult dreams appear during this phase, such as the sudden shifting of the dream setting in time and space, and the condensing of dream characters into images with composite meaning. They are dreams we have all experienced, which take the following form: "I was in place A, but it was also place B," or "I was with a person who was a mixture of so-and-so and what's-his-name." Prince Frederick's dream described in Chapter 3—"I was sometimes at Rome, and sometimes at Wittemberg"—is an explicit example of this phenomenon.

A dream space begins to be delineated that emulates certain aspects of the real world virtually, creating a world in miniature inhabited by characters capable of acting, pursuing goals, and even reflecting on the consequences of the actions they carry out in the dream. The continuity between dreaming and waking is heightened, with a frequent recurrence of actions and situations experienced while awake. The dream accounts that come from this age band see the emergence of a genuine curiosity about the multiplicity of people, objects, and relationships from the real world, with less of a focus on the dreamer's own physiological states.

But even if the dreams reported between the ages of five and seven are structurally comparable to the dreams of adults, the dream activity of children does not completely integrate the dreamed scenarios with the representation of the dreaming self—a dominant imagined being, capable of voluntary actions, emotions, and reason during the dream. For Foulkes, these dreams reflect a period of transition in the child's development, marked by a quick reorientation of focus, a shifting from oneself to the outside world. Unlike the dreams of adults, children's dreams in this age band often have animals or family members as their main protagonist. It is as though the representation of the dreaming self were left on the back burner during this specific period in the maturing of dreaming.

While in general the dream content is similar for boys and girls, some specific differences appear during this stage. Girls aged five to seven report more dreams involving social interaction than boys, with happier outcomes and conflicts resolved. Meanwhile, boys report more dreams with unknown male characters than girls do. They also report more dreams with animals. While we do not yet have exhaustive studies comparing gender differences in the dreams collected in different cultures, research suggests that the differences between the sexes described by Foulkes might be widespread.[7] Still, it is likely that the differences reflect similar discrepancies between boys' and girls' experiences across different cultures, rather than any absolute biological distinction.[8]

Children aged between seven and nine have already been through a large part of the process of attaining full dream competence. This is typically the phase in which the active representation of the dreaming self is established. Dreams in the first person become prevalent. There is a significant increase in the proportion of dream accounts obtained after being woken from REM sleep, but curiously, children of this age are also able to dream during slow-wave sleep. Narrative structure becomes more complex and starts to evoke specific affects, with a slight preponderance of pleasant dreams. Dean, the boy who at the age of four described only two dreams on waking from REM sleep, produced eleven dream accounts at the age of nine. The increased complexity is striking:

> DEAN: We were tree planters and we went up to this place and we planted a tree. And the next day we came back and the tree was already grown. So we planted more and they all grew and there was a forest fire and they wouldn't burn down. So we made forests out of them, and then some men were chopping them down for firewood and when they chopped them down the fire wouldn't work. So they told it to the state police and [the mayor] said that they planted trees and they won't burn.[9]

EXPANDING REPERTOIRES

Between the ages of nine and eleven, few qualitative changes occur in the dream elements, but many quantitative changes, with an expansion of the symbolic repertoire, an increase in the capacity for remembering dreams, and a consolidating of the dreamer's own role in the narratives. The dreams of girls and boys continue to diverge gradually, with an increased prevalence of motor behaviors in the male narratives. The dreams become more idiosyncratic at the start of preadolescence, starting to offer less reflection of the general features of the age band and more of the specific personality of each dreamer. The dreamer's own emotions and expectations start to gain in importance, too.

And then, at the end of preadolescence, between the ages of eleven and thirteen, dreams go through their second major phase of maturation. The capacity to remember dreams and the prevalence of dreams during REM sleep stabilize at levels similar to those of adults. During this age band, the individual differences in personality, intellectual repertoire, and social skills become obvious. We observe a greater balance in the emotions that are dreamed, with positive and negative affects appearing at equivalent levels. The plots of dreams become richer in variety and subtlety, parallel to the construction of dream contexts that are better balanced between self-representation of the dreaming self and the different characters and objects of the world of dreams. In general, dreams become less centered on relatives and more on other participants from the child's social milieu, such as schoolmates or neighbors. The differences between girls' and boys' dreams become clearer, with a great distinction in dream narratives according to the social roles that are more typical of each gender. Girls have more dreams about female characters, while boys have more dreams about male ones. Boys' dreams present with more sensory activity than girls' dreams do. They also include more narrative conflicts with unfavorable outcomes, including attacks carried out by other boys.

ADOLESCENCE AND MATURITY

The most decisive influence on dream content during adolescence in Foulkes's sample was each dreamer's intellectual and emotional development, which reflects their own specificities and individual features during a time of fast hormonal transformation. While romantic relationships are normally very important at this age, it is actually striking that dreams reported by adolescents are not totally dominated by sexual elements—though they do reflect increased curiosity about their bodies and the differences between men and women, the greater differentiation between sexual roles, and the maturing of the reproductive system.

The brain that dreams is the same brain that lives through waking experience, and so the more complex the mental tissue, the more complex the dreams will be, too. At age fifteen, the dreamer is an active character in the dream's reality, desiring and choosing and acting in a virtual setting that is multifaceted and full of nuance. The dream of a sixteen-year-old female student illustrates the importance of social relationships, and above all romantic relationships, in adolescence:

> First I dreamed that Kylie and I went to another one of Jon's parties, this time with more people from my high school. There were so many people there that they made us get into two lines, one of boys, and one of girls. Whomever we matched up with next to us was our hookup partner for that night. I got someone gross, so Kylie and I left to go home and watch a movie. I went to work out the next day . . . There were lots of boys in the gym, and I was a little intimidated. Finally, I dreamed that I was a beautiful princess and I was at the beach. I saw a Norwegian prince and we fell in love at first sight.

Considering Foulkes's experiment in its entirety, the profound transformation that dreams express, from infancy to adulthood, is quite clear. The psychological maturing of dreams follows a course parallel to the mental development experienced in waking life. Between the passive, static dream of a three-year-old child and the dramatic, cinematic dream of a fifteen-year-old girl there is a huge cognitive distance, which is bridged through the countless experiences that have been lived through and dreamed over the course of the more than four thousand nights in the intervening period. More recent studies have confirmed this general pattern of dream development, but observations made outside of the laboratory context, in volunteers' homes, for example, show that small children are also capable of producing dream accounts that are rich in movement, social interaction, emotion, variety of characters, and an active representation of the self, so long as there is a greater familiarity with their environment.[10]

Despite its broad scope, Foulkes's research had the significant limitation that it sampled almost exclusively children coming from middle-class U.S. families, with a reasonable degree of education, an adequate meeting of the children's material needs, and a social context that was peaceful. These biases probably explain the low rate of nightmares in the dream series that Foulkes collected. Another piece of research found a prevalence of nightmares in the dream accounts produced in the 1990s by children in the Gaza Strip and in Kurdistan. With their daily exposure to the high levels of stress that characterize the experience of war, the children sampled by the Finnish psychologists Antti Revonsuo and Katja Valli, of the University of Turku, showed significant continuity between waking time and dreams, with nightmares that were frequent, intense, violent, and even warlike. This was a sharp contrast with the dreams of Jordanian and Finnish children, who dreamed peaceful narratives in the normal safety of countries without war.[11]

CONFLICTS, EMOTIONS, AND AUTONOMY

Along with physical violence, economic violence tends to have a great impact on the quality of sleep. Many studies have demonstrated the existence of sleep problems in low-income communities. The adverse conditions that cause this include stress, anxiety, unsafe environments, overcrowded rooms, and conditions made uncomfortable by noise, temperature, and damp, among others. Research into more than eleven thousand adolescents aged between ten and eighteen showed that exposure to violence produces a negative impact on sleep, of the girls in particular.[12] Low-income families tend to live in small homes where beds are shared and sleep is regularly disturbed, owing to the differences in the working and studying schedules of different family members. Research into more than three thousand three-year-olds showed that sleep deficits are associated with low levels of maternal schooling as well as with domestic overcrowding and poverty.[13] Another piece of research using 1,400 Finnish adults showed that sleep quality was not much affected during the serious economic crisis of 1990, except in low-income individuals, who presented worse-quality sleep, more insomnia, and a greater use of substances taken for sleeping.[14] The reduction in sleep time is generally more pronounced among individuals of lower socioeconomic status, reaching an unbelievable 3.8 hours a day in some jobs.[15]

A large-scale study of sleep problems in underdeveloped countries, carried out on more than forty-three thousand people from Ghana, Tanzania, South Africa, India, Bangladesh, Vietnam, Indonesia, and Kenya,[16] revealed problems that were serious or extreme in almost 17 percent of participants, with considerable variety among countries— from 4 percent in Kenya to 40 percent in Bangladesh. The study found a consistent connection between the high prevalence of sleep problems, on the one hand, and low levels of schooling and poor quality of life on the other. The social component can affect sleep directly, as children in poor families often need to work to supplement the family income. A study of the effects of work on the sleep of students aged between fourteen and eighteen showed that young people who studied and worked ended up waking earlier on weekdays than those young people who only studied, leading to a significant reduction in their total duration of nighttime sleep.[17] The correlation between the quality of sleep and academic performance is also found among medical students.[18] Whether from an excess of work or studying or other sources of tension, the stress is harmful to their sleep. A recent study of the sleep habits of more than fifty-five thousand college students in the United States suggests that sleep deficits can be even more detrimental to academic performance than excessive alcohol or marijuana use, anxiety, or depression. Every night with poor sleep per week was accompanied by a 10 percent greater chance of quitting a course and a fall of 0.02 in grade point average.[19]

DISRUPTIONS AND JOURNEYS

Disruptions to sleep in childhood, when they occur under conditions that are comfortable and safe, are generally not very serious, and are easily resolved. Difficulties in getting to sleep and nighttime awakenings are common but transitory. The production of nightmares can be substantial between the ages of three and ten,[20] tending to fade away after this age band. The most common narratives in children's nightmares include the death of relatives, dangerous falls, or being chased by people known or unknown. There is a significant link in children between sleep deficit and irritability, with tantrums and bursts of bad temper. In addition, anxious children tend to have more nightmares, as one might expect.

In situations that are very stressful in real life, children tend to develop recurring nightmares, with sad and frightening narratives that can return almost identically night after night, leading to feelings of fear when the time comes to go to sleep. Meanwhile, children whose anxiety levels are low, raised in environments of care and protection and without stressful disruptions, tend to report positive dreams in which their desires are pursued and frequently fulfilled. But nightmares and uneasy insomnia can occur in happy homes, too: what is irrelevant to some people can be frightening or painful to others. Between the two extremes, across the whole range of possibilities in the wheel of fortune, children's dreams tend to reflect the situations that the dreamers have experienced, both on the affective plane and the symbolic one.

Dreaming is a slow, gradual learning process that probably begins in the maternal uterus, with the formation of the first sensorial representations at the body's border with the outside world. These impressions, which are still only diffuse, are imagined reflections of the outside world, flickering shadows in the depths of a cave in which bit by bit we find ourselves living. Over the course of childhood and adolescence, dreams reflect the experiences of youth, filled with novelty and expectation. In adulthood, people become accustomed to routine and sometimes forget about themselves, but even in old age they remain capable of transporting their minds away. First experiences, first desires, first dreams are the foundations of this ability. It is for this very reason that elderly people dream and get so emotional about their tender, almost eternal childhoods.

We do not know for sure what mental experiences accompany our endings, but it is striking how many religions sustain a belief in life after death. Richard Linklater's movie *Waking Life*, an extraordinary philosophical narrative made up of monologues and dialogues about dreams, life and its end, suggests that the neuronal processing that occurs during the transition toward death generates sequences of dreams that are ever more abstract, occuring in a state of altered brain activity in which time becomes elastic. This period would be dominated by the emotions and memories experienced throughout life, taking just a few seconds to create a sense of eternity in the hell, purgatory, or heaven constructed by the dreamers themselves on their own personal journey. This daring artistic conception of death received some unexpected scientific backing in 2013, when researchers from the University of Michigan reported high levels of neural activity in the brains of rats subjected to cardiac arrest, around thirty seconds after their heartbeats stopped.[21]

Whatever the end point may be, the maturing of dreams is an important corollary to the development of a well-defined personal identity. If children's dreams are often impoverished in emotions and images, static, and even contemplative, the maturing of dreams up to adulthood leads to a rich dream process in which the dreamer becomes the main agent of events, that is, an active operator immersed in their internal virtual setting—which they normally do not control, but inhabit. How this mental state came to evolve is the subject of the next chapter.

6

The Evolution of Dreaming

Because sleep is very ancient, it evolved to have a great variety of psychobiological functions, the generating of dreams being just one of these. The properties of sleep developed at very different times, under quite different evolutionary pressures. Determining a point of departure for sleep requires that we go back 4.5 billion years and imagine the conditions in which the first self-replicating molecules appeared. The planet was volcanic, with quite a lot of water and an atmosphere that did not yet have oxygen in it. The first unicellular organisms, dated to between 4.28 and 3.77 billion years ago, resembled bacteria from hydrothermal vents, which feed on iron oxidation.[1]

When the sun was in the sky, the temperature rose, facilitating molecular diffusion and accelerating chemical reactions. From the beginning of time, the setting of the sun over the horizon has been accompanied by a drop in the temperature on Earth—and a slowing of chemical reactions. This alternation, which has been almost unchanging for over 1.6 trillion days and nights, is the basis of the coupling of Earth's rotation to the behavioral cycles of almost all forms of life that have ever existed on the planet. With the exception of those environments that are very deep down, all planetary life has evolved under an alternation of dark and light approximately every twelve hours. For this reason, very similar circadian rhythms are found in almost all living beings on the planet.

Nearly 1.5 billion years had to pass before the first multicellular beings emerged. These were bacteria, which were capable of photosynthesis and the formation of cell colonies. These ancestors of today's cyanobacteria spread widely across the ocean and raised the concentration of oxygen in the atmosphere so much that they wiped out a large proportion of the life that existed 2.4 billion years ago. The cyanobacteria destroyed almost all the anaerobic creatures and led to the photosynthetic capacity of algae and plants, making the planet into a thriving producer of biomass through solar energy. This created the bases for the evolution of herbivores, which in turn created the bases for the evolution of carnivores.

ARCANE RHYTHMS

Sunlight, originally an accelerator of chemical reactions and later the energy base of the food chain, began from a certain point to be used by living organisms to detect and act upon environmental changes. Cilia and flagella began to evolve that were able to produce movement toward the surface of water, where photosynthesis is possible.[2] Biological mechanisms appeared that were capable of "turning on" and "turning off" behaviors according to the availability of light, which diversified into many other derivative mechanisms, at both molecular and cellular levels. Countless unicellular organisms demonstrated a circadian rhythm of activity and rest.[3]

In 2017, the U.S. biologists Michael Young, of Rockefeller University, and Jeffrey Hall and Michael Rosbash, of Brandeis University, received the Nobel Prize in Physiology or Medicine for their discoveries relating to the molecular clocks that determine the circadian rhythm, a period close to twenty-four hours. Studying the fruit fly *Drosophila melanogaster*, these researchers showed that the circadian clock involves a periodic variation in the levels of molecules codified by a particular group of genes, whose mutations can shorten, expand, or even eliminate the circadian period, affecting behavioral, physiological, and molecular rhythms.

The demonstration of the occurrence of periodic quiescence even in jellyfish shows that sleeping does not require a brain, and can occur even in a very primitive nervous system.[4] Melatonin, a sleep-inducing hormone produced in the first half of the night by the human pineal gland, would appear to have originated 700 million years ago, when animals resembling marine worms evolved cells capable of capturing light and moving around with the beating of cilia during the day, but not at night.[5] The mechanism for this dichotomy was the nighttime production of melatonin, which in the absence of light stimulates neurons to stop the movement of the cilia. Sinking slowly during the nighttime quiescence and swimming upward in the frenzied daytime, our arcane ancestors embodied the yin and yang of the solar cycle in what we see today as two fundamental states of the body: sleep and waking.

And then, around 540 million years ago, the first structures appeared that resembled eyes. Today, eyes are found in all animals with bilateral symmetry, possessing heads and tails as well as backs and fronts. In all these animals, eye formation in the embryo is controlled by the same genes. These are also very like the genes that regulate the circadian clock, which in vertebrates involves an important group of neurons, called the suprachiasmatic nucleus. This small cluster of neurons, of the order of merely twenty thousand cells, is responsible for the communication between the light-sensitive retinal cells and the melatonin-producing cells.

The signals of the presence or absence of light are transformed several times, from photons to structural alterations of small and large

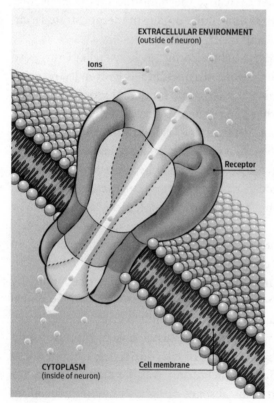

Figure 6. Receptor anchored in the cellular membrane, functioning as an ion channel. When ion channels open, ions such as sodium, potassium, chlorine, and calcium can cross from the extracellular space into the intracellular space, or vice versa.

molecules, which open and close channels that are anchored in the membranes of neurons (fig. 6). These channels open to allow the flow of ions, which lead to the release of chemical substances, which in turn activate more structural changes to molecules in other cells, and so forth, generating short-, medium-, and long-term consequences for the whole nervous system. Between us and the worms, the huge evolutionary distance notwithstanding, the ancestral function of many molecular mechanisms has remained, such as the role of melatonin in the regulation of sleep.

SLEEPING IS NOT ONLY RESTING

While rest is opportunistic, happening only when it is necessary or possible, actual sleep has a time for starting and finishing, and needs to be replaced when it is lacking. When people are exposed to the natural succession of day and night, they normally complete a cycle in a period of twenty-three hours and fifty-six minutes, as the light restarts the pacemaker every day. People who are experimentally isolated from the light-dark cycle, however, whether in caves or other closed-off environments, exhibit a sleep/wake cycle with an average period that is twenty-four hours and eleven minutes long.

The slightly longer duration of the cycle observed in the absence of physical clues to indicate the passage of time, compared to a normal light-dark cycle, indicates the evolution of a mechanism that, in the event of a delay in the morning brightness, allows the animal to keep sleeping for a little while longer, reducing the risks of predation.[6] If the sun is taking its time to rise, best to wait nice and quiet in the burrow.

In evolution, novelties that work out well tend to remain, and spread, and finally become very old. To judge by its prevalence in both invertebrates and vertebrates, sleep is extremely ancient, possibly predating the Cambrian explosion, when most animal groups originated. Today it is believed that fish date back 500 million years, followed by insects at 400 million years, reptiles at 340 million years, mammals at 225 million years, and birds at 150 million years. By way of comparison, human beings of the *Homo sapiens* species seem to have appeared a mere 315,000 years ago.[7]

The opinion of science as to which animals sleep has changed significantly in recent decades. Good and reliable quantitative behavioral studies have benefited from the advent of the computer and precise tiny-scale movement sensors to show that bees, scorpions, and cockroaches present regular periods of quiescence with a low sensitivity to sensorial stimuli.[8] Because of its importance to genetic research, the fruit fly has been well studied as regarding the question of the occurrence of sleep. Careful behavioral records have demonstrated well-defined cycles of activity and quiescence. Another indication that these flies sleep is the fact that a period of enforced sleep deprivation is followed by the characteristic rebound that is observed in mammals, that is, a compensatory increase in the time of sleep coming later.[9] In spite of the convincing demonstration that flies do in fact sleep, there seems to be no anatomical correspondence between the parts of the nervous system that are necessary for sleeping and the areas of the human brain involved in the creation of sleep.[10] All the same, these flies do share with mammals some of the important cognitive benefits of sleep, as we will see below.

THE SIMPLE SLEEP OF FISH AND AMPHIBIANS

For a long time there was discussion about whether sleep as such might not exist among fish and amphibians, in the sense of there not being a period of behavioral and physiological quiescence corresponding to some fixed phase in the day-night cycle. It is true that fish and amphibians can be seen to rest, in a fleeting fashion and without predetermined periodicity, as a function of satiety and the lack of a risk of predation. These unpredictable, momentary conditions mean that the resting we see in fish and amphibians is opportunistic. Fish often live in murky or deep waters, environments in which the difference in brightness between day and night simply doesn't exist. In the absence of light, fish depend very little on their sight for avoiding predators and finding food and sexual partners. Instead, they depend crucially on scent and on navigation using electromagnetic fields.

Those few species of fish that have already been studied in a laboratory have shown a sleep behavior evidenced by periodic quiescence, increased movement after the administering of caffeine, and behavioral deficits following sleep deprivation.[11] Nevertheless, sleep deprivation is less stressful and there is less rebound. In fish that live in and around coral reefs, we see continuous swimming day and night, but these animals are suspected to be capable of swimming in their sleep.

In amphibians, the scientific information is scarcer still. The American bullfrog, a large amphibian of diurnal habits, does present circadian variations in behavior, but its responsiveness to sensorial stimuli is greater in the period of quiescence than in the more active period. This information suggested that sleep might be suppressed in amphibians owing to their greater physical vulnerability, but a subsequent study of the small tree frog *Hyla septentrionalis* demonstrated the existence of sleep in that species—perhaps owing to its occupying a less dangerous ecological niche.[12] No sign of REM sleep has been recorded either in fish or in amphibians.

THE COMPLEX SLEEP OF LAND VERTEBRATES

Unlike fish and amphibians, reptiles, birds, and mammals have a body surface that is dry and impermeable. They also have an amniotic sac containing the embryo during its development, keeping it warm, damp, and well padded. These adaptations allow land vertebrates to occupy habitats that are very remote from the water of rivers, lakes, swamps, and seas.

Three hundred fifteen million years ago, the landmass of the planet formed a vast single continent named Pangaea. While the water environment was infested with terrible carnivorous vertebrates and invertebrates, the vast land environment explored by the first reptiles, including the lands that make up Antarctica today, was an edible Eden filled with plants and insects. To judge by the fossilized sets of teeth found in Nova Scotia, Canada, the first reptiles were insectivores quite similar to today's lizards. Initially without any natural predators and with food in abundance, these animals quickly spread into different species, occupying ecological niches that were very different, but all marked by the light-dark cycle.

The transparency of the air and the periodic abundance of light favored the evolution of an increasingly sophisticated and powerful visual system in land vertebrates. The main advantage of sight is the ability to perceive other creatures and objects at a distance. On the other hand, the major disadvantage is the periodic lack of light, which makes feeding more difficult and greatly increases the risk of nighttime predation. To this day, for almost all herbivorous land vertebrates, it is standard behavior to hide and group together to sleep at night. It is therefore possible that slow-wave sleep, which reduces the rate of metabolism and which already existed in a primitive form in fish and amphibians, developed in land vertebrates as a collateral effect of the need to hide from daytime predation. Since it is impossible to search for food inside a burrow, it became adaptive to remain immobile, reducing temperature and energy consumption, and even reaching a state of torpor.

As sleep progresses, there is a reduction in the frequency of brain waves of up to 50 percent, that is, there is a deceleration of these waves with a corresponding increase in their "size" or amplitude. We call this state slow-wave sleep. The quiescence of the whole body co-evolved with slow brain waves that make the cells momentarily silence their functioning with each wave cycle.

It is very hard to conjecture about the events that led to the evolution of REM sleep, the phase during which the most vivid dreams take place. For a time, it was believed that the origin of the difference between slow-wave sleep and REM sleep could be dated to the Triassic period, 225 million years ago, which saw the evolution of those ancestors common to all mammals. These were nocturnal insectivores, physically similar to small rodents.[13] This narrative was based on the premise that reptiles and birds do not have REM sleep, an opinion that was predominant among experts for some decades.

However, this state of high brain activation with minimal corporeal activation, which until recently has been held to be exclusive to mammals, is today well documented in several bird species and some reptile species.[14] For a time there was controversy over the existence of REM sleep in the echidna, an odd insectivorous mammal from Australia and New Guinea, endowed with defensive spines and a proboscis specialized for the extraction of ants, termites, worms, and larvae. The supposed absence of REM sleep in this species would have been no more than a curiosity were it not for the fact that the echidna is an animal of the Monotremata order, primitive mammals that possess certain reptilian characteristics, such as the absence of placentas and reproduction with egg-laying. If REM sleep is absent from those animals closest to the mammals' common ancestor, it is likely that REM sleep evolved independently in mammals, birds, and reptiles. However, more recent electrophysiological studies have demonstrated the presence of REM sleep both in echidnas[15] and another monotreme, the platypus,[16] which can spend a full eight hours a day in this state, the greatest ever observed in any species. The ostrich, one of the birds that most closely resembles the common ancestor of all birds, demonstrates a sleep pattern very similar to that of the platypus.[17] The results strengthen the hypothesis of a single origin for REM sleep common to land vertebrates. The separation of slow-wave sleep from REM sleep could therefore have originated 75 million years before the Triassic, in the Carboniferous period, the time of the invasion of the land by the ancestors of amphibians and reptiles.

THE END OF THE SLEEPING DRAGONS

If the assertion above is correct, it is very likely that the dinosaurs, reptiles of all sizes that dominated the planet from 230 million years ago,[18] slept and dreamed in a way very similar to that adopted by their closest relatives that still exist on the planet: birds. This would have been a cyclical sleep pattern, marked by the quick, irregular alternating of slow-wave sleep and REM sleep. There is obviously no fossil in existence that might corroborate this hypothesis, but there was nonetheless an intriguing discovery in western China of the remains of two troodontidae, dinosaurs from the start of the Cretaceous period that had feathers and were phylogenetically close to birds.[19] The fossils were found in a similar position to that taken by birds when asleep, with their neck bent downward so as to tuck their heads beneath their forelimbs. It would appear that they were asleep when surprised suddenly by death. It is tempting to imagine that REM sleep played an important role in the dinosaurs' domination of the planet. The presence of REM sleep in so many different species suggests a feature of great physiological significance. What important function could this have fulfilled, and what pressures of selection shaped it?

One interesting hypothesis posits that REM sleep appeared as a preparation for waking up, raising the activity of the neurons in the cerebral cortex to a level close to waking, after the long period of low activity that characterizes slow-wave sleep. The main argument in favor of this hypothesis is the fact that individuals woken from slow-wave sleep show sensorial, motor, and cognitive deficits that take some minutes to dissipate. Besides, awakenings normally happen after REM sleep, suggesting that it functions as a facilitator of the transition between slow-wave sleep and waking. It is possible to imagine how waking with the capacity to be fully alert would provide those vertebrates able to enter REM sleep with important competitive advantages. Another possibility is the importance of REM sleep for the establishment of a proper correspondence between muscular cells and neurons during development. The cerebral activation of the motor areas during REM sleep, which stimulates brief contractions all around the body, makes it possible to calibrate movements and actions even in newborns, well before they are carried out in the real world.[20]

It is tempting to imagine the advantages that REM sleep might have given the dinosaurs. Could this state have contributed to the large reptiles' ecological hegemony over 170 million years? What role did REM sleep play in the hard-fought struggle for survival during their long reign? The subject is a fascinating one, though necessarily speculative. The fact is, with or without REM sleep, 66 million years ago chance intervened and the dinosaurs were wiped off the map. By an incredible coincidence of uncommon events, an asteroid fell onto what is today the Yucatán peninsula in Mexico and totally changed the course of life on Earth.[21] The fall at 45,000 miles per hour of a rock of between six and nine miles wide and weighing between 10^{12} and 10^{14} tons onto an area of shallow seas with large reserves of gypsum, a mineral rich in sulphur, caused a vast emission of toxic gases immediately following the impact. Unusual seismic shocks and an increase in volcanic activity were followed by intense changes in climate. The gas emissions were so great that a thick layer of clouds blocked out the sun for months, possibly years.[22] After the intense heat of the explosion, equivalent to an atomic bomb 10 billion times more powerful than the one at Hiroshima, what followed was a profound, persistent winter. Photosynthesis on land and in the water was interrupted. Altogether, these changes wiped out 75 percent of animal and plant species in a brief space of time. All the dinosaurs disappeared except for those that were the ancestors of birds, as did countless species of mammals, fish, mollusks, plants, and even plankton. If the asteroid had fallen just a little earlier or a little later, it would have struck deep water and the consequences of the impact would have been far less significant. Once again, the dynamics arising from the planet's rotation on its axis had a decisive influence on the evolution of life on its surface.

CRISIS IS OPPORTUNITY

The mass extinction at the end of the Cretaceous period allowed the radiation of species on a large scale, marked by an accelerating divergence of the morphological characteristics of those groups of animals that survived the catastrophe. Species previously constrained within ecological niches full of competition and predation suddenly found themselves in niches that had been vacated on various levels of the food chain, especially at the top: all the large predators had disappeared. The new evolutionary pressures meant that after the extinction, many new species of mammals appeared, as well as birds, lizards, and fish, in a process of gradual adaptation to the new available niches. Primates and cetaceans, the two groups of mammals with the greatest cognitive capacity, both spread globally after the extinction of the dinosaurs.

The drop in temperature after the sun sets makes nighttime activity almost inviable for reptiles, which depend on external heat to warm themselves up and thus activate their metabolism. It was precisely their capacity to generate body heat that allowed the mammals to occupy ecological niches during the night, even with great variations in ambient temperature, at different times of the year. Meanwhile, a comparison of the behavior of almost 2,500 different mammal species suggests that the strictly diurnal habits of mammals arose only after the extinction of the dinosaurs, with the appearance of simian primates between 50 and 30 million years ago.[23] The need to conserve energy during sleep favored the evolution of the behavior of grouping several individuals together to sleep at night, which is so typical among mammals and birds, especially in cold environments.

The advent of an internal control for heat production favored REM sleep, since the maintaining of body temperatures at a suitable level is essential for this kind of sleep to occur. The echidna, for example, only exhibits REM sleep when the ambient temperature is around 77°F.[24] Meanwhile, REM sleep co-evolved with an almost complete relaxation of the muscles, which makes it possible to maintain strong cortical activation without any great motor repercussions. The almost complete relaxation of the body during REM sleep allows dreaming to reach a high level of vividness without waking the animal or causing undesirable behaviors, which might attract predators.

THE LONG REM SLEEP OF MAMMALS

One of the main differences in the patterns of REM sleep in the different species of vertebrates concerns its duration. While reptiles and birds have short sleep cycles, with episodes of REM sleep that are just a few seconds long, REM sleep in mammals frequently lasts dozens of minutes, and can reach up to an hour or more in some species. In general the quantity of REM sleep is in inverse proportion to body weight, such that smaller animals tend to have more REM sleep. However, if the effect of body weight is excluded from the analysis, we see a strong correlation with the degree of immaturity of the body at the moment of birth. Animals that are relatively mature at the moment of birth, such as sheep and giraffes, which demonstrate quite a lot of autonomy soon after they are born, are characterized by small amounts of REM sleep (around one hour per day in total). Meanwhile, those mammals that are relatively immature at their time of birth, such as humans and platypuses, have amounts of REM sleep that are enormous, especially in the earlier phases of life.

A human baby cannot feed itself, move about, defend itself, or clean itself. A baby platypus does none of these things either, and cannot even regulate its own temperature without contact with its mother. Curiously, both of these spend around eight hours a day in REM sleep. In newborn mammals, which have not yet opened their eyes, the high levels of electrical activity prompted by the large amount of REM sleep protect the brain from atrophy caused by the absence of stimuli. The important role of REM sleep during the development of the embryo and in extrauterine learning is related to its role in the regulation of those genes that are used by the neurons for maintaining and modifying their connections.

In short, REM sleep has a central role in the development of the fetus and the newborn, especially in animals that are more immature at birth, which need a large amount of modification to reach their adult phase. Immaturity is a disadvantage at the beginning of life, since neonatal fragility requires constant parental care. However, this characteristic becomes a great advantage in the long term, when the successful individual, who has escaped from any lethal dangers in childhood and had the chance to develop to adulthood in the care of good guardians, learns how to optimize its occupation of its ecological niche as a result of the extensive collection of memories and skills it has acquired.[25] As we will see below, REM sleep plays a crucial role in the consolidating of learning in the long term. For anybody who needs to learn a lot, REM sleep is vital.

SWIMMING, FLYING, MIGRATING

The adaptation of mammals, birds, and reptiles to aquatic and aerial niches, as well as to migrations, is associated with profound changes in sleep patterns. Elephant seals migrating between Alaska and California spend up to eight months in the sea without being able to rest on terra firma. During their migration, these animals periodically dive to depths of more than nine hundred feet. In some of these dives, they stop swimming and just allow themselves to sink, spinning gracefully, probably asleep. This circular movement allows the elephant seals significantly to reduce the speed at which they sink toward the bottom of the Pacific Ocean.[26]

Eight thousand miles away, off the Seychelles, experiments tracking sea turtles equipped with devices for recording time and depth have shown long dives without taking a break for breath, reaching sixty-five feet in depth and fifty minutes in duration. Sensors on their jaws show that while carrying out these deep dives, the sea turtles cease to make the mouth movements that pump water, a behavior that is necessary to enable an olfactory sensing of the environment. The results suggest that sea turtles sleep submerged right in the middle of the Indian Ocean.[27] The strategy of sleeping far from the surface is adaptive, as animals swimming close to the surface have their silhouettes very visible to predators approaching from below. Besides, the surface limits possibilities of escape.

Since it is not guided, but rather the product of chance, evolution will often result in quite different solutions to the same problems. Unlike elephant seals and the turtles of the Seychelles, cetaceans such as whales and dolphins do not sleep submerged in the water, but they exhibit unihemispheric sleep, that is, they are capable of sleeping with just one hemisphere of their brain at a time.[28] This allows them to remain in constant motion, emerging periodically to breathe. The absence of REM sleep in these animals has been interpreted as evidence that retaining a part of the brain at high levels of electrical activity, so as to maintain continual motor activity, is enough to meet the demand generated by the lack of REM sleep.

For cetaceans, unihemispheric sleep might be the only way of sleeping, but in birds, episodes of unihemispheric sleep are mixed with bihemispheric episodes that include REM sleep.[29] The high risks and energy costs involved in long-distance migration can lead to surprising adaptations. White-crowned sparrows, which make an annual migration of more than 2,500 miles from Alaska to California, show a reduction in sleep of almost 70 percent during the migration period, even when they are unable to fly because they are trapped inside cages. Intriguingly, in this period they show none of the behavioral deficits typical of sleep deprivation.[30] Genetically and hormonally programmed to carry out their impressive long-distance migration every year, the white-crowned sparrows simply skip their sleep during this period—and show no signs of missing it.

Decades ago, it was suggested that unihemispheric sleep while flying might explain those flights that last days and even weeks without a break. In 2016, a team led by the ethologist Niels Rattenborg, of the Max Planck Institute for Ornithology, published the first evidence of unihemispheric sleep during flight. In collaboration with researchers from the Swiss Federal Institute of Technology and the University of Zurich, Rattenborg implanted small electronic devices in the skulls of frigate birds, seabirds that make their nests on the Galápagos Islands. The miniaturized sensors recorded movements of the head but also the brain waves produced by the electrical activity beneath the skull, called electroencephalographic waves (EEG). Frigate birds have the largest wing surface-area relative to their body weight of any birds, and are able to fly over the ocean for weeks, without once coming to rest. When the researchers recovered the devices and analyzed the data, they confirmed that in ten days the frigate birds had flown nearly two thousand miles without landing, alternating long periods awake with short periods of sleep. During the day the frigate birds remained alert, foraging actively, but once the sun had set, they began to fly at a higher altitude, going into a state of unihemispheric sleep for several minutes, circling on the rising air currents and keeping one eye open, turned to the direction in which they were flying.[31]

SLEEPING IS DANGEROUS

In addition to the need to remain in motion, unihemispheric sleep seems to be connected to the maintaining of high levels of alertness, which

can reduce the risk of predation. In order to investigate this phenomenon, a group of researchers under Rattenborg's leadership recorded
brain waves during sleep in groups of four ducks simultaneously, lining
the animals up side by side, so that the two in the middle were in a safer
position—flanked on both sides—while the animals on the edges found
themselves with only one side garrisoned and therefore less safe. The
results showed a substantial increase in the quantity of unihemispheric
sleep in the animals on the edges, which were potentially more exposed
to predators. The open eye in each episode tended to be the eye pointing toward the unprotected side.[32] The animals placed in the center
exhibited normal sleep in both cerebral hemispheres.

The high levels of predation in sub-Saharan Africa and the long migratory journeys also impose tough restrictions on the sleep of mammals on the African savannas. The penalty for sleeping too much is
loss of one's offspring or even one's own life. Fitting elephants' trunks
with actimeters, small devices capable of continuously recording movement, allowed the South African ethologists to demonstrate that these
animals sleep standing up, actively protecting their children through
the night. Adult elephants sleep only two hours a night, in fragmented
episodes.[33] Among baboons, the socially dominant animals are more
alert, with fewer episodes of relaxed sleep, suggesting that sleep is reduced by social stress.[34]

The historical references to human sleep in antiquity and the Middle
Ages occurring in two consecutive nighttime phases converge with the
observation of the same phenomenon in modern agricultural populations without access to electric light.[35] Would this be different in groups
of hunter-gatherers? In order to investigate this question, researchers from the University of California–Los Angeles equipped hunter-
gatherers from Tanzania, Namibia, and Bolivia with actimeters. To the
researchers' surprise, the results confirmed that these people's sleep
took place in a single nighttime phase, with a duration very similar to
that experienced by adults in the world's large industrial metropolises.[36]
However, another study on hunter-gatherers in Tanzania showed that
the adults in the group hardly ever all slept at the same time: while
the oldest went to bed earlier and woke earlier, the younger ones slept
and woke later. As a result, at any one moment at least one-third of
the group was awake.[37] Since older people tend to sleep less, the study
suggests that our ancestors' grandparents fulfilled a crucial role in keep-

ing watch at night, as was required to reduce the risk of predation. The briefer, more superficial and flexible sleep of the hunter-gatherers allows for a refined harmony with environmental changes, risks and opportunities alike. They don't punch a clock at the office, nor do they have scheduled times to plow or harvest, but they need to be very alert to any arrhythmias of nature.

When the first hominids spread across Africa millions of years ago, they were as well equipped to sleep and dream as any other mammal. These were dreams of dangerous escapes and hunts that our ancestors transported out of the African continent countless times, via successive migrations, until a group of about a thousand people left east Africa around seventy thousand years ago and spread their descendants across Asia, Oceania, Europe, and finally America in the subsequent millennia.[38] What followed, the long trajectory of our ancestors across the whole planet, gradually moved us away from the natural world toward the cultural world, changing the way in which we sleep[39] and creating a dream space filled with symbols for designating all creatures and things, even those that were merely imagined. In order to understand this transition, we need to delve into the biochemistry that governs the figments of our imagination.

7

The Biochemistry of Dreams

Nighttime has arrived. After many hours of intense motion and alert reasoning, we assume a horizontal position and embark on a radical journey of consciousness alteration. When we lay our head on our pillow and close our eyes to sleep, profound changes occur in our brain waves, and chemical substances are released by our nervous systems in many different ways. First, we experience the surrender to the darkness with eyelids closed, beginning a reversible disconnection between our body and the outside world. Next come the transitory dreams from the start of sleep, which then give way to dreamless (or nearly dreamless) sleep, a state of abandoned quiescence and a great reduction in sensorial reactiveness. Finally, nearly two hours later, the intense, vivid dreams begin, which are the ones we sometimes remember on waking.

In the mid-twentieth century, the ancient conception of sleep as a state of homogenous calm induced by an absence of stimuli suffered a terminal shock. The discoveries that refuted the theory of sleep as a passive process originated in the revolutionary studies of the sleep/wake cycle at the University of Chicago, carried out by the U.S. physiologist Nathaniel Kleitman and his then doctoral student Eugene Aserinsky. With careful observation of eye motility during the sleep of twenty adult volunteers, the researchers discovered that the periods of quiescence alternated with a more agitated sleep, REM sleep,[1] which involved rapid movements in both eyes, choppy breathing, irregular heartbeat, and fast brain waves (fig. 7)—all this in spite of a general relaxation of the body. The publication of this incredible discovery in the journal *Science* in 1953 gave a huge impetus to the characterizing of the different phases of the sleep/wake cycle.

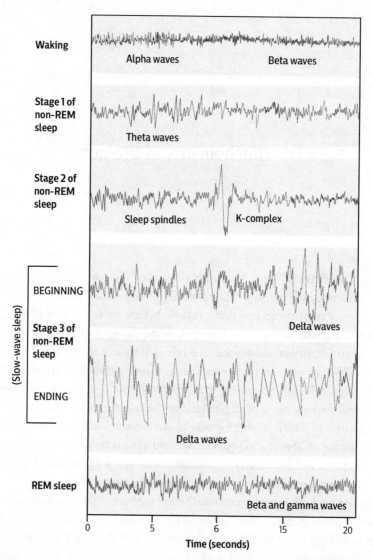

Figure 7. *The brain waves recorded by electroencephalography (EEG) vary signifi-cantly in different phases of the sleep / wake cycle. Each phase is marked by distinc-tive brain waves, characterized by different speeds (frequency) and sizes (amplitude). One complete cycle passes in sequence through all of the phases described, from top to bottom. The transition into sleep involves the occurrence of a number of specific brain waves, beginning with large, very slow electrical oscillations called K-complexes, which occur as a single wave with duration typically under one sec-ond, and is often followed by oscillatory bursts of a higher frequency (approximately ten cycles per second) called sleep spindles. As the sleep gets deeper, we see the appearance of consecutive slow waves in tandem, under four cycles per second, called delta waves, which become slower and increase in size as the sleep goes on.*[2]

THE PARADOX OF REM SLEEP

Following the identification of REM sleep in Kleitman's lab, William Dement, then a doctoral student, decided to carry out a deeper investigation of something Kleitman and Aserinsky had observed regarding a possible increase in the frequency of dreams during REM sleep. Upon waking research volunteers from this exact phase in their sleep, Dement and Kleitman reported in 1957 that about 80 percent of the episodes of REM sleep coincided with dreams—much more than in non-REM sleep, in which dreams had been occurring in fewer than 10 percent of episodes.[3]

Two years later, the French neuroscientist Michel Jouvet, from Claude Bernard University in Lyon, began to publish important studies on the physiological properties of REM sleep—which he named "paradoxical sleep," owing to its being a state of heightened cortical activity but almost complete corporeal quiescence. This quiescence originates from a small group of neurons whose activation, specifically during REM sleep, secretes neurotransmitters that inhibit motor neurons directly involved in the muscular control of posture. Among other discoveries, Jouvet showed that cats subjected to a lesion in this group of neurons start to move about vigorously during REM sleep, performing—though asleep—various behaviors typical of the species, such as attacking, exploring, or mewing.[4]

Jouvet interpreted these behaviors as evidence that cats dream during REM sleep. Though the cerebral regions associated with vision and preparing movements are quite well activated during this state, sleep is not interrupted. This is only possible because REM sleep happens in a condition of almost complete suppression of motor responses, as Jouvet discovered. However turbulent the plot of a dream might be, the dreamer's behavioral reactions are almost entirely inhibited.

Taken altogether, the experiments of Kleitman, Aserinsky, Dement, and Jouvet wiped from the map the idea of sleep as a state of brain inactivity, instead bringing into play the notion of sleep as an active state, during which the brain processes information as intensely as it does during waking. The discovery that dream activity occurs during a well-defined neurophysiological state—REM sleep—began to tame a phenomenon that up until that moment had remained elusive. It became possible to determine with precision at what moment a person is dreaming. This opened the way to understanding the functions of sleep and dreams.

In spite of their enormous influence, the original observations made by Kleitman and his team were not totally validated by subsequent research. Using a broader definition of dreaming, which can encapsulate the whole spectrum of mental content that can occur during sleep, David Foulkes showed as early as the 1960s that at least 50 percent of wakings that happen outside REM sleep are concomitant with dream activity of some kind.[5] Dreams comprise a vast range of different but related experiences, from the scenes that accompany the period of falling asleep, through the fragments of thoughts and feelings that occur during slow-wave sleep, culminating in the vivid, intense dreams of REM sleep, with their very well-structured narratives.

THE STRUCTURE OF SLEEP

We know today that in general the sleep of mammals is in two stages, characterized by striking differences in the levels of brain activity. The first major stage of sleep occurs mainly in the first half of the night and is subdivided into three sub-stages of an increasingly deep sleep, collectively called non-REM sleep (NREM). The second main stage—REM sleep—occurs mostly in the latter part of the night. One complete cycle of human sleep lasts around ninety minutes and consists of a fixed sequence of successive states: N1 → N2 → N3 → REM sleep. This cycle is repeated four or five times per night, until the person wakes up.

But let us return to the start of sleep in order to better understand its dynamics. The process of falling asleep begins with the disappearance of the alpha waves, which are typical of the state of being awake with closed eyes, and the appearance of what are called theta waves, typical of the N1 state. The first dream images occur in this initial stage of sleep and continue through the following stage, N2, apart from when the brain waves called K-complexes are occurring (fig. 7). These quite slow, isolated waves, which are typical of N2, provoke a mental shutdown, a sudden loss of consciousness that precedes N3, dominated by waves that are equally slow but that occur in series, called delta waves.

Stages N1 and N2 are very short, almost always between five and twenty minutes in duration. The N3 state lasts longer, but its episodes get shorter as the night progresses. REM sleep, meanwhile, happens in short episodes at the start of the night, getting progressively longer, with the longest episodes at the arrival of morning. While the first episode of REM sleep of the night lasts only a few minutes, the duration of the final one can exceed an hour.

The episodes of REM sleep do not only get longer over the course of the night, but also progressively more intense. There is an increase in eye movement, localized muscular spasms, and the vividness of the reports of the dreams, as well as vaginal blood flow[6] and penile erections.[7] REM sleep reaches maximum duration when the body temperature reaches its lowest level. In spite of the lack of thermoregulation of the body during REM sleep, certain areas of the brain show an increase in temperature.[8] In situations of thermal discomfort, whether this is above or below the normal band of body temperature, REM sleep is substantially reduced, while NREM sleep persists.

NEUROTRANSMITTERS AND MENTAL STATES

These huge variations in mental content in the different phases of sleep are related to corresponding variations in the levels of neurotransmitters. When a person is awake, their brain releases large quantities of noradrenaline, serotonin, dopamine, and acetylcholine neurotransmitters, whose origins go all the way back to the first animals, more than 500 million years ago. These neurotransmitters play important roles in the modulating of attention, emotion, motility, and motivated behaviors in general.

When the eyes are closed and the body relaxes to sleep, there is a reduction in sensorial stimuli and a shift in the balance between the different neurotransmitters. During slow-wave sleep, there is a small reduction in the levels of dopamine, while acetylcholine levels begin to oscillate strongly. At the same time, there is a reduction in the levels of three neurotransmitters that are very important for brain function: noradrenaline, serotonin, and histamine. This happens because the production centers for these neurotransmitters are inhibited by the intermittent release of acetylcholine, as slow-wave sleep gets progressively deeper. Finally, in the shift to REM sleep, the levels of acetylcholine rise sharply, the levels of dopamine increase slightly, and the levels of noradrenaline and serotonin plummet to practically zero. What have these chemical changes got to do with the experience of dreaming?

In 1977, the Harvard University psychiatrists J. Allan Hobson and Robert McCarley proposed the theory that the great change in subjective experience in the transition to REM sleep can be explained by the activation of those cells that produce acetylcholine and the deactivation of those cells that produce serotonin and noradrenaline. Variations in the levels of these neurotransmitters are enough to explain five fundamental characteristics of dreams: while (1) the intense emotions and (2) the strong sensorial impressions derive from the high levels of acetylcholine, (3) the illogical content, (4) the uncritical acceptance of the dream events, and (5) the difficulty in remembering them upon waking are the result of the almost nonexistent levels of noradrenaline and serotonin. Since it was first formulated, Hobson and McCarley's theory has influenced several generations of neuroscientists, propelling them toward the search for pharmacological and anatomical explanations for dreams. This is not a matter of reducing a psychological phenomenon to biology, but of trying to understand how it is that the chemical interaction of totally unconscious cells generates the somewhat conscious experience of dreaming.

DETOXING AND RESTORATION

Since Hobson and McCarley first proposed their theory, many other facts relating to sleep have been discovered, which have made the explanation of the phenomenon much more complex. As an evolutionarily very ancient behavioral state, sleep serves many biological functions, based on mechanisms that evolved at different moments and have distinctly different synergistic benefits. Only in the last five years has it become clear that one of the most important functions of sleep is the detoxing of the brain. Neural functioning while awake generates unwanted molecular by-products, such as the proteins called beta-amyloids, the accumulating of which is related to Alzheimer's disease. Experiments with dyes and radioactively tagged beta-amyloid show that falling asleep considerably expands the tiny space between the cells in such a way as to increase the spread of toxins through the cerebrospinal fluid, the transparent liquid produced by the brain that communicates with the circulation of the blood and allows the exchange of substances with the rest of the body.[9] It is possible that this effect is due more to the position of the body than to sleep per se,[10] but since human sleep almost always occurs in a horizontal position, in practice sleeping provides a quick and efficient cleaning out of the molecular trash accumulated by the brain while awake. It is therefore not surprising that a quick nap can have such restorative effects on one's attention, nor that sleep deprivation is a risk factor for Alzheimer's. A study of 177 French adolescents showed that a reduction in the duration of sleep correlates to worse school performance and to a smaller volume of gray matter in various areas of the cortex.[11]

SLEEP ILLNESSES, SLEEP PILLS

Being a biological function that is complex and essential for life, sleep is also subject to countless physiological or psychological disorders. The main pathologies that are directly or indirectly associated with sleep are nocturnal apnea, West syndrome, epilepsy, night terrors, somnambulism, restless leg syndrome, narcolepsy, cataplexy, and the recurrent nightmares that are characteristic of post-traumatic stress disorder, which will be dealt with below. While episodes of somnambulism and night terrors happen during slow-wave sleep at the start of the night, those well-formed nightmares with high levels of anxiety typical of REM sleep occur in the second half of the night. Disturbances in both phases of sleep are linked to anxiety, depression, and psychosis. Patients with neurological damage similar to those of Michel Jouvet's dreaming cats can develop a behavioral disorder of their REM sleep, characterized by an explicit acting out of their dreams.

A number of different substances inhibit the coming of sleep, such as the orexin naturally produced by the brain, the lack of which causes narcolepsy, marked by excessive drowsiness, the sudden onset of REM sleep, and, frequently, by the abrupt loss of muscle tone, called cataplexy. Among those substances that inhibit sleep are caffeine, amphetamines, methylphenidate (Ritalin), and cocaine, which are either extracted from plants or synthesized in the lab. Other substances encourage drowsiness, some of them produced by the body itself (adenosine, melatonin, and leptin) and others made industrially, such as alcohol, barbiturates, benzodiazepines (such as Diazepam and Rivotril), and Z-drugs (such as Zolpidem). In the latter cases, the quality of sleep can suffer, representing something that is more a temporary cerebral shutdown than a natural period of quiescence and memory processing.

If sleep is made up of different physiological states with specific profiles of neurotransmitters, it is unsurprising that it would be altered by these substances, by their analogs, and even their biochemical precursors (the raw materials used in their production). People with Parkinson's disease, who have low dopamine production, are normally medicated with L-dopa, a molecule that is the basis for the synthesizing of dopamine. This treatment can cause powerful dream experiences, which patients describe as real hallucinations.[12]

SLEEPING AND SPORTS

One of the areas in which sleep science has the greatest application is in high-performance sports. Intensive exercise causes a loss of fluids, damage to muscle fibers, and the exhaustion of biochemical sources of energy, such as glycogen. The adequate restoration of the cells and tissues by sleep is essential if an athlete is to maintain strength, precision, stamina, and speed.[13] On average, an athlete who is aged eighteen has a reaction time to stimuli that is much lower than an athlete who is forty, but this difference can be eliminated if the younger athlete is deprived of sleep and the older one has slept well. Sleep deprivation also has a negative impact on the production of testosterone, which increases muscle mass in men as well as women and is released mostly during sleep.[14]

High-performance sports trainers almost always use special sleep regimens as part of their preparation of their athletes, after a competition as well as before it, aiming for a reduction in reaction times, a sharpening of motor coordination, and the replenishment of metabolites.[15] In Formula One car racing, a good part of Ayrton Senna's exceptional performance as a three-time world champion was credited by his trainer Nuno Cobra to a strict adherence to a regime of going to sleep early. In American football, the use of power naps has become common, and stars like Tom Brady stop all activity at nine p.m. in order to get nine hours of uninterrupted sleep.

NEUROGENESIS AND HORMONE REGULATION

One of the earliest functions of sleep relates to its contribution to neurogenesis: the creation of new neurons. In humans, neurogenesis continues until the start of adolescence[16] and is located in the dentate gyrus, a layer of neurons that serves as the entrance to the hippocampus for sensorial information of various kinds. Sleep deprivation causes neural inflammation and a reduction in neurogenesis in the dentate gyrus, two effects that are associated with depression.

Another of the essential roles of sleep is the control of the levels of some of our bodies' most important metabolic regulators, such as the growth hormone that is needed for cell reproduction and development, and the hormone cortisol, which is essential for responding to stress. In the first half of the night, which is mostly filled with slow-wave sleep, levels of the growth hormone reach their peak, while cortisol levels are at their lowest. In the second half of the night, which is mostly spent in REM sleep, the hormone profile is reversed: the release of growth hormone stops and that of cortisol increases, reaching its peak at the beginning of waking. In normal conditions, cortisol levels will then remain low for the rest of the day,[17] but stressful situations can increase the level of this hormone at any moment. One of the many consequences of this increase is a weakening of the synapses in the hippocampus,[18] which is damaging to learning and harmful to previously acquired memories.

Sleep has a close relationship to the regulating of the appetite, too. People who sleep little see increases in the level of the hormone ghrelin and a reduction in leptin, which increases the ingestion of food and the tendency toward obesity. Chronic sleep deprivation causes a devastating combination of metabolic, hormonal, emotional, and cognitive damage, and is a risk factor for illnesses as varied as cerebrovascular accidents, multiple sclerosis, headaches, epilepsy, somnambulism, Alzheimer's, and psychosis.

MICROBIOTA, SLEEP, AND MOOD

If sleep is susceptible to significant modification by chemical agents, it is not surprising that it should also be affected by the vast array of bacteria, viruses, yeasts, and protozoa that make up our microbiota. The discovery of this relationship goes back to 1907, when the French psychophysiologists René Legendre and Henri Piéron began to carry out pioneering experiments transfusing cerebrospinal fluid between pairs of dogs, in which one of them—the "donor"—had previously been deprived of sleep for up to ten days. The results showed that the "recipient" animal, which had not been subjected to the sleep deprivation, fell into a deep sleep about an hour after the transfusion. Legendre and Piéron interpreted these results as evidence of the accumulation in the waking brain of a sleep-inducing substance.[19] During the same period, the Japanese physiologist Kuniomi Ishimori carried out similar research and came to the same conclusion. In 1967 this substance was isolated, and in 1982 it was finally identified as the peptide muramyl, which originates in the cell wall of bacteria and induces a slowing of brain waves—which can explain the increase in slow-wave sleep and the reduction in REM sleep when a person is ill due to an infection.[20]

It is calculated that a typical adult, in their usual flora, has 50 percent more microorganisms than cells of their own body. Intestinal micro-organisms alter the quantity of serotonin produced by the roughly 500 million neurons located in the walls of the alimentary canal, a digestive nervous system that sends far more axons to the brain than it receives. While this system is not directly involved in the making of decisions or in the planning of actions, it can have no small amount of influence on these processes. Serotonin has a decisive impact on digestion, but it also produces powerful effects on the mind, altering the mood. Most of the serotonin produced by the body is found in the alimentary canal, which explains the link between strong emotions and gastrointestinal disturbance. Indeed, depression is affected by microbiota through a number of mechanisms, including changes to sleep patterns.

Interestingly, fasting has been used and continues to be used in the world's major religions—Christianity, Islam, Hinduism, Buddhism, and Judaism—to attain transformative visions. Amerindian peoples are known for the use of fasts to induce meaningful dream revelations, as also happened abundantly in ancient Egypt, Greece, and Rome.[21] One contemporary study of the relationship between feeding and dreams looked at almost four hundred people in Canada and confirmed that long periods of fasting are associated with more vivid dreams.[22]

THE CHEMISTRY OF DELUSIONS

Even if a number of substances can induce sleep, there are not many that can convincingly emulate the experience of dreaming. The drugs that have come closest to this are psychedelic substances, which can provoke a wide range of effects from subtle alterations of perception and emotion to truly dreamlike hallucinatory experiences. If the brain is a pharmacy, then an understanding of the chemistry of delusions can allow us to mimic their natural processes; by using certain plant, fungal, and animal extracts, we may pay a visit to the dispensary of dreaming. Endocannabinoid neurotransmitters have analogs in plants, such as the delta-9-tetrahydrocannabinol molecule (THC) and the cannabidiol molecule (CBD), two of the more than a hundred cannabinoid molecules found in cannabis.[23] Among the analogs of serotonin are N,N-DMT from the leaves of *Psychotria viridis* used in ayahuasca, 5-MeO-DMT from the Amazonian snuff made from the bark of *Virola theiodora* and from the secretions of the Sonoran Desert toad *Bufo alvarius*, the mescaline present in the peyote cactus *Lophophora williamsii*, the psilocybin of the *Psilocybe cubensis* mushroom, and lysergic acid diethylamide (LSD), synthesized from an alkaloid of the ergot fungus.[24] The root of the African iboga plant used in the Bwiti religion contains a potent psychedelic alkaloid called ibogaine. Leaves from the Mexican *Salvia divinorum* contain salvinorin, a substance capable of inducing rapid and intense dissociative trances. It is fascinating to imagine the very long process of discovery of the pharmacological properties of these fungi, animals, and plants of power, pored over for many millennia in bold self-experimentation, in search of the modes of use and the doses that distinguish venom from remedy.[25] Now the body had become the laboratory.

All of the molecules described above act via receptors (proteins capable of altering their shape following a connection to specific molecules) anchored in the cell membrane of neurons. Often these receptors are channels that open up when they change shape, allowing the passage of ions such as sodium and calcium through to the inside of the cell (fig. 6). In other cases, these receptors turn into enzymes when they change shape, prompting chemical reactions inside the cells. In the case of LSD and 5-MeO-DMT, the main receptors activated are the serotonin receptors. In the case of cannabinoids, the main receptor activated in the brain is called CB1.

CANNABIS, SLEEP, AND ECSTASY

The first cannabinoid to be discovered in the brain itself was called anandamide, a fusion of the amide chemical structure with the word *ananda,* which means happiness in Sanskrit. Anandamide is a powerful inducer of slow-wave sleep and REM sleep, leading to a reduction in waking time. Other significant endocannabinoids, such as 2-araquidonoyl-glycerol, are also sleep inducers.

The similarity between dreams and the effects caused by cannabis is partial but undeniable, especially for the diffuse cognitive change that reduces short-term memory while at the same time increasing creativity. Several pieces of research into the effects of cannabinoids attest to the complexity of the changes they cause. The cannabinoid THC is a stimulant—it accelerates thought and provokes the imagination. In small doses it can increase the duration of slow-wave sleep, but in high doses it is an anxiogenic, causing an increase in wakefulness and a reduction in REM sleep. The cannabinoid CBD is anxiolytic, protecting against short-term memory deficits, increasing the time spent awake, and reducing the duration of REM sleep. In excessively high doses, both of these induce sleep.

It is probably for this reason, and because of the residual amnesiac effects of cannabis consumed before sleeping, that its users report finding it harder to remember a dream after using the plant. As a result, cannabis and its constituents can be effective for treating the recurring nightmares that are typical of post-traumatic stress disorder.[26]

If the reduction in REM sleep caused by cannabis effectively reduces the possibility of dreaming and of remembering dreams, the effects of cannabis consumed while awake are themselves oneiric. Perception is enriched, the boundaries between things seem less fixed, logical connections are loosened, remote ideas become connected, and thoughts become more interesting. It is as if cannabis reduced nighttime dreaming (nightdreams) in favor of daytime reverie (daydreams).

SEROTONIN AND PSYCHEDELIA

There is a striking likeness between the dream state and the effect produced by psychedelic substances similar to serotonin, such as LSD and dimethyltryptamines (DMT) like N,N-DMT and 5-MeO-DMT.[27] The powerful effects of these molecules on mental functioning were initially posited by psychiatry in the 1950s as models of psychosis. In 2017, a pharmacological study by the team of Swiss psychiatrist Franz Vollenweider, of the University of Zurich, showed that the activation of serotonin receptor 5-HT_{2A}, a molecule that is found in our brain, is fundamentally necessary for LSD to cause subjective effects similar to dreaming, such as an increase in cognitive bizarreness and the dissolving of the boundaries of one's body. Despite their potent psychic effects, these substances do not cause dependency and have low toxicity.[28]

The ingestion or inhalation of DMT causes powerful visual imagery with eyes closed, frequently in two distinct phases. At first the field of vision is overtaken by vibrant colorful patterns, a real kaleidoscope of colors and geometric shapes in fractal repetition. Then there are complex shapes of animals, plants, or objects, which come to occupy the entirety of the field of vision, in a dizzying overlapping of shapes in motion. The first phase is nothing like dreaming, nor like any other state of consciousness that is normally experienced. Its abstract content might perhaps correspond to the effect of the N,N-DMT on the actual retina, through the activating of geometric patterns that are characteristic of the network of photoreceptor cells itself. The second phase, however, has the intensity, the shapes, and the textures that are characteristic of dreaming, since it is filled with complex objects in strong colors and movement. However, it is not common for this second phase to have plots or narratives, except in the case of very high doses, which can trigger lengthy and profound experiences that greatly resemble dreams, with complex social interactions, fantastical scene-setting, and even oceanic feelings. In 1988, the U.S. researcher J. C. Callaway raised the hypothesis that the N,N-DMT actually produced by the brain is directly involved in the generating of visual images during REM sleep, but so far there has been no convincing evidence of this.

From a scientific point of view, the best-researched preparation of N,N-DMT is the brew called ayahuasca, which means "vine of the spirits" or "vine of the dead" in Quechua. In addition to N,N-DMT, ayahuasca contains inhibitors of enzymes that degrade neurotransmitters, leading to a rise in serotonin, dopamine, and noradrenaline levels. Also known as *hoasca, daime, yagé,* or simply *vegetal* (plant), ayahuasca is used for curative and divination purposes by indigenous groups in the Amazon and Orinoco basins, as well as by syncretic churches that have spread their revelation-filled sacrament across the world.

One of the most typical effects of ayahuasca (while still not a common one) is *miração,* a vision-action state dominated by powerful visual experiences actively navigated in spite of the eyes being closed, with images that are as vivid as actual reality but fantastical, full of symbolism and of the profound, bright, colorful presence of animals, plants, zoomorphic creatures, ancestral spirits, and divinities whose purpose is to advise and to cure.

Even when vivid visions do not take place, the consumption of ayahuasca provokes a spiritual or psychic purging, which includes the revisiting of past actions and severe self-criticism. This mental purge is frequently in parallel with (and might perhaps be triggered by) a physiological purge, in the form of vomiting and diarrhea, often followed by an ecstatic feeling of redemption. These effects are not surprising, given that almost all serotonin receptors are located in the gastrointestinal tract. The dynamic of ingesting the brew and the subsequent purging, combined with the religious syncretism of indigenous and African beliefs with Christianity, makes the religions that use ayahuasca a powerful cultural space for representing—right here in the twenty-first century—the cycle of death and rebirth that the species has always craved.

VISION WITH CLOSED EYES

The striking similarity between the visual experience of dreaming and the vision induced by ayahuasca led the Catalan pharmacologist Jordi Riba, then at the Sant Pau Biomedical Research Institute in Barcelona, later at the University of Maastricht, to carry out pioneering experiments into the trance that the brew induces. Using electroencephalography to record brain waves before and after the consumption of the ayahuasca, Riba and his team showed an increase in the power of rapid brain waves in parallel with a reduced power of slow brain waves.[29] Comparing it to the stages of sleep, the cerebral state induced by ayahuasca is closer to REM sleep than to slow-wave sleep. This fact, which is consistent with dreams' similarity to the *miração*, raises a few basic questions: which areas of the brain are activated after the consumption of ayahuasca? Does it make a difference if the eyes are kept open or closed? Does ayahuasca increase the power of the imagination?

Motivated by these questions, the Brazilian neuroscientist Dráulio de Araújo, my colleague at the Federal University of Rio Grande do Norte, coordinated some research into brain activity under the effects of ayahuasca with a focus on the ability to imagine visual objects. Brain activity was measured by functional magnetic resonance imaging during the carrying out of two consecutive tasks: visual perception with open eyes and visual imagination with closed eyes. The protocol was inspired by a classic study carried out by U.S. neuroscientist Stephen Kosslyn, then at Harvard University, in which he showed that the imagining of visual objects activates the primary visual cortex in proportion to the mental effort being made.[30]

Before presenting the results of this study, one caveat must be mentioned. I took part in the experiment design and the first gatherings of data for this research at the hospital of the University of São Paulo in Ribeirão Preto, where Araújo was then professor. I can attest to how hard it is to bring the ayahuasca experience into a magnetic resonance scanner set up inside a hospital. This is both because of the physiological changes described above and because of the beliefs of the volunteers, who considered the scanning sessions a particularly difficult passage through the portal to the spiritual world. The volunteers were followers of Santo Daime, one of the main syncretic religions to use ayahuasca as a sacrament, along with the União do Vegetal and Barquinha. To those people who practice this syncretic cult rooted in symbols from the Amazon rainforest, the hospital environment, where it is believed that souls are suffering and are frequently disembodied, is particularly challenging.

When we compare the data before and after the consumption of the ayahuasca, we see an increase in brain activity in various areas of the cerebral cortex that are related to vision, to the recovery of episodic memories, and to intentional and prospective imagination. Not only did the visual areas correspond to the zones activated during dreams or psychotic hallucinations, but the activity in the primary visual area, the cortical region anatomically closest to the retina, showed a strong correlation with symptoms that resembled psychosis, which were experienced after the consumption of ayahuasca. In addition, there were significant changes in the relationships between the different parts of the brain, revealing a major functional reorganization of brain activity.[31] The results suggest that the active effort made to see with closed eyes—intending to imagine—actually produces, under the influence of ayahuasca, the feeling of seeing an imaginary scene quite clearly. Four years later, similar results were obtained with LSD by the group led by English pharmacologist David Nutt, of Imperial College London, which showed a powerful activation of the visual system even when the eyes are closed.[32]

Building on the research on ayahuasca, Araújo collaborated with the Indian physicist Gandhi Viswanathan, along with then doctoral student Aline Viol and other researchers from the Federal University of Rio Grande do Norte, to show that the consumption of the brew increases the degree of cerebral connectivity.[33] With this increase in entropy, the mind is effectively able to become more "open," reaching a more flexible state in which thoughts about the future or the past are no longer mentally identified with the reality they represent, but rather freely associated. Similar phenomena were observed with other psychedelic substances, such as psilocybin and LSD.[34] It is therefore easy to understand what led the Neolithic shamans to use psychedelics to provoke divinatory visions. These substances are called entheogens, or internal manifestations of the divine, with the same Greek root as the word *enthusiasm:* bringing God inside.

The relation between dreams and entheogens is an intimate and complex one. In the words of the Brazilian anthropologist Beatriz Labate, "In traditional societies, the state of wakefulness is not considered the 'normal' or 'superior' way of being in the world and coming to know reality. Dreams and other altered states of consciousness are considered totally legitimate means of learning and revelation." Societies like these divide reality into two or more planes, one visible and the other invisible. In order to access the "other side," the invisible plane of souls and divinities, it is necessary to dream or to make ritual use of entheogens so as to bring the numinous dimension of existence into perception. It is believed that it is only in these border states that it is possible to see through the surface of the people, animals, plants, and things in the world, deepening one's familiarity beyond mere appearances. This invisible dimension, which is experienced as non-human, is at least in part the reason for what happens in the world on this side, on the plane of the visible.

The Kaxinawá people, who live in the Amazon rainforest between Brazil and Peru, drink ayahuasca infusions to attain visions and access the spiritual world.[35] This consciousness altering is directly related to dreams, febrile delusions, and even comas, all states considered to be at the extreme frontier of being—and which for this very reason are capable of producing authentic revelations of the invisible reality where the spirits live. The dream work, just like the work done by ayahuasca, functions among the Kaxinawá to reveal the hidden side of the world. By raising the vividness of the waking imagination with closed eyes to the level attained in dreams, even reaching the level of reality as perceived when eyes are open, entheogens assign the visions concreteness and likelihood, making the encounter with one's own memories a brave and moving discovery. Is this what controlled madness is? What is madness?

8

Madness Is a Dream One Dreams Alone

A young man, C.S., suffered from paranoid schizophrenia. Afflicted prematurely with dementia, he was interned in a public hospital at the age of twenty-one. He was experiencing incessant hallucinations of a female voice who insulted him and threatened him with death, and he started having visual hallucinations and seeing threatening figures. His psychiatrist prescribed him the medication risperidone, a powerful blocker of dopamine and serotonin receptors, used preferentially for delusional psychosis. However, even when treated with the maximum daily dose, the patient went on believing in his delusions and hallucinations. He heard voices every day and felt impulses to disappear into the scrubland like a wild animal.

After several months, C.S. was discharged and taken home on medication, but he continued to suffer, constructing excruciating persecutory delusions of imaginary slanders and intrusive threats. He went on having impulses to run away to the scrubland, but he never went. His impulse was restrained, present but impotent. During this period of fragile normality, he described a dream in which the voice, as in his waking day, was threatening to kill him. Then he went out of the house, he saw a man attacking his mother, he killed him, he was arrested, he claimed to be sick, and he was freed from prison. On his release he felt very well, and the dream ended. This same dream was repeated several times. The patient considered it a pleasant one "because it lets the anger out and everything is OK in the end." When the effects of the dopamine were reduced, the medication inhibited the motor impulse to obey the voices while awake, but did not contain it during dreaming, where everything can be resolved with no negative consequences for the dreamer. In the parallel universe of dream activity, the patient had total freedom to express his psychotic symptoms, and sleep turned into a perfect escape from the social restrictions of waking life. It is not surprising that one of the collateral effects of risperidone should be drowsiness, as the medication mimics the fall in the levels of dopamine and serotonin that occurs naturally when we fall asleep.

DEMENTIA PRAECOX AND CHILDHOOD FABULATION

Despite advances in science, the prognosis for cases such as C.S. remains challenging. Schizophrenia is a potentially devastating illness, with complex causes that are both genetic and environmental. On the one hand, there are clear but diffuse signs of the illness's genetic inheritance, with its prevalence in certain families and many genes somewhat related to the symptoms. On the other hand, the long-term psychic damage caused by a lack of care from the mother and father, or worse, from parental interactions that are outright negative, seem to play a role in the development of the illness. Schizophrenia is characterized—among other symptoms—by the appearance, in adolescence or at the start of adulthood, of perceptual hallucinations and delusional beliefs, combined with affective blunting, a loosening of logic, and the disordering of thought. Paranoia is frequently also a part of this picture, causing a progressive deterioration in social relations.

Curiously, hallucinations, delusions, and a loosening of logic also occur in the dreams of adults and children who are healthy, as well as in the fabulation that is quite normal in the waking life of children. Consider, for example, a child's nightmare recounted by the actual psychiatrist of the patient described above, an oneiric plot of which Stephen King would be jealous. It is a long account, but it provides a detailed illustration of the anxiogenic script typical of the nightmare, which is marked by rising suspense and a more powerful multisensoriality than in any horror movie.

The characters in the dream were various relatives of the dreamer's, and the setting was her family's summer home, surrounded by a dense forest that did not exist in waking reality. Women of all ages were arriving, excited about the vacation, but her father did not seem satisfied. He kept to himself, cleaning the knives and rifle; he stocked up with large-caliber cartridges, arranged his backpack, and went off to go hunting. At first the women were all glad about the vacation, but bit by bit, one by one, they started to disappear. Somebody would go out to use the bathroom and she wouldn't come back. Another would go off to find the first and then she wouldn't come back either. The dreamer called for her father, but he didn't show up; she started to be suspicious about him, but she seemed to be the only person to have this suspicion. The disappearances continued with increasing frequency, but still her mother insisted there was no cause for worry.

The dream reached its first climax with some terrifying visual details when the dreamer walked past a room and saw her aunt suspended from the ceiling, hanged, bug-eyed. She ran to find her mom, but when they came back there was no trace of the body or of the rope. She insisted to her mom that they were in danger, and her mom finally agreed to leave, though reluctantly. Her mom asked, "Where's your sister?" And then they realized that she, too, had disappeared. The hyperrealist suspense was intensified when the dreamer found traces of blood on the floor, mixed with a strong smell of decomposition that led to the bathroom. The trickle of blood went right up to the laundry basket.

The second climax came when the dreamer opened the laundry basket, where she found just half of her sister. On seeing her mutilated daughter, their mother became desperate and decided to run away, but the half-dead sister leaped sobbing out of the laundry basket and, dragging herself over, begging them: "Don't leave me here, don't leave me here!" They picked her up, and they could see all her internal anatomy, her organs, muscles, and bones, sectioned, a scene so intense that it reversed the perspective of the lucid dream: the dreamer even thought, "This isn't a dream, it's real!" Mother and daughters fled the house, but although they kept on running, the setting didn't change, the house seemed endless. Then came the third and most distressing climax: the dreamer looked back, she saw some writing that was moving like the final credits of a movie, and concluded in despair that they would be trapped in that setting forever.

And then the girl who would become a psychiatrist woke up. Just another normal dream of an equally normal child? A typical family, an inexorable career? Some real trauma, or an excess of TV? How is it possible for a small child to experience such detailed horrors and yet still retain her sanity to the point of choosing to take professional care of people who are experiencing psychic suffering? Put in perspective, medicine, biology, and history suggest that the functions and dysfunctions of dreams are at the core of the human mind. From a qualitative point of view, psychotic hallucinations and delusions do not differ greatly from the dreams reported by most people.

In fact, the conception of madness as pathological disconnection from the external world—a world whose accurate perception would be shared by "normal people"—is a very recent one. As we saw at the start of the book, in different cultures in antiquity, delusions and hallucinations that today we associate with psychosis were interpreted as instances of contact between the worlds of the living and the dead, a capacity to address the gods, bestowing the ability to predict the future, interpret other dreams, reveal omens, and pronounce prophecies. Madness has an incomparable importance in the connection between men and gods, whether through the mysterious Pythia in Delphi or the megalomanic pharaoh capable of moving mountains and multitudes. But the development of Christian civilization progressively segregated the pagan madman, divesting him of divinatory powers, which were now the exclusive privilege of the sanctified and beatified of the Church.

At the end of the Middle Ages, the social exclusion of people afflicted with madness reached a degrading level. If it is likely that there were psychotics among those who burned in the fires of the Inquisition, then those who ordered the cruelty were brutal psychopaths. The *Malleus Maleficarum,* a fifteenth-century manual for persecuting witches, prescribed violent death for women afflicted with what we would today call delusions and hallucinations. Tortured and executed in Germany, France, and especially Spain, for being possessed by the devil, individuals experiencing a psychotic episode and other destitute people suffered the consequences of their social inadequacy in their flesh.

FROM THE SHIP OF FOOLS TO PSYCHIATRIC WARDS

The waxing and waning of the Inquisition and the population shift toward the towns dispersed gregarious bands of travelers with psychosis who were kept apart from society, wandering with no rest and no destination, sailing the great rivers of Europe on rustic rafts to beg for alms from city to city, without being accepted as residents anywhere. This was the *Ship of Fools* painted by the Dutch artist Hieronymus Bosch and studied by the French philosopher Michel Foucault, a vessel that remained off the coast of normality but very close to it, tolerated without being attacked. This type of exclusion went on for centuries and lasts to this day in the figure of the crazy beggar, completely disassociated from productive activities, free to fully live out all the delights and horrors of his condition.

With the end of the Renaissance, a different vision of the madman became increasingly dominant, which was reflected in the creation of the first public mental asylums. Though the first institutions devoted to the treating of lunatics appeared in the Arab world in the ninth century, it was the Christian Europe of the seventeenth century that saw the spread of institutions specializing in the admission of psychiatric patients, with illnesses defined by specific behavioral symptoms.

Even if it was motivated by the state's need to contain, exclude, and punish people considered crazy, the establishment of asylums inadvertently benefited the study of madness and the search for methods to treat it. Bringing mentally ill people together into an environment under the control of doctors created a previously unknown space for clinical examination, launching the empirical bases for a medical discipline geared toward mental illnesses. A madman was now no longer the augur of antiquity nor the monster of the Middle Ages, but the host of a natural phenomenon, studied by somebody who was "normal," who was not mad.

In the second half of the nineteenth century, psychiatry developed out of the identification and classification of various kinds of mental illness. Unlike neurology, which was already successfully cataloguing the close correspondences between brain damage and perceptual, motor, or cognitive deficits, psychiatry was dealing—as it does to this day—with much more subtle disturbances, whose causes were not revealed by mere neuroanatomical examination. From that time onward, an understanding was developed that there are at least two general types of psychiatric illness. Psychoses would be those mental illnesses whose origins are "organic" and whose prognoses are poor, since their physiological and/or anatomical causes are hard to access therapeutically, while neuroses have cultural origins, causing disturbances that are much more easily treatable through different kinds of therapies.

Seen by Freud as particularly useful in the psychotherapy of neuroses, dreams at the end of the nineteenth century were widely considered a phenomenon similar to psychosis but not pathological. This was the belief of Emil Kraepelin and Eugen Bleuler, founders of psychiatry and the first scientists to describe schizophrenia. After all, people going through a psychotic break behaved as if they were inhabiting an intense dream despite being awake, as if they were immersed in an intimate reality that was more real than social reality itself. One corollary of this reasoning is that dreams are a normal moment of psychosis in all people, even those who experience no psychotic symptoms when awake. While Kraepelin and Bleuler disagreed with Freud in many regards, they agreed that dreams were clearly akin to psychosis, probably with common mechanisms and with great therapeutic power.

This opinion spread through medical thinking in the first half of the twentieth century and came to have quite a lot of influence in Europe and America, but the discovery in the 1950s of the first antipsychotic drugs, all of them to some degree dopamine D_2 receptor antagonists, led to a decrease in interest in the psychosis-dream relationship. There was no longer any reason to investigate the dreams of psychotic patients, nor to try to understand the relationship between oneiric fantasies and schizophrenic delusions. The place of subjectivity in the treatment of psychosis gave way to something much more concrete, simple, and objective: drugs that were capable of reducing the action of dopamine on the brain.

From the point of view of a patient's relatives, pharmacological therapy was a real miracle, because it cut off at the root those antisocial behaviors that were so alarming in psychosis. From the patients' own point of view, the success was more debatable, as the inadequate handling of the dosage often neutered emotions and dulled movement. Decades later, the newest state-of-the-art antipsychotics no longer target the dopamine receptor exclusively, but also seek out the serotonin, noradrenaline, and glutamate receptors. Psychiatric drugs possess a wide range of chemical affinities for many receptors, leading to complex pharmacological effects that modulate different aspects of the mind, such as mood, cognition, and social interactions.

While the dream-psychosis relationship fell off the radar in psychopharmacology, neuroimaging studies did show a striking similarity between REM sleep and psychosis. In both states the dorsolateral prefrontal cortex is deactivated, generating a negative feedback that suppresses its many important functions still further: memory of work; planning, inhibition, and voluntary control of motor actions; decision-making; logical and abstract reasoning; and subtle social tuning. This cortical deactivation leads to a disinhibition of subcortical structures involved with emotion, such as the nucleus accumbens and the amygdala, related to a positive or negative valuing of stimuli, respectively. The combination of the deactivating of the dorsolateral prefrontal cortex and the activation of these subcortical structures has the potential to explain the appearance of the bizarre thoughts, affective disorders, hallucinations, and delusions that characterize both psychosis and dreaming. Interestingly, schizophrenic patients present a greater frequency of nightmares than healthy individuals,[1] with more hostile content, a larger proportion of strangers among the dream characters, and a lower frequency of dreams in the first person.[2]

WITHOUT DOPAMINE THERE IS NO REM SLEEP

Curiously, it was an investigation into the electrophysiological effects of dopamine on rodents that ended up bringing psychosis close to dreams again in the field of psychopharmacology, through research I carried out with the U.S. psychiatrist Kafui Dzirasa and the Portuguese neuroscientist Rui Costa at the lab of Brazilian neuroscientist Miguel Nicolelis at Duke University. We were inspired by an anecdotal polysomnography observation made by the Austrian psychiatrist Ernest Hartmann, of Tufts University, who in 1967 recorded the case of a non-medicated schizophrenic patient whose psychotic break was preceded by fragmented sleep, with many short episodes of REM sleep (fig. 8). The data suggested that psychosis is related to an intrusion of REM sleep into waking life.

While fascinating, what Hartmann found was not replicated in the decades that followed. Perhaps he made a mistake, or perhaps the particular case does not represent a phenomenon that is regularly observed in a sufficiently large sample of patients. The most likely reason for this discrepancy is simply the adoption of stricter ethical procedures for research from the 1970s onward, which prevent the carrying out of research into non-medicated patients.

Be that as it may, interest in the subject went into hibernation until, one fine autumn afternoon, we got excited about the possibility of testing on mice the hypothesis Hartmann had suggested. In the building next door to the neurobiology department of Duke University Medical Center, biologists Marc Caron and Raul Gainetdinov had raised a number of strains of transgenic mice, among which was a strain with artificially high levels of dopamine in their synapses. Mice of this

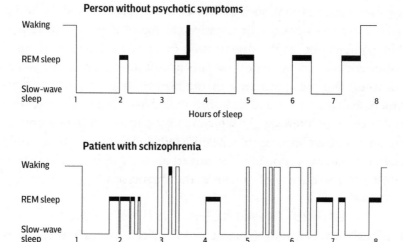

Figure 8. Polysomnographic records collected in the 1960s suggest that patients with schizophrenia suffer from an excessive intrusion of REM sleep. The increase in the number of episodes over the course of a night was accompanied by a reduction in the duration of each episode.

kind, which present erratic behavior, are considered an animal model of psychosis. Through various behavioral, electrophysiological, and pharmacological experiments, we discovered that the neural oscillations in these animals when awake were strangely similar to those found during REM sleep.[3]

However, when we administered an antagonist drug, which inhibits dopamine D_2 receptors, similar to those first antipsychotic drugs of the 1950s, the abnormal intrusion of REM sleep into waking life was reduced. When the animals were treated with an enzyme capable of totally interrupting the production of dopamine, REM sleep was completely eliminated. It was then possible to restore REM sleep in these animals using a dopamine D_2 agonist. Altogether these experiments produced the first direct evidence that dopamine is absolutely necessary for REM sleep, and corroborated the idea that psychosis mixes waking with REM sleep. A finding that would have brought a gleam to the eyes of Kraepelin, Bleuler, and Freud, over their solemn mustaches.

It is therefore possible that the mental disturbances of psychosis make the differentiation between fantasy and reality difficult precisely because they are the result of an invasion of waking by sleep. Even if delusions and hallucinations can involve any combination of sensory modalities, such as sight, touch, and even smell and taste, the breaching of the boundary happens, mostly and crucially, in the domain of language. The great majority of psychotic symptoms are auditory, typically taking the form of sarcastic, demeaning, accusatory, or imperative voices, which are sometimes incessant and which sound convincingly "inside the head" and seem totally real. Moments of relaxation are conducive to their expression, as is exemplified by a traditional song from Capoeira de Angola, which repeats like a mantra: "I'm sleeping, I'm dreaming, they're speaking ill of me . . ."

What is confusing and frightening in this situation is the vivid feeling that these are *other people's* voices, since an internal dialogue is a healthy mental fact of life, whether in the form of a reflexive monologue or the evoking of clichés and expressions to suit the moment. The French psychoanalyst Jacques Lacan, agreeing with Freud, observed that the basis for internal mental dialogue is the voices of one's parents, the first and most important auditory expressions in the social world, codified so strongly that they are reactivated and reverberate throughout one's life, making up the foundation for the social norm that is expressed as the superego.[4] In a very concrete way, we are verbally formed by our direct ancestors. Their representations speak within us and even for us—they remain even after their owners have disappeared. As in the play *Waiting for Godot*, by the Irish playwright Samuel Beckett, the dead insist on talking:

ESTRAGON: All the dead voices.
VLADIMIR: They make a noise like wings.
ESTRAGON: Like leaves.
VLADIMIR: Like sand.
ESTRAGON: Like leaves.

(Silence.)

VLADIMIR: They all speak together.
ESTRAGON: Each one to itself.

(Silence.)

VLADIMIR: Rather they whisper.
ESTRAGON: They rustle.
VLADIMIR: They whisper.
ESTRAGON: They rustle.

(Silence.)

VLADIMIR: What do they say?
ESTRAGON: They talk about their lives.
VLADIMIR: To have lived is not enough for them.
ESTRAGON: They have to talk about it.
VLADIMIR: To be dead is not enough for them.
ESTRAGON: It is not sufficient.[5]

This conversation echoes Jaynes's hypothesis regarding dreams about dead ancestors. In his bold conjecture, Jaynes stated that today's psychotics represent the socially maladjusted persistence of an ancient mentality, a memory of a time when hearing voices was common. Psychotics would be living fossils of a kind of human consciousness that was born in the Paleolithic period, flourished in the Neolithic, expanded in the Bronze Age, and resoundingly collapsed in the Iron Age, around three thousand years ago.

In order to construct this theory, Jaynes built directly on countless archaeological finds, but he also got indirect support from the Jungian[6] and Freudian[7] idea that psychiatric illnesses can resemble the mental functioning of children, of today's hunter-gatherer people, or of our ancestors. To Freud, "Primitive men and neurotics . . . attach a high valuation—in our eyes an *over*-valuation—to psychical acts. This attitude may plausibly be brought into relation with narcissism and regarded as an essential component of it."[8] In his conception, religions are illusions that obey instinctive desires and seek to take control of reality:[9] "Religion is comparable to a childhood neurosis."[10] Melanie Klein, the Austrian pioneer of the psychoanalytic study of children, proposed a related concept. To Klein, the perversion and fantasies of the first decade of life find a transient correspondence in psychosis.[11] She proposed that the mental world is constructed from the internalization of objects: parts of people (breasts), people, animals, and things.[12] Over the course of their normal development, children frequently experience disturbing dreams in which their parents, who are so familiar to them, stop being trustworthy protectors, to become threatening, unfamiliar, unpredictable adults. The reverberating voices of these distorted fathers and mothers echo the psychotic voice of Norman Bates's mother in the Alfred Hitchcock movie *Psycho,* that wicked and cynical speech heard so often in childhood, in moments of self-pity and unsettled quiet: imaginings when awake or dreams when asleep that represent the worst nightmare of any young mammal, predation arising out of the very archetypes of parental care—a father or mother trying to kill their child.

The persistence of these fantasies is an echo of our ancestral past. The Bible tells us, in Genesis 22, the story of Abraham, who at God's command decided to kill his own son, Isaac. The patriarch tied the boy to the altar and was already preparing to execute him when an angel from the Lord appeared to dissuade him, pointing out a ram that might replace the child in the sacrifice. In the Quranic version of this story, the divine order to kill his own child came to the patriarch in a dream. From Medea to Herod, the texts of antiquity are filled with infanticides. In classic paranoid schizophrenia, self-referential, persecutory delusions are a frequent occurrence, and are characterized by hearing voices that are commanding, seductive, or threatening, and often cynical, scathing, and sarcastic as well. The dream narratives reported by patients with schizophrenia include a higher than usual proportion of characters who are strangers, who are male, and who appear in a pack.[13] The desire to escape society, to vanish into the forest, disappear into the mountains, is typical of those with schizophrenia. They would rather be abandoned to nature than harried by the evil of culture.

QUANTIFYING PSYCHOTIC LANGUAGE

If psychosis is a psychologically archaic state, which was common in our historical past and persists today in the early stages of development, it ought to be possible to find linguistic traces that are common to children, psychotics, and texts written at the time of the pharaohs. Prompted by this interesting and somewhat outlandish mission, I teamed up with physicist Mauro Copelli from the Pernambuco Federal University to embark upon a mathematical analysis of the structure of the language of adults and children, both healthy and psychotic, in order to compare it to the structure of Bronze Age texts.

This research began in 2006 when Natália Mota, a young medical student who would later become a psychiatrist and complete a master's and doctorate on the subject, started to record dream narratives of patients undergoing psychosis. In order to quantify structural dif-

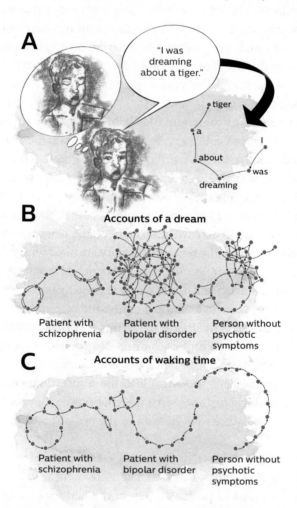

Figure 9. Dreams represented as graphs can assist in the diagnosis of schizophrenia. (A) Each word corresponds to a node (circle), and the temporal order of two consecutive words is represented by an edge (arrow). Accounts of dreams (B)—but not of waking time (C)—allow us to distinguish between schizophrenic patients, patients with bipolar disorder, and persons without psychotic symptoms.

ferences among the accounts, we decided to transform each one into a graph made of words (fig. 9A). A graph is a simple mathematical structure for representing any network of elements, such as a city's bus routes, the metabolic pathways on the inside of a cell, or a social network on the Internet. Using this representation to analyze reports produced by people of different ages, we discovered that the structure of an account of a dream is highly informative about the patient's psychiatric state.[14] Figure 9B shows representative examples of dream accounts produced by two kinds of patients undergoing psychosis, those with schizophrenia and those with bipolar disorder, compared to the dream account of a person without psychotic symptoms. There is a striking difference between the graphs, which are short and simplified in patients with schizophrenia but long and involved in bipolar patients when they are in a manic phase, full of escapes and loops. Persons without psychotic symptoms present an intermediate pattern that falls between the two types of patient. It is as though "normal" people were halfway between the linguistic poverty of schizophrenia and the discursive richness of mania. Curiously, none of this happens in accounts of waking experiences lived during the day, which in all these groups is presented chronologically, directly, with few loops (fig. 9C).

Taken altogether, these linguistic phenomena allow us to use the reports of dreams to quantify and make an early diagnosis of schizophrenia, in a way that is quick, cheap, and noninvasive. Dreams are therefore clinically useful, because dreams give us a sharper picture of the structure of the dreamer's mind than we can draw from daytime questioning. In psychoanalytic terms, this corroborates the idea that dreams are indeed a royal road for accessing the mind's deepest structures.

When we carried out a structural comparison between the dream accounts of children and psychotics and the texts of antiquity, especially those from Sumer, Babylon, and Egypt, the similarity was obvious: low levels of lexical diversity, small word networks, many short-range loops, and few long-range loops. The maturing of the structure of language follows similar paths during individual development and over the course of history, with an increase in lexical diversity and in the size of the word networks, as well as in long-range loops. Intriguingly, this maturing process undergoes an abrupt transition between 1,000 and 800 B.C.E., that is, between the civilizational collapse at the end of the

Bronze Age, when the Trojan War took place, and the period of cultural renaissance at the start of the Axial Age, the time when *The Iliad* and *The Odyssey* were transposed from the oral register into written form. The structural similarity between texts written in the Bronze Age and dreams reported today by healthy children or adults with psychotic symptoms establishes a connection between psychology and history, a bridge to the recent past when people dreamed while awake and didn't know they were doing it.

Obviously, all these ideas are based on examining subjective reports of human mental experience, both anonymous and with known authorship. In the next chapter we will see how these accounts are constructed in the waking and the sleeping brain.

9

Sleeping and Remembering

In human beings, it is possible to draw a clear distinction between memories that can be consciously declared and those that cannot. Once the mind has been trained and has matured, it is generally quick and easy to acquire memories of the first kind. What is Bob Dylan's actual surname? In which century did Queen Nzinga of Angola fight and reign in Africa? Who consolidated the Quilombo of Palmares in Brazil, the largest maroon community ever to exist in the Americas? What is the correct proportion of water and rice for making a good risotto? The answers to these trivia questions—Zimmerman, the seventeenth century, Aqualtune (daughter of the King of Kongo), one part rice to three of water and half of wine, adjusting to taste—depend on these so-called declarative memories. They are very different from those memories that we need to ride a bicycle, surf, or play Capoeira. Memories of this latter type tend to take a while to learn, since they demand countless repetitions that can reconfigure the vast neural circuits responsible for the representation of highly complex sensory-motor behaviors. It is not possible to teach somebody to surf merely by translating the movements into words. Riding a bicycle is not the same as reporting on it orally. Playing Capoeira is an ineffable learning of the body, even if reading about this Afro-Brazilian art might help someone to understand it.

SLEEP, REMEMBERING, AND FORGETTING

If in our waking lives we acquire new memories, it is mostly while asleep that they reverberate and are transformed. While the reverberation of memories is implicit in the psychoanalytic concept of the day residue, there is no mention in Freud's work of the role of sleep in learning. Carl Jung came closer when he said that dreams prepare the dreamer for the day to come. However, the first experimental approach to the relationship between sleep and learning did not happen in Europe, the undisputed center of scientific knowledge in the nineteenth century, but in the United States, whose university tradition at that time was still very young.

At the start of the 1920s, the researchers John Jenkins and Karl Dallenbach from Cornell University attempted to replicate a classic experiment, carried out decades earlier by one of the founders of modern psychology, the German Hermann Ebbinghaus. The experiment consisted of teaching volunteer participants a list of syllables that did not exist in their language, and to then measure how these syllables were retained over time. With this simple procedure, Ebbinghaus had discovered forty years earlier that memories, once they have been acquired, decline in an exponential fashion as time passes, defining a "forgetting curve" that characterizes the dynamics of memory in countless different species. Jenkins and Dallenbach's innovation was to ask the participants to go to sleep right after they had learned the syllables.[1] For the purposes of comparison, they repeated the experiment but kept these volunteers awake. Surprisingly, for identical intervals of time, retention after sleeping was far greater than retention after remaining awake. The participants in the waking group were undergraduate students, and so their post-training time was spent in regular classes. This fact prompted a joke that persists to this day among scientists in this field: Jenkins and Dallenbach proved that sleeping is better for learning than going to school.

Joking apart, we know today that the relevant variable for the low retention among the waking group is sensorial and cognitive interference. While awake, the brain is constantly bombarded by stimuli of all kinds, which greatly interfere with the mnemonic process. We find a good example of this phenomenon when we try to hum a song while listening to a different piece of music. The effort required to do this simple task is in proportion to the volume of the interfering music, demonstrating the difficulty the waking brain experiences in isolating itself from contact with reality.

For some mysterious reason, Jenkins and Dallenbach's discovery was not replicated by their contemporaries, and instead remained obscure and without any consequences for decades. With the exception of a couple of minor studies in the 1940s, the discovery passed through the Second World War and the start of the Cold War unnoticed. Those were pre-Internet days, in which information still flowed very viscously, slow and capricious, not obliged to be known. While in the 1950s the United States became the epicenter of the research into REM sleep and its relationship to dreaming, at first nobody followed up the cognitive aspect. Jenkins and Dallenbach's results, published in 1924, had to wait forty years to be revisited in greater depth.

JOUVET AND THE FLOWERPOT

There was an energetic resurgence of interest in the subject in France and the United States toward the end of the 1960s, when a new generation of researchers influenced by Michel Jouvet turned their attention to the cognitive importance of sleep. The experimental design of these studies had as their common denominator post-training sleep deprivation in rodents. The flowerpot method invented by Jouvet proved to be so simple, efficient, and cheap that it quickly spread across many laboratories that were interested in studying the biological effects of sleep deprivation. It was simply a matter of putting the animal on a small platform, an upside-down flowerpot, surrounded by water. The method is based on the fact that slow-wave sleep is accompanied by a drop in muscle tone, which is reduced still further with the beginning of REM sleep. If the platform is small enough, the animal will fall into the water when it loses muscle tone, waking immediately. Using a platform of the right diameter it is possible to deprive an animal of sleep completely, or deprive it only of REM sleep. The first experiments to use this method showed that rats subjected to a range of tasks—spatial learning, acquired fear, operant conditioning—present a deficit in the evocation of memories after being completely deprived of sleep, or specifically deprived of REM sleep.

Sleep that has been missed needs to be replaced or compensated for. This need applies especially to REM sleep, since being deprived of this is invariably followed by a positive rebound in proportion to the amount of sleep missed. Curiously, the opposite is not true: while it is possible significantly to increase the duration of REM sleep simply by increasing the total duration of sleep, this does not cause a negative rebound the following day, that is, there is no subsequent reduction in the duration of REM sleep. Emotions have a lot of impact on these dynamics. Situ-ations of moderate anxiety lead to a reduction in the total time spent in REM sleep, but situations of extreme stress, such as a life-or-death emergency, can result in a great increase in the amount of REM sleep as soon as the danger has passed. All this suggests that REM sleep fulfills a vital role for an individual's cognitive health.

Over the course of the 1970s, a number of researchers showed con-clusively that sleep deprivation is harmful to learning.[2] There was a real burst of excitement on the subject, in the context of international competition and collaboration whose focus was REM sleep, which was taken to be the most interesting stage of sleep because of its intimate relationship with dreaming. The passing of time, however, saw the formation of a movement of growing resistance to the idea that REM sleep possesses some cognitive value.

STRESS OR A LACK OF SLEEP?

The harshest criticism made by the skeptics focused on the experi-ments' weak point: the method used to produce the sleep deprivation. The flowerpot that Michel Jouvet devised and popularized, used to cre-ate the deprivation, is inherently stressful. If it is very small, it causes the animal to fall into the surrounding water at the first sign of falling asleep. If it is slightly larger, it allows them to sink into deeper sleep up until the moment when muscle tone is so low that the animal rolls off the platform and falls into the water. The enforced bath following a certain threshold of atonia, which is intrinsic to the method, invari-ably causes a great shock. It is obvious that the situation faced by the animals when these kinds of experiments are carried out is stressful and unnatural.

Besides having their sleep interrupted in such an abrupt way by fall-ing into the cold water, rats subjected to the flowerpot method experi-ence a severe limiting of their body movements, being prevented from carrying out the majority of their natural movements. Rats need to move around, and after a few hours of deprivation they end up walk-ing freely around the flooded cage, becoming incapable of remaining dry. As a result, the animals show irritation and generalized metabolic changes, including the release in the hippocampus of the glucocorti-coid stress hormone, which can have deleterious effects on memory.

With so many collateral effects, assigning the cause of memory deficits to the lack of sleep alone would be arbitrary at the very least.

This argument was a legitimate one, and it prompted new experiments with less stressful methods for causing sleep deprivation. In order to get around the problem of stress, the then doctoral student William Fishbein and his supervisor William Dement took advantage of the important behavioral differences between rodent species. Unlike rats, which when they are adult weigh more than half a pound, small mice weighing just one ounce are able to hang from the bars that serve as the roof of their cage for a long time. They can spend hours walking upside down, clinging to the cage bars spontaneously in a movement that is so free that it even allows them to consume water and feed. When the flowerpot method was used to deprive mice of sleep, the animals presented much lower levels of stress than the rats, as they were able to remain on the platform for only the stretch of time when they wanted actually to sleep. Still, the experiments into the sleep deprivation of mice after learning did lead to deficits in the evocation of memories, strengthening the theory that sleep helps with their consolidation.

All the same, the argument about the stress inherent in the method continued to be made. Another alternative used was the gentle but effective interference of the researcher, which was done each time the animal was about to fall asleep. It is obvious that this method varies depending on the researcher's attentiveness, which weakens the experimental data and makes their interpretation inconclusive. At this point the storm clouds of dissent were already thickening. Two U.S. researchers, psychiatrist Jerome Siegel from the University of California–Los Angeles, and neuroanatomist Robert Vertes from Florida Atlantic University, became known for their forceful arguments against the cognitive hypothesis of sleep.

SKEPTICS VS. LONE RANGER

Their questions spread. If REM sleep is so important for cognition, why do reptiles, birds, and even mammals like the echidna not have REM sleep? If REM sleep is used for learning, why does an animal as intelligent as a dolphin not have REM sleep, while other less intelligent animals, such as the armadillo, have it in abundance? Why do people whose REM sleep is reduced as a result of treatment with antidepressants not present learning deficits? Why isn't there a strong correlation between the time spent in REM sleep and the capacity to learn in humans?

The defenders of the theory counterargued that we are not totally sure that dolphins do not experience REM sleep, as their episodes might simply be too short to be recorded. Besides, the dolphin is descended from a land mammal that arrived late into the water environment. It is likely that REM sleep was reduced or eliminated in cetaceans so as to prevent total atonia in an aquatic environment, which could lead to drowning. In the context of a specialization to occupy such a different new environment, the cognitive functions of REM sleep could have been replaced by other metabolically equivalent processes. The armadillo, on the other hand, spends long periods burrowed underground. Evidence from the last two decades has shown, contrary to what had been previously believed, that REM sleep does occur in echidnas, in birds, and even reptiles. In addition, treatment with antidepressant drugs raises the levels of the neurotransmitters that are important for the formation of memories, such as noradrenaline, dopamine, and serotonin. It is therefore likely that the consolidation during waking hours is increased, making up for the effects of a reduced amount of time spent in REM sleep.

The 1980s saw the intensifying of the debate and a hardening of its tone. Battle lines were drawn, and the territory came to be clearly divided over the cognitive properties of sleep. For a time, it was like a dialogue of the deaf. Discouraged by the now acrimonious atmosphere of the sleep research conferences and by the ever-more-aggressive anonymous revisions to articles submitted for publication, the veterans began gradually to quit the field. For more than a decade, scientific interest in the relationship between sleep and learning declined considerably.

Over the course of this turbulent period, a hefty Canadian maverick from Trent University, the psychologist Carlyle Smith, was the almost totally lone ranger defending the cognitive role of REM sleep. In a number of experiments conducted on rodents, he showed positive effects of REM sleep during specific windows of time following learning, revealed by periods of greater vulnerability of memories to sleep deprivation.[3] But Smith in isolation was unable to change the minds of the critics of the cognitive theory of sleep, and the impasse persisted until the early 1990s, when an unlikely new actor appeared, one who tipped the scales with experiments carried out directly on humans.

THE MANY LIVES OF STICKGOLD

The U.S. psychologist Robert Stickgold had three independent careers connected to science. His interest was awakened for the first time in sixth grade, when a teacher performed a simple experiment: he walked a long distance away on a lawn and crashed a pair of cymbals together. Stickgold was far enough away for there to be a perceptible difference between what he saw and what he heard. Light is faster than sound! There and then, he decided to be a scientist.

Years later, by which time he was in the first year of medical school, he was fascinated by an article by Francis Crick in the magazine *Scientific American* about the genetic code, which had just been discovered. After spending the whole night tenaciously unraveling the text, Stickgold decided to be a biochemist. The following summer he managed to get a job as an assistant at Francis Neuhaus's laboratory at Northwestern University, cultivating bacteria for him in five-gallon glass carboys. The four months he spent in this lab allowed him his first publication, a paper in the *Journal of Biological Chemistry* about the biosynthesis of the bacterial cell wall, which also served as his final course paper.

Stickgold graduated from Harvard and did his doctorate in biochemistry at the University of Wisconsin–Madison. As a graduate student, he began to take an interest in the relationship between the mind and the brain, but opted to postpone his interests when he took a course in physiological psychology, which led him to conclude that this was not yet a science. That was the distant year of 1965, and the field of research that one day would come to be called cognitive neuroscience was still only crawling.

For a while, Stickgold gave up on the brain. In the 1970s he started to write science fiction, achieving some success in the field. And then he returned to the study of the brain, and this time it was for good. The shift in direction began in 1977 during his postdoc, when somebody pointed him toward an article by the British neuroscientist David Marr, who though still very young had developed widely accepted theories about the functioning of the cerebellum, the neocortex, and the hippocampus. Marr's influential theories were based on the connectionist assumption, according to which behaviors and thoughts are emergent properties of a network composed of interconnected elementary units, a system that is locally simple but capable of generating global complexity owing to the vast range of collective patterns it can take on. Any resemblance to neuronal networks is no mere coincidence. While Marr's seminal ideas did not manage to transform Stickgold into a connectionism fundamentalist, they did definitively alter his way of conceiving of the brain.

But it was the impact of J. Allan Hobson and Robert McCarley's activation-synthesis theory that really steered Stickgold toward research into sleep and dreams. From the 1990s, having by now entered middle age, the biochemist and writer got himself a job as a technician in Hobson's laboratory at Harvard, where he began a new academic career, starting again from scratch in the fields of psychology and neuroscience. From then on, his trajectory was meteoric. Following an altogether unconventional path, Stickgold was soon promoted to assistant professor and finally made full professor and director of the Center for Sleep and Cognition at Harvard Medical School.

Among his various fundamental discoveries, Stickgold provided the first demonstration of the oneiric reverberation of images coming from a computer game. The effect was detected in the transitional state called hypnagogic sleep, which includes the two first phases of sleep (fig. 7).[4] The game that was used, a classic video game called Tetris, requires the player to act upon differently shaped blocks that drop down from the top of the screen. As the blocks fall, the player needs to turn them so that they slot into the virtual floor, which rises as the game proceeds. As the slotted-in blocks accumulate, the task gets progressively harder, which keeps the attention and the emotions engaged. The oneiric reverberation of Tetris detected by Stickgold's team was so strong that it happened even in patients suffering from amnesia caused by extensive bilateral lesions in the hippocampus area. Though the patients did not even remember playing the game, they reported striking dream images in geometric shapes falling incessantly. These experiments, which were published in 2000, showed that human dreams do indeed contain elements that are linked to waking experiences, that is to say, Freudian day residue. The study marked the return of dreams to the pages of *Science* for the first time since 1968.

THE GREAT DUEL OF CHICAGO

One of the most exciting moments in this dream renaissance happened in 2003, at the annual gathering of the Associated Professional Sleep Societies (APSS), which on that occasion was celebrating fifty years since the discovery of REM sleep. Researchers from across the world came together enthusiastically for six days in Chicago, in a real flurry of excitement: interest in dreams was experiencing a powerful resurgence both in the scientific realm and among the general public. The gathering took place in the same city where REM sleep had been discovered, which happened also to be the birthplace of Stickgold, the researcher who was most shaking things up in the field at that time.

One of the panels included in APSS's commemorative program was a potentially explosive debate on the relationship between sleep and learning. For thirty years, the opponents of this theory had been keeping it in check, with a number of circumstantial arguments and indirect pieces of evidence. For decades the theory didn't develop, it merely defended itself. However, the group of results published by Carlyle Smith in the 1980s had just lately, at the end of the 1990s, received the heavyweight backing of Stickgold. Even so, a look at the pages of the main scientific journals showed that the cognitive theory of sleep was always being confronted with opinion pieces enumerating evolutionary, neurological, and psychiatric reasons why the theory had to be wrong. The main authors of these articles of entrenched resistance were Robert Vertes and Jerome Siegel.

On that June day the hundred-seat auditorium was crammed to double its capacity, an audience filled with students but also including the presence of the leading sleep researchers in the world. Scientists famous and unknown alike were sitting on the floor, down the side aisles, and even outside the room, while Smith and Stickgold were seated on one side of the stage, for the defense, against Vertes and Siegel, on the attack. The audience's nerves and elation were palpable. There was a vague dissatisfaction in the air at the two stubborn opponents of the cognitive theory of sleep. Many young scientists carrying out doctoral or postdoctoral work, of whom I was one myself, wanted to make progress investigating the mechanisms that might explain the theory, but the truth was that the mood in that area of research was still heavy, echoing the seismic clashes of previous decades.

It was an epic duel between two totally opposed views. In this famous clash, which is vividly recalled by those colleagues with whom I shared the excitement of witnessing it in person, the fate of a whole area of research was at stake. Smith gave a robust presentation of the many pieces of evidence he had produced in support of the theory. The proof that sleep deprivation is more harmful during certain windows of time post-learning than in others did suggest—albeit indirectly— that the stress caused by this deprivation should not be the explanation for the memory deficits, since the period of deprivation was the same in all the groups. Vertes was scathing and came across as being much less interested in the evidence that connected REM sleep to the processing of memories than in the familiar litany of hypothetical obstacles, such as the supposed absence of REM sleep in echidnas. Siegel followed the same path, making it clear that he did not recognize any validity to Smith's position.

Stickgold hit back. He called the argument that quite intelligent animals such as dolphins necessarily must have a lot of REM sleep simplistic, and he offered an analogy that made the audience buzz: "After all, the fact that feet are used for locomotion does not make the centipede the fastest animal." Then Stickgold said that it would be best to leave these indirect arguments to one side and focus on the direct evidence obtained empirically in the lab. Then he projected results of his now-classic studies showing that the learning of visual patterns is crucially dependent on preserving sleep in the first post-learning night. To deal with the argument that deprivation is very stressful and harms performance for reasons that are independent of sleep, as his opponents maintained, Stickgold did not measure the learning in the first post-learning day but after four days had gone by. While the participants in the research had slept adequately in the subsequent nights, and had not presented any tiredness or sleepiness at the time of testing, they still presented deficits in performance.[5]

Siegel counterattacked with great mental agility, disregarding every-thing Stickgold had just shown and insisting that nothing yet proved the theory, as there was a series of theoretical obstacles—and he repeated everything all over again, as if he were deaf to the new empirical results. Then Stickgold raised his eyebrows, like a man who wasn't going to take any crap from anyone. Even from a distance it was possible to see his blue eyes glittering with annoyance. His breathing became choppy, the tension rose, and finally Stickgold burst out with a bit of scientific jargon that brought the house down: "What part of *p* less than 0.05 corrected for Bonferroni do you not understand?"

This line, referring to the Italian mathematician Carlo Bonferroni, meant that the empirical data showing the consolidation of memories by sleep are supported by very rigorous statistical tests, calibrated using a method that is considered a gold standard, based on Bonferroni's contributions to probability theory. Using a technical term, Stickgold was stating that there was an incredibly low probability that the results had been a product of chance. It was a devastating counterattack, a technical *ippon* that asserted the supremacy of empirical evidence over authoritative opinion.

The releasing of tension caused by those words was cathartic. The hall was filled with applause and whistles. Vertes, looking as though he was just about to walk away from the panel, said that he had become disenchanted at the path things were taking, and that he would not be taking part in any more debates on this subject. Those in attendance responded with more applause, shouts, and laughter. Many had stood up, and there was no way for things to continue. Nor was there any need. Vertes and Siegel had already thrown in the towel. To general delight, Stickgold and Smith had turned the tide.

THE REDISCOVERY OF THE COGNITIVE ROLE OF SLEEP

The start of the 2000s was a time of great flourishing of scientific interest in the cognitive role of sleep. The wave of demonstrations of the sleep-memory relationship begun by Stickgold led Sara Mednick, then a doctoral student in his lab and today professor of psychology at the University of California–Riverside, to look into the cognitive effects of napping, a short episode of sleep experienced during the day.

The powerful restorative abilities of naps have been known for a long time and have been crystallized in traditional practices, such as the Spanish and Mexican siesta. There are historical records of the great Renaissance painter and polymath Leonardo da Vinci—as famous for his brilliance as for his eccentricity—having slept for various half-hour spells over the course of the day, in order to make better use of his time for working and creating. Owing to their restorative abilities, such naps have come to be called "power naps" in the United States. Sara Mednick's experiment consisted of comparing the performances of people carrying out a task of visually differentiating between textures, both before and after taking a nap. The task required locating a pattern of three diagonal lines against a background of horizontal lines. In general, performance improves over the course of the first session of carrying out the task, but after two or three sessions repeated on the same day, performance declines owing to fatigue in those areas of the brain involved in processing the stimuli.

In a first study, it was shown that a brief nap of thirty to sixty minutes, which allows the development of the initial stages N1 and N2 and of slow-wave sleep, but not the appearance of REM sleep, restores performance to pre-fatigue levels. In a second study, Mednick, Stickgold, and the psychologist Ken Nakayama showed that longer naps, between sixty and ninety minutes in duration, which include REM sleep as well as slow-wave sleep, make it possible not only to compensate for fatigue but also significantly to improve performance of the task. While a short nap was enough to restore the capacity for sensorial processing,[6] only a longer nap actually boosted the learning.[7] The effects of these longer naps were so powerful that they supplied the same benefits as a whole night of sleep. Another U.S. psychologist from Stickgold's group, Matthew Walker, today a professor at the University of California–Berkeley, demonstrated the importance of pre-learning sleep for the acquisition of memories.[8] As we have seen above, sleep deprivation causes an accumulation of toxins in the brain, which likely explains these findings.

Another scientist who began to lead the field at the end of the 1990s was the German neuroscientist Jan Born, who considerably expanded Stickgold's findings. Both men's trajectories showed how detours in a scientist's path can lead to fundamental discoveries. Jan Born is a native of Celle, a town in the north of Germany. As legend would have it, some centuries ago, in order to preserve the honor of their daughters, the virtuous inhabitants of foggy Celle had opted to build a prison rather than having a university.

When Born completed high school, his father—who was a judge—suggested a career in the military, as he didn't consider his son brilliant enough to devote himself to the law. The boy's response was quite unlike what his father had expected: he studied psychology. Shortly before receiving his diploma in psychology, Born decided to steer toward behavioral neuroscience (called biological psychology in those days). This happened because Born, despite finding psychoanalysis positively stimulating, considered that most of what he had learned in his psychology studies had lacked a solid basis in experimentation.

After his doctorate, Born made one final crucial adjustment to his trajectory, prompted by his curiosity and by the desire to increase the efficiency of his small laboratory at the University of Ulm, which was lying idle at night. Born decided to launch some research into sleep, which at that point felt like an ill-defined smudge in the middle of the lovely neuroscientific landscape that was growing so rapidly. He knew that the two halves of sleep differed in their cortisol levels, but also in their amounts of slow-wave sleep and REM sleep. Might the suppressing of the release of cortisol during slow-wave sleep be necessary for this stage's role in consolidating declarative memories? And what was to be said of the effects of sleep on non-declarative memories—also called procedural memories—which are necessary for carrying out coordinated motor actions, such as riding a bike or playing soccer? Might the different phases of sleep be important for different kinds of memory?

Born and the then doctoral student Werner Plihal carried out studies to answer these questions, using polysomnographic records, psychological tests, and the administering of drugs. The results showed that slow-wave sleep is necessary for consolidating declarative memories, while procedural memories have a greater need for REM sleep.[9] In addition, administering a cortisol analog during slow-wave sleep harms the consolidating of declarative memories, but leaves the consolidation of procedural memories intact. In the last decade, the U.S. psychologist Ken Paller of Northwestern University broadened out this landscape, showing that REM sleep also plays an important role in the consolidation of declarative memories when the task is more difficult and demands greater cortical integration, such as in the acquisition of new vocabulary.[10]

SLEEP AT SCHOOL

In the last twenty years, the scientists mentioned above and a number of others have demonstrated the role played by sleep in the consolidation and restructuring of memories, and even in the selective forgetting of content. These findings have implications both for everyday lives and for the progress of pure science. Similar effects were demonstrated in monkeys, rats, and flies. From a practical perspective, the main social utility of this research is the optimization of the sleep regimen in accordance with cognitive or metabolic goals, with educational or therapeutic aims in mind. Among the various alternatives, the most promising is the siesta—that is, the use of naps in a school environment to increase academic performance.

The first studies in this field were published recently. In 2013, Rebecca Spencer and her team at the University of Massachusetts–Amherst showed that a post-training nap increases the learning from a memory game in children at kindergarten, in proportion to the quantity of sleep spindles. In 2014, research I carried out with then master's student Nathália Lemos and the linguist Janaina Weissheimer, all of us then at the Federal University of Rio Grande do Norte, showed that a post-learning nap increases the duration of declarative memories acquired in the classroom by sixth-grade students.[11] The biologist Thiago Cabral completed master's research under my guidance in which he showed an increase in declarative learning when a normal class is followed by a period of thirty to sixty minutes of sleep.[12] And with doctoral student Ana Raquel Torres and the Brazilian neuroscientist Felipe Pegado, of Aix-Marseille University, I am currently investigating the effectiveness of training followed by sleep for providing a lasting consolidation of letter discrimination in children aged five to seven who are in the process of learning to read and write. The results show that post-training naps completely preserve the learning of letter discrimination over time, doubling reading speed, while in the absence of this sleep there is a significant fall in performance after four months.

The use of sleep for optimizing school learning is still at an early stage, but its progressive adoption seems inevitable. The creation of napping rooms or siesta clubs, as well as the adopting of individual pods for sleeping, are proposals for an education that is biologically more intelligent.[13] Delaying the start of classes also seems to help, especially among adolescents. The physiological changes of the start of puberty push back the times for going to sleep and waking up, causing young people to arrive at school even sleepier.[14] In 2016 and 2017, high schools in Seattle delayed the start of classes by almost an hour. The change was associated with a significant increase in the length of sleep and a 4.5 percent rise in their students' grades.[15]

In the area of basic science, meanwhile, the clear demonstration of the cognitive role of sleep granted free access to a deeper layer of questions, which had not been able to make it onto the agenda until the impasses of the past had been overcome: Which biological mechanisms are responsible for this very benign psychological effect? What alterations to the electrical activity of neurons help to explain the formation of memories? Which molecular and cellular changes make it possible to understand their storage over the course of a whole life? In the next three chapters we will be devoting ourselves to considering these questions in detail, dealing with the role in the reverberation of memories played by genes, proteins, electrical oscillations, and neuronal circuits activated during sleep. In Chapter 13, we will resume the main themes of the book toward its central argument. Now would be a good time to begin or resume your dream diary.

10

The Reverberation of Memories

The research into the mechanisms that lead to the reverberation of memories during sleep is a quixotic saga of idealists who dared to cross the bridge between biology and psychology at a time when this path was still only an imaginary one. It is also the tale of some scientific heavyweights whose brilliance was equaled only by the scale of their stubbornness. The beginnings of this story go back to the pioneering studies of the electrical activation in recurrent neural circuits, carried out in the 1930s by the Spaniard Rafael Lorente de Nó. This young and brilliant disciple of Santiago Ramón y Cajal emigrated to the United States in 1931 and five years later settled in New York, hired by the important biomedical research institute that just a few years later would be transformed into Rockefeller University.

By this time, Lorente de Nó was a scientific celebrity, a prodigy whose precocious talents had already added a number of important discoveries to his résumé. It was he who had first described the cellular structure of the cerebral cortex, characterized by the organization of the neurons in sets of vertical cylinders that function as elementary processing modules. Lorente de Nó was also a pioneer in the detailed description of the internal structure of the hippocampus, a region of the brain whose evolutionary origin is very old, as is evidenced by its presence in birds and reptiles as well as mammals, but whose function was unknown at the time.

A solid reputation acquired in neuroanatomy gave Lorente de Nó the prestige he needed to occupy his new academic position with enough resources and freedom to attempt a big methodological leap, using a technique—then still incipient but very powerful—for measuring neurons' electrical activity: electrophysiology. The Rockefeller Institute didn't skimp on funds when it came to setting up Lorente de Nó with the best equipment of the day. Inside the spacious laboratory, which had very high ceilings and was completely lined in copper in an attempt to isolate electrical noise, Lorente de Nó became the first person to try to understand how the activity induced in a specific neuron could be propagated to other cells in such a way that it would return to its origin after a time, via recurrent connections.

CLOSED CIRCUITS AND RECURRENT ACTIVATION

Based on the anatomical observation that certain cerebral circuits form closed loops capable of feeding electrical activity back to the original place of stimulation, Lorente de Nó decided to investigate the trajectory followed by periodic pulses of electrical activity applied to loop circuits like the hippocampus. Based on a lot of anatomy and some electrophysiology, he posited the idea that neural circuits that include closed loops are capable of reverberating electrical activation for some time after the interruption of the stimulus, creating cycles of activation that are only dissipated after several repetitions.

The idea of the reactivation of closed neural circuits proposed by Lorente de Nó seduced neuroscientists right through the twentieth century, because recursive processes like these can be—and indeed are—the basis for different kinds of rhythms, oscillators, clocks, or physiological pacemakers. From Lorente de Nó's trailblazing research onward, countless specialized neuronal connections were discovered in the brain, which generate recurrent waves of activity via multiple brain structures. The combination of closed-loop brain architecture, inhibitory neurons capable of temporarily eliminating electrical activity, and the release of different neurotransmitters (such as acetylcholine) gives rise to rhythms of variable duration, which are characteristic of global cerebral states like waking, slow-wave sleep, and REM sleep (fig. 7).

OSCILLATIONS, RHYTHMS, AND MEMORIES

Different sub-states occur within each of the main states, in the form of long episodes of electrical oscillations in specific cerebral regions. As we will see below, these oscillations coexist in time and space, creating harmonies that are established at specific moments in such a way as to optimize communication between areas of the brain. However, the complex syntax of neural oscillations was still totally unknown when the fertile imagination of Canadian psychologist Donald Hebb was gripped by the idea of a reverberating circuit. In February 1944, when he learned of Lorente de Nó's recent discoveries, Hebb had an epiphany. All of a sudden, he was able to see in the reverberation of electrical activity a natural way of storing memories.

Might reverberating circuits be our blocks for building and assembling memories, fundamental elements in the construction of mental representations of events and objects? Might electrical reverberation be the fundamental process capable of sustaining cumulative learning in our vast neuronal mesh? Maybe neural reverberation really is the key to unlocking our incredible capacity for acquiring new representations of the world around us without losing (too many) previously stored representations.

HEBB ASKS FOR AN INTERNSHIP

Excited about the potential of these ideas and eager to collaborate with the great Spanish master, Hebb wrote to Lorente de Nó on April 28, 1944, volunteering for a one-month internship in his lab. The man offering himself to work for free was no beginner. Having completed his doctorate under the guidance of Karl Lashley, first at the University of Chicago and then at Harvard, Hebb had learned from the cream of the physiologists and psychologists of his day. In 1936, Hebb had defended his thesis and accepted a position as research assistant at the Montreal Neurological Institute, under the direction of neurosurgeon Wilder Penfield, who would become known for carrying out pioneering experiments of electrical stimulation of the cerebral cortex. One account by Penfield exemplifies the amazing discoveries made possible by his method, as he describes an instance in which electrical stimulation "somehow summoned a past experience" in the patient in his care:

A patient . . . was complaining of seizures during which she sometimes fell unconscious to the ground in an epileptic convulsion. But, immediately preceding such an episode, she was aware of what seemed to be a hallucination. It was always the same, an experience came to her from childhood.

The original experience was as follows: She was walking through a meadow. Her brothers had run on ahead along the path before her. A man following her said to her that he had snakes in the bag he was carrying. And she was frightened and ran after the brothers. This had been a true experience. Her brothers remembered, and her mother remembered hearing of it.

Afterward, for some years, the experience came back to her in her sleep, and she was said to have a nightmare. Finally, it was recognized that this little dream was a preliminary to an epileptic seizure that might come on at any time, day or night. And the dream sometimes constituted all there was of the attack.

At operation, under local anesthesia, I mapped out the somatic sensory and motor areas for purposes of orientation, and I applied the stimulator to the temporal cortex. "Wait a minute," she said, "and I will tell you." I removed the electrode from the cortex. After a pause, she said: "I saw someone coming toward me, as though he was going to hit me." It was obvious also that she was suddenly frightened.

Stimulation at a point farther forward caused her to say, "I imagine I hear a lot of people shouting at me." Three times, at intervals and without her knowledge, this second point was stimulated again. Each time she broke off our conversation, hearing the voices of her brothers and her mother. And on each occasion she was frightened. She did not remember hearing these voices in any of her epileptic attacks.

Thus the stimulating electrode had recalled the familiar experience that ushered in each of her habitual attacks. But stimulation at other points had recalled to her other experiences of the past, and it had also produced the emotion of fear. Our astonishment was great, for we had produced phenomena that were neither motor nor sensory, and yet the responses seemed to be physiological, not epileptic.

Penfield's experiments demonstrated that mere cortical activation can trigger dream experiences, in the form of chains of memories that can maintain unity and coherence even after multiple repetitions of their activation. The account continues:

A young woman (N.C.) said, when her left temporal lobe was stimulated anteriorly . . . , "I had a dream, I had a book under

my arm. I was talking to a man. The man was trying to reas-
sure me not to worry about the book." At a point 1 cm. distant,
stimulation . . . caused her to say: "Mother is talking to me." Fifteen minutes later, the same point was stimulated: The patient
laughed aloud while the electrode was held in place. After withdrawal of the electrode, she was asked to explain. "Well," she
said, "it is a kind of long story, but I will tell you . . ."[1]

Hebb was an outstanding member of Penfield's team, and there he
made important discoveries about the psychological effects of brain
lesions. Therefore, when he volunteered his services to Lorente de Nó,
he was offering a vast wealth of empirical and theoretical knowledge
about the mind and its biological basis. Yet Lorente de Nó still showed
no interest, categorically refusing the request for an internship in a
letter dated May 1, 1944: "At present my work is concerned with the
relationship between the production of the nerve impulse and the
metabolism of the nerve, a problem that is of little immediate interest
to a psychologist."

The disappointment did not stop Hebb. Alongside his empirical
research, he devoted himself to creating a theory that would forever
change the understanding of the neural bases of psychology. Speculat-
ing freely about the possible biological mechanisms for the formation
of memories, Hebb identified a series of phenomena that even today
continue to be at the lively experimental frontiers of neuroscience. His
book The Organization of Behavior, published in 1949, would come to
occupy a place as the most influential of all neuropsychological theories
to this day.[2] Hebb correctly predicted that the acquisition of memories
demands—at the level of individual neurons—the summation of multiple activations originating from different upstream neurons, which
can lead to the strengthening of the connections between them. One
popular phrasing of the Hebbian hypothesis is that "neurons that fire
together wire together." Hebb proposed that the consolidation of a
memory starts with its electrical reverberation through the recurrent
neural circuit, which then causes a group of neurons to start working
together, in synchrony. This in turn increases the excitability of this
group of neurons, eventually corresponding to a physiological representation of the place, object, or event that has been memorized.

In order to understand the enormous progress represented by this neuronal conception of what it means to learn, it is necessary to consider that until the end of the 1940s, the different branches of psychology continued to advance in just the same way they had done since the nineteenth century: with no unifying theory, violently warring and without any contact with neurobiology. Behaviorism, then the most successful branch of psychology, provided a very detailed quantification of animal behavior in controlled laboratory conditions, but was not prepared to open the cerebral "black box" that produces the mind. On the other hand, neurophysiology, which was starting to understand the simpler aspects of the nervous system, had no aspiration to get anywhere near to the mental phenomena. Among those few people who did dare to take a walk on the wild side, ignorance of the mechanisms of thought was so complete that even a renowned neurophysiologist like the Nobel laureate Roger Sperry devoted years of his life to investigating whether consciousness was caused by electromagnetic fields, rather than neuronal firings, a possibility that has been completely dropped today. It was therefore with no small amount of daring that Hebb wrote again to Lorente de Nó, this time to present his work to him. At the age of forty-four, Hebb predicted: "I believe that my book will be able to show that modern ideas in neurophysiology, and particularly some of those that you have developed, have a revolutionary significance for psychological theory."[3]

WINSON ASKS FOR AN INTERNSHIP

He couldn't have been more right. Fifteen years went by and another unlikely scientist came onto the scene: the New Yorker Jonathan Winson, an old-style gentleman with an unpredictable trajectory. The story begins with a prematurely aborted career in technology. After completing a master's in aeronautical engineering at the California Institute of Technology and gaining a doctorate in mathematics from Columbia University, Winson married and moved to Puerto Rico to manage the family's successful shoemaking business. Goodbye, science, theaters, and refined restaurants; hello, palm trees and blue waves!

Nearly twenty years later, after the death of his father and the lucrative sale of the business, Winson and his wife, Judith, decided to return to New York, wanting access to its intense cultural life. They were cultured, sophisticated, and hungry for the concerts, exhibitions, and lectures that were lacking in San Juan but existed in abundance in New

York. In particular, they were seeking access to the psychoanalytic circles that were flourishing in the Big Apple of the 1960s. But Winson,
in addition to being a humanist and a Freudian, also had significant
technological and scientific inclinations. Aged forty-four, well set up in
life and far from the ideal age for beginning a career in experimental
science, Winson knocked on the door of the laboratory of Professor
Neil Miller, from Rockefeller University, and offered his services to
work for free as an apprentice.

   This took no small amount of boldness. By 1967, that small university, occupying a single block on the Upper East Side, already had one
of the greatest concentrations of Nobel Prizes per square meter in the
world. It was also a bastion of independent, unconventional attitudes.
Winson was not only accepted as a researcher, but over the years he
was promoted successively to technician, assistant professor, associate
professor, and emeritus, even being granted the honor of occupying
Lorente de Nó's copper-lined laboratory for carrying out his research.

UNCOVERING THE FUNCTION OF THE THETA RHYTHM

Winson's first major contributions were related to the theta rhythm,
formed by quite regular brain waves that totally overtake the hippocampus during certain states, lasting several minutes. Discovered in rabbits
in the 1950s and later observed in rats, cats, monkeys, and humans, the
theta rhythm was a great mystery until the mid-1970s, when Winson
started to decipher it. The paradox is that the same rhythm appears in
completely different situations depending on the species being studied (fig. 10). While the theta rhythm in rats is largely proportionate to
speed, in rabbits it only occurs when the animal is stationary. To make
things more complicated, the theta rhythm in cats can occur both when
the animal is immobile and when it is in motion. The cherry on the
cake of this scientific enigma is REM sleep, which coincides with hippocampal theta rhythm in all of these species.

   Winson saw that the key to understanding the theta rhythm is identifying, depending on the ecological niche of each species, what kind
of behavior requires a high degree of attentiveness to the environment.
Rats are prey from the point of view of some species, such as cats,
but they are predators according to other species, such as mice. They
are characterized by being excellent explorers of their environments,
which they navigate with agility and great attentiveness in search of

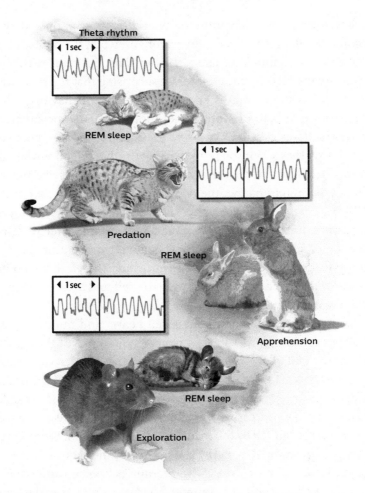

Figure 10. In mammals, the theta rhythm of the hippocampus happens when awake and alert, and, even more intensely, during REM sleep. The theta rhythm is present in different species during the carrying out of behaviors that are essential to the survival of each species, with a frequency of between four and nine waves per second.

food. It is therefore during the spatial exploration of new environments that the theta rhythm appears most strongly in rats. Rabbits, on the other hand, are prey par excellence, and they freeze when placed in a new environment, assuming a posture up on two legs with their ears raised, in a fearful search for predators. In rabbits, the theta rhythm occurs during their alert immobility and disappears once they have become familiarized with the environment, when the animals start to forage slowly on four legs. The cat, meanwhile, is a feline with obvious predatory instincts. It is therefore not surprising that the theta rhythm occurs when the animal is engaged in hunting behaviors, whether against mice or balls of wool, either motionless in anticipation of an imminent attack or running toward their target, about to pounce.

In his synthesis, Winson proposed that the theta rhythm while awake is explained by attentional behaviors typical of each species (fig. 10). By analogy, he suggested that the occurrence of the theta rhythm during REM sleep would indicate a physiological state capable of processing those memories acquired while awake in the most protected sensorial isolation of sleep, but with the same heightened attention that is used during waking experience. REM sleep would therefore be a reflexive state in which the brain pays attention to itself and to its representation of the world it knows.

Winson's interpretation conquered hearts and minds within the small but growing community of neurophysiologists of the hippocampus. At the end of the 1970s, he discovered that the destruction of the hippocampal theta rhythm by damage to another part of the brain, the medial septum, causes a sharp loss of spatial memory in rats. Published in the journal *Science,* this was the first direct demonstration of the important role played by the theta rhythm in cognition. We know today that the hippocampal theta rhythm is vital for the acquisition, processing, and evocation of declarative memories, which can be narrated verbally, such as one's last summer holidays, a best friend's wedding party, or a recent dream.

ELECTROPHYSIOLOGICAL DAY RESIDUE

Inspired by the Freudian idea of the day residue, Winson and the Greek neuroscientist Constantine Pavlides, who was then his doctoral student, decided to ascertain whether the neurons that are most stimulated while awake are also the ones most activated during sleep. In order to test this hypothesis, they made use of one specific property of certain neurons in the hippocampus, called pyramidal because of the conical shape of their cell bodies. These neurons are selectively activated only when the animal passes through a very specific location in space, such that each neuron corresponds to a restricted spatial field, within which the neuron is activated and outside of which it is not. It was the discovery of such place neurons as a mechanism for the mapping of space that earned the American John O'Keefe and the Norwegians Edvard and May-Britt Moser—Hebb's scientific grandson and great-grandchildren—the 2014 Nobel Prize in Physiology or Medicine.

In Pavlides and Winson's experimental design, the activation of the pyramidal neurons restricted to specific spatial fields made it possible to compare well-activated neurons with others that were kept almost silent. After surgically implanting electrodes into the hippocampus, the researchers identified and recorded pairs of pyramidal neurons with spatial fields that did not overlap—that is, each neuron had a preference for a different place. Then, using transparent acrylic domes to restrict the animals' position to the field of one of the neurons, and taking care not to remove the visual cues that allowed for spatial positioning, the researchers made this neuron activate repeatedly during the whole duration of the recording, while the other neuron remained deactivated.

After twenty minutes of this, the researchers positioned the animal in the recording cage, outside the spatial fields of both neurons, and allowed the rat to sleep spontaneously for several hours. The results, which were published in 1989, were revelatory: neurons that had been more activated while awake were specifically reactivated during the subsequent sleep, both during slow-wave sleep and REM sleep (fig. 11).[4] The study provided empirical support for the idea that neuronal activity during sleep reverberates the experiences of the waking mind. It was nothing less than the first electrophysiological evidence of the day residue that had been put forward by Freud.

A few years later, this discovery was deepened by the U.S. neuroscientist Matthew Wilson during his postdoctoral work at the University of Arizona. Wilson made a quantitative analysis not only of the changes in the rates of activity of the neurons of the hippocampus, but of the changes in the synchrony between the moments of different neurons' activation. In other words, he measured not only how much each neuron increased or decreased its activity, but also the proportion in which any two neurons are activated together, in synchrony.

It is no coincidence that Wilson studied under the Canadian neurophysiologist Bruce McNaughton, who in the late 1970s was close to Donald Hebb, and shared his enthusiasm for studying neuronal

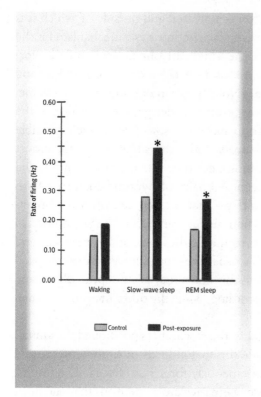

Figure 11. Impressions from the day: electrophysiological day residue. Hippocampal neurons activated while awake by persistent exposure to their spatial fields present more neuronal firings during sleep than those neurons that had not been exposed (control).

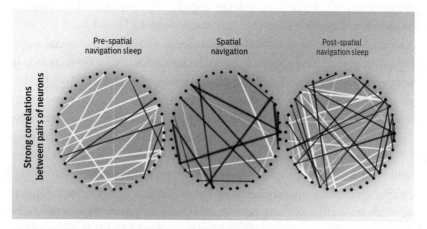

Figure 12. Patterns of synchronized activity between pairs of neurons in the hippocampus arise during spatial navigation and are retained during the sleep that follows. Only the strongest correlations—those with a high degree of synchrony—are shown. Each of the forty-two points corresponds to one hippocampal neuron: the darker the line, the stronger the correlation.

synchrony. In 1994, Wilson and McNaughton published results that became instant classics (fig. 12). First, they showed that during waking, when the rat moves along a particular trajectory and the theta rhythm predominates in the hippocampus, new patterns of synchrony arise between pairs of hippocampal neurons. Then, they showed that the same patterns reverberate, with a certain amount of background noise, during the slow-wave sleep that followed. In 2001, by which time he was a professor at the Massachusetts Institute of Technology, Wilson and doctoral student Kenway Louie demonstrated similar effects in REM sleep.

In order to understand the difference between Pavlides and Winson's initial findings and Wilson's subsequent findings, imagine that each action potential of each neuron is a musical note on a score. Pavlides and Winson's fundamental discovery was like saying that the notes most often played while awake are heard again during sleep. Wilson's results showed that it is not only the notes observed while awake that get repeated in sleep, but also their combinations in chords and melodic phrases. The metaphor of memory as a musical score allows us to use these findings to imagine how it is that something remembered from the daytime can resurface in dreams.

REVERBERATION OR REACTIVATION?

The subsequent exploration of these discoveries by different research groups, of which mine was one, managed over the last twenty years to establish that the electrical reverberation of memories is at its greatest during slow-wave sleep, becomes variable during REM sleep, and decreases significantly while awake.[5] Given the consistent increase in neuronal reactivation during NREM sleep, the greater variance during REM sleep, and the reduced duration of REM sleep compared to NREM sleep (about 1:4 in rats and humans), one should conclude that NREM sleep plays a dominant role in neuronal reverberation, while REM sleep plays a secondary role in this phenomenon. In practical terms, this means that the first half of the night, which is dominated by NREM sleep, is essential for the reverberation of those memories acquired while awake.

Proposed by Donald Hebb more than sixty years ago, the term *reverberation* has been replaced in recent decades by the word *reactivation*, but this term does not fully describe the phenomenon in question.

Although the reverberation of the activity of neuronal networks does reduce substantially during waking, it does not disappear completely when animals are awake. The strength of mnemonic reverberation while awake is in inverse proportion to the degree of interfering sensorial stimulation. Since traces of memory can be continuously detected in the post-acquisition period across all behavioral states, the correct thing to say is that the relevant sensorial experiences cause sustained reverberations, rather than discrete reactivations. Why do we not see that there are dreams there in the background when we are awake? The explanation for this is the torrent of stimulations coming to us from our five senses. In scientific jargon, the reverberating patterns of neural activity associated with past experiences are to a large part—though not entirely—masked during waking life by the sensory stimuli received. To paraphrase Freud, *dreams are like stars: they are always there, but we can only see them at night.*

And yet some people do manage to perceive them even when awake, such as in the creative reveries attributed to the Austrian composer Wolfgang Amadeus Mozart by his first biographer:

> Mozart wrote everything easily and quickly, which might at first sight have looked like carelessness or haste. He did not even go to the piano as he wrote. His imagination presented the entire work as it appeared to him, clearly and vividly . . . In the tranquil silence of the night, when no obstacle hindered his soul, the power of his imagination turned incandescent . . . [6]

HIGH OR LOW FIDELITY?

Another term that gained in popularity, related to the word *reverberation*, is what is called memory replay, the high-fidelity repetition of past patterns of neural activity. However, the reactivation of memories during sleep does not evoke them perfectly, like playing a recording, repeating content from the daytime. Rather it is a noise reactivation, like a band playing live and from memory. For this reason, the sound is "dirtier," with the clatter of the reverberations that compete for the neuronal activity produced during sleep. The final result is more a jam session than an exact copy, more vinyl than MP3.

This dirty reverberation is probably related to the fact that vast portions of the mammalian brain are devoted to the simultaneous representation of a range of perceptions and actions. As a result, individual neurons are recruited to participate in different synchronized neuronal groups, combining multiple pieces of information in such a way as to make the detection by the researchers of any single specific memory difficult. The same single musical note used countless times in different scores produces very different effects on the hearer depending on the context in which each note occurs. Now imagine that these scores could be performed in parallel, simultaneously, and this phenomenon becomes easier to understand.

One notable counterexample that proves the rule is found in an Australian bird that is often used in scientific research, the zebra finch. During sleep, a group of neurons that are involved in the motor production of birdsong present a truly faithful repetition of the neuronal activity observed when the animal sings: an almost perfect copy. This rare case of replay derives from the highly specialized neural processing that is carried out by these neurons, whose activity is required for controlling the vocal muscles responsible for singing. All of these neurons are fully devoted to the sequential codifying of one single memory, repeated right through the course of their life in a fixed, unchanging form: the bird's own song. During the zebra finch's sleep, high-fidelity replay does in fact exist.

STRENGTHENING AND WEAKENING OF SYNAPSES

The great importance that neuroscientists assign to neuronal reactivation comes from its connection to the consolidating of memories and to a phenomenon of the greatest importance to the neurobiology of learning: long-term potentiation. Although Hebb could see in 1949 that the simultaneous activation of multiple neurons must persistently change their connections to other neurons that are further downstream, one or more synapses away from those neurons activated, the phenomenon remained purely theoretical for nearly two decades. It was not until 1966 that the first empirical evidence appeared that an electrical stimulus can strengthen a group of synaptic connections in a lasting way. Studying stimuli and electrical responses in the hippocampi of anesthetized rabbits, the Norwegian Terje Lømo, then a doctoral student, managed for the first time to electrically induce an artificial memory, making neurons "remember" stimuli they had received. Doing research at the lab of Norwegian neurophysiologist Per

Andersen at the University of Oslo, Lømo first alone and then with his British colleague Timothy Bliss published the first evidence of long-term potentiation.[7] The discovery is the cellular analogy of the operation of addition, essential to the functioning of the biocomputer we carry with us inside our skulls.

In 1982, the Japanese neurophysiologist Masao Ito published the first evidence of the opposite phenomenon, a reduction in synaptic strength obtained by low-frequency stimuli, called long-term depression—the neural equivalent of the operation of subtraction. Since then, the study of the potentiation and depression of synapses has become one of the most dynamic fields of research in neuroscience.

Inevitably, there were criticisms: the stimuli were applied at too high or low a frequency, creating a situation that was somewhat artificial. The argument was made that memories that were acquired naturally probably depended on different mechanisms, but with time it became clear that the potentiation and depression of synapses also occur under stimulation at frequencies close to those observed in the brain. Progress in the research ended up showing that the mechanisms triggered in these experiments are exactly the same as those used for "natural" learning.[8]

If the Mount Olympus of science is inhabited by just gods, then Lømo, Bliss, and Ito are yet to share a Nobel Prize for discovering how the synaptic landscape is shaped. Until that day comes, those students who have fallen in love with an understanding of the biological mechanisms of learning will still have the chance, at international congresses and courses, to share some good beers with the jolly Tim Bliss, an enthusiastic, enjoyable narrator of his own essential discovery.

ENCRYPTING MEMORIES

As late as the end of the 1980s, Pavlides and Winson made another surprising discovery: stimuli at identical frequencies can cause opposite effects when they are administered at different points in the theta rhythm cycle.[9] At the peaks of this brain wave, when the neurons are depolarized and therefore easily excited, stimulation causes a potentiation of the connections. In the troughs of the wave, when the neurons are hyperpolarized and therefore hard to excite, the same stimulation causes a depression of connections. Since then, these findings have been replicated by other research groups,[10] and have come to be seen as a

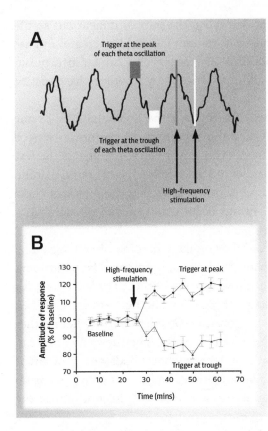

Figure 13. The phase of the stimulation in relation to the theta waves determines whether the neuronal connections will be potentiated or depressed. (A) Peaks and troughs in the theta oscillations were detected in real time to trigger the high-frequency stimulation in the following cycle. (B) Stimulation at the top of the theta oscillation increases the amplitude of the response, causing long-term potentiation. Stimulation at the trough of the theta oscillation provokes a reduction in the amplitude of the response, causing long-term depression.

central element in the process of memory acquisition (fig. 13). This dependency on the phase of the theta waves allows stimuli of identical frequency to produce diametrically opposed effects, strengthening or weakening the connections between the neurons.

We know today that the acquisition of any memory requires both the strengthening and the weakening of synapses, selectively increasing and reducing the strength of the connection between small subsets of the human brain's total synaptic network, which has many hundreds of trillions of synapses altogether. We also know that the selection of these synapses depends on the attention paid to the stimuli, an attention that coincides with the theta oscillations in the hippocampus.

Starting with this discovery, the neural melody began also to benefit from the beginnings of a harmony. It is believed today that the theta rhythm functions as a score for the occurrence of the notes, that is, of the higher frequency oscillations and neuronal firings. The theta

rhythm muffles those notes that fall at the start of the bar but amplifies those that fall at the end, so its temporal upholding creates a phase
space for the notes' distribution.[11] This is a mechanism for incorporating novelties, which shifts old memories away to other phase spaces,
to other areas of the bar.

The importance of these findings for the understanding of the
relationship between sleep and the processing of memories started
to become clearer when the neuroscientist Gina Poe, one of the few
voices in neuroscience to explicitly acknowledge Freud's seminal influence on the field, showed for the first time that the phase of theta
rhythm at which neuronal firings occur could encode the familiarity
of a memory.

Poe's story makes her another unlikely heroine of our saga. She
was born in Los Angeles to a very poor family, with no father around.
Two years later, her mother moved to San Diego, in order to look for
work and affordable housing, taking Gina and her brother with her.
The family was dependent on government aid, as the jobs her mother
managed to get only paid the minimum wage. They never owned a car,
they only had a TV for a while, and the neighborhoods where they lived
were violent, too. Though she sometimes had to go hungry to ensure
her kids got fed, Gina's mother was a firm believer in education as the
key to her kids' being able to escape poverty.

And so it proved. In fifth grade, the clever, curious girl fell in love
with the activities led by her science teacher. Whether it was dissecting an ox eye or measuring the color preference in invertebrates, at age
eleven young Gina could tell for the first time that she wanted to be a
scientist. Less than a decade later, in 1983, she managed to get admitted
to the prestigious Stanford University. In her neurophysiology course,
she saw the neurobiologist Craig Heller recount the scientific steps that
had led to his discovery that mammals do not thermoregulate during
REM sleep.[12] He said that this made mammals even more vulnerable
in this state—it was a risk to the organism and yet quite necessary, as
indicated by its widespread occurrence across species. And most important of all: nobody knew why! Gina got excited and thought how much
fun it would be to discover some fundamental phenomenon in an age
when so much was already known, or seemed to be. Except that her
initial motivation dissipated when she found she needed more money
to pay the university fees. Since she didn't need credits from that course

in order to graduate, she gave up her classes and increased the hours she was working as a cook at a restaurant.

The story could have ended there, but fortunately it did not. Some years later, Gina managed to get a research assistant job at a veterans' hospital, to study the brain activity in Air Force pilots while they were flying at low altitude and under high gravitational forces, equivalent to several times the gravitational acceleration experienced on the surface of Earth (9.8 m/s^2). The aim of the research was to determine reliably whether pilots were losing consciousness, so that planes might move onto automatic pilot without the need for a human command. As part of this research, Gina went to a conference of sleep specialists and realized that this was a field with absolute oceans of important unanswered questions—a field therefore where it is possible to aim to make a great difference, to discover some really important thing, or in short, to dream big. But Gina only learned the irony when she was officially admitted into the doctoral program in neurosciences at the University of California–Los Angeles: a doctoral scholarship in the United States pays better than any research assistant's job. Gina loved her doctorate and never looked back.

Gina, the heir to the scientific lineage of the legendary Per Andersen, John O'Keefe, and Donald Hebb, was a postdoc in the lab of Bruce McNaughton and Carol Barnes at the University of Arizona when she made her important discovery, which was published in 2000.[13] In order to understand her breakthrough, it is important first to remember that it is at the peak of the theta waves that neurons have the greatest chance of being activated. As we have seen above, Pavlides and Winson had discovered in 1988 that the stimulation of the hippocampus at the peaks of the theta rhythm produces long-term potentiation, while this same stimulation when the theta rhythm is in a trough produces long-term depression.

Combining the pieces of the jigsaw, Gina Poe formulated the hypothesis that new memories had to be codified at the peaks of the theta rhythm, while old memories destined to be forgotten must be codified in the troughs of the theta rhythm. Gina implanted electrodes into the hippocampus of rats and started to record the activity of the place neurons, whose activation was selective for the specific areas of the box in which the experiment was carried out. After a first block of data had been collected for a while, the wall on one side of the box was removed, creating a new, much larger space. This prompted a remapping of many of the place neurons, which began to respond selectively to the areas in the new space.

Comparing the firing phase of the neurons that had been remapped for new places and that of the neurons that still mapped old ones, Gina was able to confirm the separation of phases that she had anticipated. When a rat visited the new environment, the neuronal firings occurred at the peak of the theta rhythm, both when awake and during the subsequent REM sleep. However, when the same rat visited an environment with which it was familiar, the firings that happened at the peak during waking started to occur in the troughs when it was in REM sleep.

It is as though the already known past was being represented in the negative phase of the theta rhythm, which induces long-term synaptic depression and therefore forgetting. The representation of novelties, meanwhile, was concentrated in the positive phase of the theta rhythm, leading to a strengthening of connections and therefore of memories. The phenomenon that Pavlides and Winson had demonstrated using artificial memories, which had been induced electrically, Poe and her mentors were now demonstrating occurring in a much more realistic situation, involving the behavior of spontaneously exploring the environment and the REM sleep immediately following this.

Although the advance in the understanding of the mechanisms responsible for the cognitive role of sleep derives mostly from the study of rodents, it was the research into humans—using EEG, positron emission tomography, and functional magnetic resonance imaging—that first established the link between learning and neural reverberation during sleep. The Belgian neuroscientists Pierre Maquet, from the University of Liège, and Philippe Peigneux, from the Free University of Brussels, showed almost twenty years ago that brain activity during post-learning REM sleep is in proportion to the acquisition of new memories.[14] This reverberation provoked local increases in the oxygenation of the blood, reflecting the increased metabolic needs of the neurons involved in the mnemonic codification. Another study, carried out during slow-wave sleep, confirmed an increase in the power of the slow oscillations (below four cycles per second) in the cortical region subjected to training, which had a significant correlation to how much was learned.[15]

ESTABLISHING CAUSALITY BETWEEN LEARNING AND SLEEP

Showing that a biological phenomenon is proportional to a psychological phenomenon does not prove that one is the cause of the other. In order to go beyond the study of correlations and try to ascertain causality, it is necessary to induce or interrupt the biological phenomenon so as to find out what happens to the psychological one. The specific importance of the theta rhythm in REM sleep for learning was demonstrated by the team of Greek neurophysiologist Antoine Adamantidis, of the University of Bern and McGill University. Using a method that interrupted the theta rhythm with great temporal precision, the researchers showed that the reduction of theta waves during REM sleep seriously harms the consolidation of memories that are initially installed in the mouse brain via the hippocampus.[16]

In humans, Jan Born's team carried out classic experiments showing that it is possible to increase learning by electrical stimulation of the brain during non-REM sleep. Using very weak and slow electrical pulses that are applied to the skull, at less than one cycle per second, it is possible to induce artificial oscillations that magnify this state's natural slow waves. The process literally amplifies the capacity to learn.[17] Surprisingly, a similar effect can be obtained using auditory stimulation in phase with the slow oscillations, a process that increases synchronization with the fastest brain waves[18] and probably results in a large accumulation of calcium inside the cortical neurons, favoring long-term potentiation and synaptic strengthening.[19]

Considered altogether, these discoveries suggest that repeating patterns of neural activity must be the cause of memory consolidation during sleep. To test this hypothesis in such a way as to convince even the most skeptical, Jan Born and the German neuroscientist Björn Rasch had the idea of using smells to reactivate memories during sleep. The capacity of smell to evoke memories, the strong association of particular scents to specific reminiscences, is well known. At some point in your life, you are bound to have experienced the sensation of being surprised by a smell that is characteristic of the past and immediately reminds you of long-distant events, transporting you completely to that atmosphere. In addition, smells are the sensorial stimuli that interfere the least with sleep. Taking advantage of these facts, the researchers decided to carry out experiments in which the participants learned the spatial positions of cards with pictures—typical of a traditional memory game—while being exposed to the scent of roses. During the sleep that followed, the participants were again exposed to the scent with the aim of making them subliminally "remember" the positions they had learned, reactivating them by multisensorial association.

The results showed that the reactivation of memories by exposure to smells was reasonably effective when carried out during sleep, but not during REM sleep, where the same levels were seen as in the experiments without the smells. This classic experiment showed that the reactivation of memories during non-REM sleep does indeed cause learning.[20] Lorente de Nó couldn't have been more wrong in his objection to Hebb: the study of neuronal reverberation is of the very greatest interest to psychology.

A SPECIFIC TRAJECTORY THROUGH THE NEURONAL MESH

But what, after all, is a memory? To begin to define this concept, let us say that it is a specific trajectory of the propagation of electrical activity across the neuronal mesh. The conscious activation of a memory is a process that stretches out in space by means of neuronal groups and which is extended in time for hundreds of milliseconds, far above the typical timescale for the activity of a single neuron, which is of the order of just one millisecond. While a single memory requires the activation of a large number of neurons, it is normally far from recruiting the entire brain, an incredibly vast three-dimensional matrix made up of hundreds of billions of cells, each one of them interconnected by axons and dendrites to thousands of other neurons. The evoking of a memory is, therefore, the propagation of electrical activity through a quite specific and limited subset of neurons and brain regions.

For each past experience evoked, there is a particular trajectory of electrical propagation through the brain, which in its latent, non-activated state, represents the memory of that experience. For procedural memories—pedaling a bicycle, playing Capoeira—the circuits mostly involve the cerebellum, the motor cortex, and the basal ganglia. Declarative memories ("What is the capital of Angola?") as well as episodic memories ("How did your trip to research Capoeira in Angola go?") require an intact hippocampus. Each trajectory has a particular likelihood of propagation, which is transformed with each new activation of the memory via mechanisms like long-term potentiation and depression. The mental repetition of the same memory is like a river that always looks the same, but is not exactly, as it follows the same bed but never with the same water and never in the same way—least of all close to the banks.

The likeliest neuronal trajectories correspond to those memories that are most reinforced, which are activated many times over the course of a life. Each time this happens, the electrical activity that passes through sculpts paths that are preferential for future neural activity, creating the impression of memorable events. The electrical activity circulates around these trajectories under the influence of reverberating networks, such as the septal-hippocampal circuit that produces the theta rhythm, and the connections of vast portions of the brain to the small and compact locus coeruleus. Situated deep in the nervous system, this "all-seeing inner eye" exercises direct control over the pupil, which dilates according to mental effort or attention, as if opening and closing the window of the soul onto the world. The locus coeruleus detects in real time everything that hurts and everything that's new, spreading this information all around the brain via the release of noradrenaline. And at night that window closes. The locus coeruleus reduces its rate of firing until it is down to a critical level. In the absence of any significant stimulus, this level is not high enough to prevent us from falling asleep.

When light gives way to darkness, the electrical activity generated spontaneously inside the brain—originally shapeless and without content—will eventually reach the threshold of activation of some particular trajectory or other, and so the first dream image of the night appears. The dream begins. The memories formed during the day are now competing with all previous memories. It is very common, even at the very start of sleep, for the memory of the previous day to disappear in the whirlwind of other reactivated memories. However, anything that was particularly striking will return, inexorably. The paths that are most deeply carved while awake have a greater chance of being reactivated than those sculpted only shallowly. And this electric reverberation of the most significant recollections is how the bank of memories we call the unconscious is woven.

ACROSS VALLEYS AND MOUNTAINS

If the brain of a newborn were a bit of topography, it would be a sandy plain furrowed only by the innate memories of the phylogenetic past. A minimal amount of software is embedded in the very form of the hardware, encoding what the child knows how to do from the moment it first tries: breastfeeding, crying, sleeping, excreting, and learning. Armed with this behavioral repertoire, the baby faces the outside world, channeling electrical activity down the neural paths that it already possesses and that are modified as it learns to perceive and to move. Continuing with our topographical metaphor, the rain that erodes the terrain corresponds to the electrical activity. And thus, with continuous alterations to the topography that is formed of a huge number of synapses, the baby begins the construction of its internal world.

As the child acquires experience, the topography is eroded. The forming of new memories reinforces small groups of specific synapses, which are useful for survival, and eliminates huge collections of synapses that are less useful. The result is that with each new bit of learning, a furrow is made and the surface is transformed, acquiring more and more indentations, valleys, and creeks. The contact with reality, like the pressure of water against the rigidity of stone, sculpts our synaptic topography until we reach old age and become a grand canyon of experiences accumulated on top of one another, a vast, deep central valley surrounded by countless other smaller valleys, each trenched and molded by autobiographical incidents. Thus the brain becomes like a palimpsest of events lived and imagined, a mental map of a whole life made up of superimposed experiences from the remotest past we can remember to the most distant future we can imagine.

On this map, the activation of each tiny furrow corresponds to the evocation of one specific memory. Traumatic experiences leave deeper furrows, as one would expect from the intense release of adrenaline and noradrenaline during extreme stress. The experience's emotional charge increases the duration and intensity of the memory, especially when the emotions in question are negative. During sleep, in the absence of external stimuli, the electrical activity generated in the depths of the nervous system forcefully reach the cerebral cortex, the hippocampus, the amygdala, and various other subcortical areas, producing vivid dream experiences. For people who have been through traumatic experiences, dreaming often results in a strengthening of the unpleasant memories, which are equivalent to revisiting the experience.

Perhaps it's true that the electrical activity that reaches the cerebral cortex during sleep is diffuse, not very specific, maybe even random, as

Francis Crick suggested. But that is not enough for us to conclude that it wipes away cortical memories the way rain erases a sandcastle on the beach. After all, once the bombardment of electrical activity reaches the cerebral cortex and starts to spread through its huge neuronal networks, the activation is propagated according to the synaptic pathways allowed by the already existing neuronal connections—which is as much as to say: by that mind's history. It might even be the case that the drops of rain fall onto the valley at random, but it is the shape of the rock that determines their course.

Returning to the comparison between different ages, the newborn has little autobiographical past, a lot of phylogenetic past, and all the future one can hope for. Anything that happens to the baby has the potential to impact its subsequent life in its entirety. In an old person, in contrast, almost nothing makes an impact any longer. The autobiographical past has become enormous, but the future is becoming ever more limited. Venerable age is often accompanied by a copious repertoire of memories, but also by a difficulty in acquiring new memories and interesting oneself in the stimuli of the world. There's nothing extraordinary anymore, there's nothing new. There is less sleep, less neural plasticity, and fewer cannabinoids produced by the brain, which are fundamental for the formation of new synapses.[21] In old age, the rock that remains is hard—and often the mind stiffens too.

By the same token, old age also brings stability. When the repertoire of accumulated experiences is vast and healthy, the oldest people become the best advisers and leaders a community can have, caring for the group with balance, panoramic vision, and zeal for both the immediate and the distant future. Among the Kalapalo and other indigenous groups from the Xingu Indigenous Park in the Brazilian Amazon, who have practiced intertribal peace for at least sixty years, "talking like a chief" requires being calm and speaking sitting down, looking at the ground with humility and pronouncing the correct repetitions of words that appease and that assert peace and respect among relatives.[22]

Genes and Memes

What are words, ideas, thoughts, concepts? In spite of their many differences, what unites these terms is the fact that they are all memories. All we perceive and all we do causes alterations to the neural circuits that act as intermediaries for our encounter with the world, constructing associations by experience, in a repeated game of making and receiving impressions. Any elderly person or anybody who lives in close contact with elderly people knows that their memories have a lot more to do with the events of their youth than with the recent past. You might have heard stories of your great-grandparents' childhoods, things they saw and heard, memorable conversations they had with extraordinary people who said unforgettable things, passed down now to their great-grandchildren like family heirlooms. How is it possible to remember one's own childhood so precisely, so vividly, and with such a wealth of detail, even after many decades have passed? And there is something more incredible still: how is it possible for a child also to start to "remember" those events, almost as though they had really lived through them themselves?

The reverberation of neuronal activity is a satisfactory explanation for the acquisition and initial retention of memories, but it is clearly inadequate for explaining how a memory can persist for days, years, decades, or a whole lifetime. It is not hard to understand why this would be ridiculous. Imagine what would happen if the retention of memories over long periods of time required that they remain continually in an active state, reverberating in the brain incessantly, all of them alive and interlinked, explosively numerous and increasingly conflicting as life went on its curving, looping, occasionally pausing way.

In this catastrophic scenario we would suffer from profound mental confusion, like that experienced by Irineo Funes, the character created by Jorge Luis Borges. The narrator of Borges's story describes an intelligent, eccentric young man who, because of a horse-riding accident, came to have a complete memory of every event he had experienced. This marvel, however, made him unable to distinguish between important events and everyday trivialities. By acquiring complete memory, Funes became a total idiot.[1]

Fortunately, our minds do not work like that. Normally we are skilled in evoking specific memories, provided that all the other memories remain deactivated—out of mind, as it were. The reasons for this can be easily intuited: just as two bodies cannot occupy the same place in space without creating distortions and breakages, two memories cannot be simultaneously activated by a person's attention without losing their identities. Memories interfere with one another, and at each moment it is necessary that one is predominant in the consciousness for thought to be able to stroll through it.

Besides, we are extremely good at forgetting almost everything that doesn't matter to our survival and comfort, as our selective attention stores only those memories to which we assign adaptive value. If it is vital that you recall the details of the first romantic dinner you shared with your other half, the menu of the lunch you had three days after that dinner will undoubtedly be a deleted file in your mind. How does the brain distinguish between memories in such a way as to store some and erase the rest? How is it possible to retain so many memories in an inactive state?

ACTIVE MEMORIES GIVE RISE TO LATENT MEMORIES

The solution to this enigma came from Donald Hebb himself, when he proposed that the consolidation of long-term memories happens in two consecutive stages. In the first, the information is immediately captured by the nervous system as electrical reverberation, creating an impression of the recent past that is instantaneous but fleeting. This reverberation declines in a few minutes, but it triggers molecular mechanisms that end up leading to modifications of the chemical composition and later the actual shape of the synapses. This second stage involves ions passing through membranes, proteins being coupled with one another, genes being activated and new proteins being constructed, in a molecular "domino effect" that occupies the seconds, minutes, and hours following the initial acquisition of the memory, until it has resulted in a remodeling of a large number of synapses.

It is this process of creation, elimination, and modification of synapses that allows for the long-term storage of a memory, perpetuating a representation that by this point no longer corresponds to the active functioning of a neuronal network but to a latent pattern of inactive synaptic connections. Days, months, or years after the memory was acquired, when a part of these connections is activated, electrical activity spreads across the neuronal network via the strongest connections, and the memory is evoked once again. Because it is able to store old memories in an inactive form, the brain can hold a huge repertoire of them without any risk of confusion. We don't get confused like Funes did because we are able, at any one moment, to remember almost nothing.

The stories that we pass on to others, the thoughts that spread, the ideas that influence one another and are replicated socially, are all strictly dependent on their ability to endure in our minds. The English biologist Richard Dawkins used the word *memes* to describe these colonizing memories, expressed as behaviors—words and other actions—that are capable of making an impression on other people and encouraging a sharing of the same ideas. The name is evocative of another replicator unit that is much better understood: the gene. In Dawkins's famous analogy, memes "are to culture what genes are to life." While it is admittedly imprecise, this analogy is tasty because without genes there would simply be no memes.

In order to understand how the synaptic modeling that perpetuates memories takes place, it is important first to understand that all of the body's cells possess the same set of genes in their nuclei. The distinctions between the different types of cells, as well as the alterations to each cell over time, depend on the changes to the subsets of genes that at any one moment are used to synthetize proteins inside each specific cell. The resemblance between genomes and libraries helps us to understand this phenomenon. Let us say that every public library on the planet is the genome of a cell, and that each gene inside each cell is a book. To complete the analogy, let us also say that the collections of all the libraries are the same.

When you walk into one of these libraries, you will see that only a small fraction of the books contained in the collection are out on loan. You could ascertain the same in other libraries, but the set of books actually being read will be different in each case and will vary over time, that is, it will be dynamic for each individual library. Very popular books will have multiple copies that can be read in parallel by different readers. In addition, each book can be read multiple times by different readers, so long as they do it one after another. Notice that, while each library has a collection that is strictly identical to the others', the books read in each one can be quite different. Some essential books will be read in all of the libraries, but the majority of the collection will only be read in certain libraries and on certain occasions. While the most-read books in some libraries will be philosophy books, in others there will be a preference for art, and in others still for biology. In each one, at each moment, a very specific set of books will be active.

Within the same body, cells in the brain, the heart, and the liver all possess the same genes but express different subsets of them, creating different repertoires of proteins that make each cell type different. Within the cell, each gene made of DNA is equivalent to one specific book, and the molecule called RNA polymerase corresponds to one of the book's readers. Each book that is read causes the creation of a complementary copy of the gene in the form of messenger RNA. This, in turn, guides the creation of a protein capable of effectively participating in cell functions, a new reading in which the information codified by the messenger RNA is translated into the sequence of amino acids that make up the protein. The complete reading of a book corresponds to the *expression* of one gene in particular. What this scientific jargon means is that the content of the book is expressed only when it is read.

IMMEDIATE EARLY GENES AND THE SLEEP/WAKE CYCLE

When a neuron is engaged in the codification of new memory, genes that codify proteins capable of remodeling synapses are quickly activated. The first genes to become involved in the process, just minutes after the electrical reverberation begins, are called immediate early genes. The expression of a specific set of these genes is indispensable for the electrical reverberation to cause synaptic modifications some time afterward.

Immediate early genes were discovered at the end of 1980s, and it quickly became evident that they are essential for learning. Given the role of sleep in the consolidating of long-term memories, the discovery of these genes clearly suggested a hypothesis: sleep must be able to induce their activation, thus causing a synaptic strengthening later on.

The first test of this hypothesis fell to an Italian team at the University of Pisa, who compared the levels of proteins codified by immediate early genes in the brains of rodents after long periods of sleep or waking. Surprisingly, the then doctoral students Chiara Cirelli and Giulio Tononi confirmed that the expression of immediate early genes was not activated during sleep but inhibited.[2] This inhibition disrupted the logical sequence that connected neuronal reverberation to the mnemonic effects of sleep, creating an undeniable paradox.

NARCOLEPTIC IN NEW YORK

This was the scientific context that I found when I came to do my doctorate in New York City, influenced by some curious personal circumstances. Because of a six-month delay, which had been necessary to finish my master's in Brazil, I arrived in New York in the depths of winter, in early January 1995. Faced with the imposing gates to 1230 York Avenue, with two heavy suitcases and a whole world of expectation, I contemplated the streets covered in falling snow and felt that nothing would be the same again. Little did I know how right I was.

I identified myself, filled in some forms, received my keys, and dragged my suitcases to one of the apartments that Rockefeller University made available to its students at a reduced rate, and which was now my home. I opened the folder I'd been given, and saw on the class timetable that a seminar for my fellow students to discuss scientific articles had just begun. I rushed out, and after some erratic navigation I walked into a large room where a few people were eating pizza. These were my new colleagues, engaged in detailed discussion of the article that had been selected for that day.

I didn't have time to feel relieved to be finally beginning my doctorate, because something very shocking happened: I simply couldn't understand a thing they were saying. It was as if all these people were speaking underwater, in vaguely familiar bubbly sounds that didn't form any words I could recognize. All of a sudden I had lost the capacity to understand English, a language I had previously been able to read and understand pretty well.

It wasn't just that I couldn't follow the discussion owing to an ignorance of recently discovered molecular mechanisms, a subject on which I was out of touch. That was true, but the situation was much more serious. I had suddenly lost the ability to understand words in English, even the most common ones, when they were spoken at that table by those people. The situation got worse when I began to feel an over-whelming drowsiness, a powerful need to shut my eyes and switch off entirely. I managed with great effort to hold out until the end. I dragged myself back to my apartment and slept like a log.

When I finally managed to wake up, I made an alarmed assess-ment of the situation, but convinced myself I would soon adapt. I never would have imagined that that mental crash would last not just a few days but the whole winter. I just lay down and surrendered to the exhaustion. I slept and dreamed, I woke up and then went back to sleep and dreamed. And dreamed and dreamed. On those white, icy nights, broken only by the ambulance sirens from the nearby hospitals, I sank into an unprecedented period of darkness, sleep, and dreams. The days didn't last long, the clouds were blocking out the sun's rays, the world out there was strange and unfriendly. Tucked up in my cozy comforter, I took to sleeping sixteen hours a day, periods that were marked by intense, vivid dreams depicting New York, the university, and the new people with whom I was now trying to interact.

While my dream life was challenging, in my waking life everything seemed headed for disaster. I still understood almost nothing that peo-ple were saying, and I was unable to make friends. To complicate mat-ters, my attempts to participate in the lab meetings of the Argentine neuroscientist Fernando Nottebohm, whose lab I had joined, invari-ably ended with my snoring embarrassingly on the meeting-room sofa. Nottebohm is a world leader in the study of the brain mechanisms of birdsong, and I was very eager to learn about this subject, but every-thing made me yawn and nothing could keep my attention. It was as if my body was deliberately undermining my scientific career.

All through January I resisted, struggling against sleepiness, but then the anxiety and tiredness gave way to sweet surrender. When February came around, in the deep silence of the snow, I gave in completely and was swallowed up into the world of Morpheus. All I wanted now was to sleep until the end of time. I even gave up on trying to spend time in the lab, so as not to harm my fledgling reputation still further. I only left the house to buy provisions and attend classes. The rest of the time,

I remained in my apartment, taking long naps interspersed with the reading of scientific articles. During this period, I began to dream in English, and my dreams became even more intense, with representations of epic narratives through unnaturally deserted New York streets on the sunny, icy morning of an endless Sunday. I even had dreams in which I realized I was conscious and managed to alter the oneiric narrative to suit my will. In these dreams at some point a swordsman opponent would appear, wanting to fight, and I sensed that I might die.

And then, just as quickly as it had come, suddenly, the sleepiness vanished. The magnificent dreams ended, and I went back to wanting to be awake. I started to come out of my den. At the beginning of April, when the days were already getting longer and the tulips were blossoming all over campus, I realized how much of a cognitive transformation I had been through. I could understand almost everything I read now, I chatted easily, and I had started to make a very special group of friends whom I cherish dearly to this day. The best novelty of that spring of adaptation was that the problems at the lab had been resolved. Under the guidance of the Brazilian neuroscientist Claudio Mello, a specialist in immediate early genes who was then assistant professor in Nottebohm's lab, I started to carry out successful experiments on the brain representation of the songs of canaries.

Claudio was the first person to show that natural stimuli cause the expression of immediate early genes, which had previously only been observed in cell cultures maintained under controlled laboratory conditions or in the brains of animals subjected to pharmacologically induced seizures. The discovery that immediate early genes are activated in animals' nervous systems while they are carrying out behaviors that really happen in nature took this area of research well beyond the world of test tubes, toward whole organisms freely carrying out complex, ecologically relevant behaviors. Claudio was a wonderful mentor, and together we published a number of studies of the expression of immediate early genes as an indicator of neuronal activity in the brains of canaries and other songbirds. Nottebohm—a generous, libertarian, and graciously grouchy man—allowed us to follow this path in his laboratory with total autonomy.

The story could easily have continued along this route—in which case this would instead be a book about vocal communication in birds—had I not become deeply intrigued by the strange phenomenon

I had experienced over the winter, the incredibly dreamy sleepiness that had kidnapped me for those first months of my doctorate. As a scientist, I couldn't help but be fascinated by that sequence of events since I'd arrived in New York, the cognitive bankruptcy with an excess of sleepiness and heightened oneiric activity lasting through the whole winter, followed in the spring by the sudden, striking adaptation in the linguistic, intellectual, and social realms.

Of course, the gradual lengthening of the days had something to do with the end of my sleepiness. As for its start, in the blizzard of early January, that was a greater mystery. While it had first seemed to me an unfortunate, awkward act of self-sabotage, which managed to wipe out all my strength just when I needed it most, that sleep ultimately proved to be a powerful processor of novelties, which was benign and undoubtedly desirable. As I allowed myself to be carried away by the internal work of sleep, truly yielding to the off-line processing of memories, in a way I managed to overcome those enormous initial difficulties, which were as much due to the stress of my new situation as to the winter reduction of the hours of natural light.

Personally intrigued, and curious to understand what had happened to me, I decided to try to understand the mechanisms of that adaptive process. When I read in an important neuroscience textbook that science knew a great deal about the causes of sleep but nothing about its consequences, I realized this was a truly important area of research. After all, the most important things are those about which we know almost nothing. I went after that "almost." On the corner of Twelfth Street and Broadway, I paid five dollars at the labyrinthine Strand secondhand bookstore for a selected works of Freud. Reading *The Interpretation of Dreams* filled me with ideas for experiments on the relationship between sleep and learning. At the same time, in the old university library, I found the many publications from the late 1960s showing that sleep deprivation causes memory deficits in rodents.[3] Soon after this, I also discovered that in the same building as Nottebohm's laboratory, one story down the big silent stairs of the old Smith Hall, there was a laboratory with a tradition of doing research on sleep in rats. It was no less than Lorente de Nó's old lab, all lined in copper, which had subsequently been inherited by Jonathan Winson and which now, following Winson's retirement, was the responsibility of Constantine Pavlides—or Gus, to his good friends.

BEREAVED IN NEW YORK

Gus Pavlides was born in Skalochori, a small village in northern Greece where the Apostle Paul once preached to the Gentiles, in the region of Macedonia, just sixty miles from Mount Olympus. When he was a kid, in the 1960s, the place had no electricity, paved roads, or running water. The village had about two hundred inhabitants, but today that number is no more than a hundred in the summer and just twenty in winter. The only elementary school has recently closed, to be replaced by a café.

At the age of about four, Pavlides started going to school with his sister. He loved it. It was a magical time of discoveries under the supervision of his grandmother, who was convinced the boy was God's gift to the world—she would say as much every day, to anybody who would listen. Throughout his early years, Pavlides lived within a twenty-kilometer radius of his village, close to nature, to love, and to Zeus.

But then, in the early 1970s, the idyll came to an end. At the age of twelve, Pavlides had to emigrate with his mother and sisters to New York, to join their father, who had been there for a decade, working in the hope of making his fortune and returning to Greece—which never happened. Pavlides's grandmother stayed behind in Skalochori and died soon afterward, leaving her grandson disconsolate.

The arrival in New York was a shock to that very timid child, who didn't have a word of English. The family rented an apartment near Fort Tryon Park, close to the northern tip of Manhattan island, where a Greek neighborhood was flourishing. Pavlides began, with difficulty, to learn English, and his studies did not go well, except in math. In moments of sadness, seeking solace, he would often visit the magnificent medieval cloisters that had been reconstructed in the park.

One day the school principal called his parents in for a meeting. The encounter between the Scottish principal and his mother, who only spoke Greek, and his father limping along in English would have been comic if it weren't so tragic. Asking Pavlides to translate his words for his parents, the principal said there was nothing more he could do: the boy was a lost cause. In his words, "He won't even be getting a job as a garbage man, as the New York Department of Sanitation requires a high school diploma, which he clearly isn't going to be able to get." It was a very tough blow, but also the impetus for Pavlides's success. Somehow he was now going to have to prove the principal wrong.

It was in high school that everything started to get better. As well as passing an exam for placement in an advanced class, Pavlides joined the tennis team and excelled. He won the New York regional tennis tournament, and the following year was admitted to the architecture program at City College. He began the course enthusiastically, but his first contact with the professors threw a bucket of cold water over the whole experience. Pavlides wanted to build skyscrapers, but the professors said that even the best student in the class would be a draftsman at most. Pavlides lost interest, transferred to the psychology course, and after a few classes decided to go into a neuropsychology lab where they were researching intracranial stimulation. He was fascinated by the degree of control the brain can exercise over behavior, and soon afterward he got a job as a technician for Neil Miller, one of the founders of research into learning and memory.

It was in Miller's lab that Pavlides met Jonathan Winson, who at the time was deeply engaged in research into sleep and memory. This period was crucial for Pavlides's development, as he regularly had lunch with the two great scientists, a fact he never tired of wondering at. In one of the thrilling discussions that characterized these lunches, the idea arose of using the properties of the hippocampus's place neurons to investigate sleep, a real "Columbus's egg" whose impressive results were described in Chapter 10.

A FAIR EXCHANGE

Six years later, excited after having read Winson's articles, I sought out the old master in order to learn how to carry out experiments on sleep and learning. He was already retired and pointed me toward his former student Pavlides, who had now been promoted to assistant professor. I knocked on the door to his room and was promptly received. In ten minutes I explained that I intended to use the same techniques from research on canary song to find out whether sleep induced the expression of immediate early genes in the brains of rats. In that first conversation, Pavlides was just as he has proved to be in all of our meetings since—practical and positive: "You start tomorrow."

Obviously I didn't know that his speed in accepting me into his laboratory was to do with a dream vision he had received not long before. In those months, Pavlides had been using a technique developed in the 1980s to radioactively mark the areas of the hippocampus activated by different stimuli. In a dream, Pavlides had pictured the hippocampus's place neurons as being organized in clusters that were responsive to the same position in space. But the radioactive method did not prove sensitive enough to produce a convincing test of this hypothesis. What was needed was a marker produced by the brain itself, which was quick and much more sensitive, something like . . . immediate early genes! Without knowing it, I was offering to bring to Pavlides's lab exactly what he needed. It was a very fair exchange, as he trained me with great care and granted me free access to his laboratory.

I threw myself in headfirst, and learned from Pavlides how to make electrodes and implant them in the hippocampi of rats, in order to monitor precisely the different phases of the sleep/wake cycle. In parallel, Mello taught me the technique for determining the level of gene expression. For three months I worked tenaciously to test out the hypothesis that REM sleep increased the expression of immediate early genes.

The result could not have been more frustrating: we confirmed that sleep *reduced* the expression of immediate early genes. I did the experiments and redid them, for months on end, and just could not believe what I was seeing. They were essentially the same results that Cirelli and Tononi had already published, but I was not yet aware of their publications. We were living through the infancy of the Internet, which would come to make the search for relevant scientific articles in electronic databases a trivial matter. Thanks to the gaps in the bibliographic tracking, I spent more than a year chasing unsuccessfully after a result that, according to those publications that existed at the time, was impossible. When I finally found those articles, I was gripped by the strange sensation that while this was certainly true, it couldn't be the whole story. An important piece of the jigsaw was missing. My ideas had been tied into a knot and I needed to untie it.

SOLVING THE PARADOX

Then, one rainy April afternoon, rummaging in the basement of the university library, I found a curious analogy that had been put forward by another Italian group, an idea that seemed as though it might be able to untie the knot and untangle the skein. According to Antonio Giuditta and his colleagues at the University of Naples Federico II, sleep is to new memories as digestion is to food.[4] According to this perspective, in order to understand how the sleeping nervous system facilitates learning, it would be necessary first to compare what happens sequentially in slow-wave sleep and REM sleep, which in the memory / digestion analogy correspond to the stomach and the intestines. However, when there is no food, the functions of these organs of the gastrointestinal system cannot be well differentiated—which is why it would also be necessary to compare what happens in the presence or in the absence of food, that is, in the presence or absence of new information.

Inspired by Giuditta's sequential hypothesis, I carried out new experiments to measure the activation levels of immediate early genes over the course of the sleep / wake cycle, but this time comparing rats exposed to a new environment for a few hours before sleeping with control rats that had not been exposed to a new environment. In addition, instead of studying periods of several hours of sleep containing all the phases mixed together, as Cirelli and Tononi had done, I opted to analyze specific episodes from each phase of sleep, carefully separating slow-wave sleep and REM sleep. The results left us buzzing. While the animals that had not been exposed to the new environment showed a low expression of immediate early genes during both phases of sleep, the expression of immediate early genes in animals that had been previously stimulated by the novel environment showed the same profile in the cerebral cortex and in the hippocampus: a fall during slow-wave sleep, but a rise during REM sleep.

This result showed that the activation of immediate early genes can indeed occur during sleep, so long as there has previously been exposure to new stimuli. It was direct evidence in support of Giuditta's sequential hypothesis, and in revealing the effect of the waking experience on the gene expression of sleep, it was also the first molecular evidence of Freud's day residue. The paradox seemed to have been resolved at last.

THE THEORY OF SYNAPTIC HOMEOSTASIS

But the connection between the Freudian concept of the day residue and some of the most fundamental mechanisms in cell biology did not happen without controversy. In the mid-1990s, Tononi and Cirelli moved to the United States to run laboratories at the University of Wisconsin–Madison. They were convinced that the fall in the expression of immediate early genes during sleep was an important phenomenon. Over the following years, they carried out various studies confirming and broadening their original findings, at a molecular[5] as well as an electrophysiological[6] and morphological[7] level. For some reason, they did not try to study specific episodes of slow-wave sleep or REM sleep, opting instead to study the results of long periods of sleep containing both phases. Nor did they use previous exposure to new stimuli. Under these restrictions, the results that came out of their labs just went on accumulating in the same direction and led them to propose a theory that would prove extremely influential.

This theory had its basis in the discovery by U.S. biologist Gina Turrigiano that synapses which have been deactivated for long periods tend to become strengthened.[8] In order to understand this finding, it is important first to consider that it is synapses—of chemical contact as well as those of electrical contact—that allow the transmission of electrical activity from one cell to another. Electrical synapses are direct connections between the membranes of two cells, which allow the free passage of ions and therefore transmit information almost instantaneously. Chemical synapses are slower, since they consist of small protuberances from the cell membrane, which are so close to the protuberances from other cells that they allow chemical contact between them. This happens through the release and spread of minuscule vesicles, real nanobubbles that contain the neurotransmitter molecules glutamate, GABA, noradrenaline, serotonin, acetylcholine, and dopamine, among others. Chemical synapses can be high- or low-efficiency depending on their size and molecular composition. In reality, there is a continuum of possible values for the strength of a synapse, between the minimum and maximum of their transmission efficiency.

Gina Turrigiano made her surprising discovery while investigating the strength of chemical synapses after pharmacologically inhibiting electrical activity for forty-eight hours. To the researcher's great surprise, the long inhibition of neural activity had made the synapses much stronger. Subsequent experience showed that after the treatment, the neurons would fire much more, becoming much more excitable. Turrigiano called this phenomenon synaptic homeostasis, using a noun whose Greek roots mean "similar" (homoios) and "static" (stasis), which is used in biology to mean "maintained in balance."

Tononi and Cirelli borrowed the idea of homeostasis to propose that the alternation between waking and sleep would result in cycling between strengthening and weakening of the synapses, respectively.[9] This theory posits the idea that the cognitive benefits of falling asleep derive from the general weakening of the synapses during sleep, which would lead to the forgetting of the weakest memories and give a comparative advantage to the stronger ones.

Over the course of two decades, the theory of synaptic homeostasis spread widely, and its proponents became increasingly influential on research into sleep and memory, publishing often in the most important scientific journals and even conquering the pages of *The New York Times*. This was a very attractive theory, simultaneously simple and general: sleep makes us forget what doesn't matter, giving relative prominence to what's important. During the day the brain "warms up," at night it "cools down."

This theory potentially explains both the weakening and the strengthening of memories, but it does not offer any mechanisms to explain the restructuring of memories in such as way as to create new ideas—a problem we will consider in detail in the next chapter. Besides, the theory depends on neural measurements obtained after long periods of sleep without differentiating between slow-wave sleep and REM sleep. This results in a significant prevalence of slow-wave sleep, leading the theory to neglect the role of REM sleep.

Incomplete theories are the essence of science, but in the case of the theory of synaptic homeostasis, the incompleteness was propagated deliberately. Over two decades, the publications by Tononi and Cirelli systematically ignored the divergent evidence produced by different laboratories in the United States, France, and Brazil. This evidence did not contradict synaptic homeostasis, but showed that it was merely the tip of the iceberg, since it had been observed in animals only in a quite particular situation, with a predominance of slow-wave sleep (to the detriment of REM sleep) and an absence of new stimuli or the learning of new tasks before sleeping.

THE THEORY OF MEMORY EMBOSSING

In several laboratories, including my own, when REM sleep was investigated in animals that had been previously exposed to novelties or to behavioral training, there was invariably an observable activation of mechanisms of synaptic strengthening during sleep, such as the activating of the expression of immediate early genes. In place of the excessive

simplicity of the synaptic homeostasis model, with synaptic strengthening exclusively during waking and synaptic weakening exclusively
during sleep, in more realistic learning situations we found a process
that is more complex, characterized by a strengthening and weakening
of complementary groups of synapses when both awake and asleep.
I called this process "memory embossing," alluding to the creation of
high and low reliefs during the embossing of wood.[10]

The theory starts from the principle that the acquisition of a new
memory demands that certain synapses be strengthened and others
weakened, while the vast majority of synapses remain as they were,
without any transformation. During post-learning sleep, the strongest
connections would be strengthened further still, while the weakest
would be further weakened. Direct and indirect evidence of this phenomenon was found in animals as varied as rats, cats, and flies, both
during the development of young individuals and in the learning of
adult animals.[11] Nonetheless, for fifteen years, the proponents of synaptic homeostasis continued to dominate the field, without admitting
there were problems with the hypothesis, unexplained anomalies, or
alternative theories.

The controversy moved toward its climax in 2014. In January, in a
review article on sleep and learning, Tononi and Cirelli admitted for
the first time that some dissent did exist.[12] They acknowledged evidence that they had previously ignored and articles they had never previously mentioned to state that the reality was more complicated than
their theory had anticipated. And not before time. Just five months
later, researchers at New York University, under the leadership of the
Chinese biologist Wenbiao Gan, published in the prestigious journal
Science a clear demonstration of synaptic strengthening during sleep
in an area of the brain subjected to learning. Using a sophisticated
microscopy technique on mice that had been genetically modified to
have fluorescent neurons, Gan and his team were able to visualize and
measure the increase in the number of synapses due to post-learning
sleep. The animals were trained to walk forward or backward on a
rotating cylinder, which caused strong synaptic changes in the motor
cortex, a region needed for the carrying out of voluntary movements.
By making detailed images of synapses before and after post-learning
sleep, the researchers showed that falling asleep is associated with the

formation of new synaptic connections. Gan and his team attributed the increase in the number of connections to slow-wave sleep, since animals deprived of REM sleep also presented with this effect.[13]

However, there are reasons to suspect that even small quantities of REM sleep are enough to strengthen preexisting synapses, analogously to feeding, which does not have to be daily, since its intermittent occurrence is sufficient to allow life to continue. Finally it was shown in rodents that a single short episode of REM sleep, lasting less than thirty seconds, is as effective in modulating the expression of immediate early genes as a long episode lasting several minutes. On the whole, the expression of immediate early genes in response to a range of stimuli is initially very robust, but it declines rapidly over time. In addition, in reptiles and birds, episodes of REM sleep do not last more than a few seconds.

If we bring together all these facts, a hypothesis arises that suggests that the most ancient function of REM sleep would have been to trigger the expression of immediate early genes immediately after slow-wave sleep. This brief, quick jolt in gene expression, which might have begun to evolve hundreds of millions of years ago in an ancestor common to all land vertebrates, has the effect of "taking a photo" of the moment, perpetuating the new synaptic connections formed between the neurons. The synaptic remodeling induced by REM sleep transforms a pattern of reverberating electrical activity in the neural circuit (an active memory) into a new pattern of synapses between cells (a latent memory). The primordial function of the regulation of genes that occurs during REM sleep seems to be the transformation of active, short-term memories into long-term latent memories that are capable not only of enduring in that particular brain but also spreading to other brains as memes: representations of people, places, events, or ideas. As they are integrated into a nervous system, these memes interact animatedly with one another, creating a simplified mental replica of the outside world, edited and filtered according to the preferences and limitations of their bearer.

THESIS, ANTITHESIS, AND SYNTHESIS

As is almost always the case in science, the controversy had not been eliminated, but had evolved. In February 2017, Tononi and Cirelli published an exhaustive study of the size and shape of almost seven thousand synapses, a truly herculean task of counting and measuring individual synapses of 0.05 micrometers squared in incredibly fine slices of brain tissue examined with electron microscopy. As in several previous studies from the same group, no attempt was made to separate slow-wave sleep from REM sleep. The study reported a reduction of about 1 percent in the average size of synapses after sleep. This is a minimal difference, but it was enough for a new entrenchment. As if this were news, *The New York Times* took advantage of the opportunity to publish one more long article on the theory of synaptic homeostasis.

However, in March 2017, Wenbiao Gan and his team published another truly revelatory study.[14] Using high-resolution images taken with two-photon microscopy, Gan presented the most complete series of experiments published to date about synaptic plasticity during REM sleep. This achievement included eleven different experimental variations on mice, focusing on different pre- and post-training moments, with important pharmacological controls and quite selective deprivation of different phases of sleep. By choosing to study the temporal evolution of live synapses under the lens of a microscope, specifically measuring the same synapse on multiple occasions over time, the researchers were able to confirm what could never have been revealed by the strategy of measuring dead synapses used by Tononi and Cirelli: that the effect of REM sleep on the synapses involves both the elimination and the strengthening of synapses, whether during the development of the mouse pups or during learning in adult animals. Whenever life requests changes to the cerebral software, it is down to sleep to do the reprogramming.

The study showed compellingly that REM sleep helps to eliminate new synapses after the massive supply of them caused by slow-wave sleep. When combined, the two main states of sleep result in a major substitution of new synapses. What is even more extraordinary is that REM sleep also acts to strengthen a select group of synapses, leading to their growth and consequently to the persistence of these connections in the long term. A huge number of synapses is generated, but almost all of them are then eliminated, allowing for the positive selection of a small number of synapses that are better adapted to the new context. In the words of Gan and his collaborators, "REM sleep is important for selectively incorporating new synapses into the existing circuits. It could be viewed as a 'selection committee' for building and maintaining the synaptic network."[15] Without REM sleep, memories would quickly disappear without a trace, unable to be accumulated for the future or transmitted from generation to generation. Without REM sleep, there would be no culture.

Sleeping to Create

Learning is a necessary condition for acquiring and propagating memes, but how are these then transformed? Replica ideas will only remain good ones if the future is the same as the past. If the strengthening of memories was the only thing that happened during sleep, we would be exaggerated versions of our parents, their characteristic features reinforced, with the same behaviors and prejudices. Fortunately, the reality is quite different: we are creatures in constant transformation, open to influence throughout our lives. How is it possible to alter memories? How are new memes invented?

Of all the mental faculties, the one that is most highly valued by entrepreneurs, artists, and scientists is creativity. The brewing of culture has always depended on imagining new forms from a recombination of old forms, and the mental construction of what does not yet exist has always benefited from dreams as a fundamental source of inspiration. While modern capitalist rationalism may have discarded dreams as an important phenomenon, oneiric ingenuity had a decisive influence on the industrial revolution. The family history of the inventor Elias Howe records the pivotal role of dreaming in his greatest innovation:

> He almost beggared himself before he discovered where the eye of the needle of the sewing machine should be located . . . His original idea was to follow the model of the ordinary needle, and have the eye at the heel. It never occurred to him that it should be placed near the point, and he might have failed altogether if he had not dreamed he was building a sewing machine for a savage king in a strange country . . . He thought the king gave him

twenty-four hours in which to complete the machine and make it sew. If not finished in that time death was to be the punishment. Howe worked and worked, and puzzled, and finally gave it up. Then he thought he was taken out to be executed. He noticed that the warriors carried spears that were pierced near the head. Instantly came the solution of the difficulty, and while the inventor was begging for time, he awoke. It was 4 o'clock in the morning. He jumped out of bed, ran to his workshop, and by 9, a needle with an eye at the point had been rudely modeled. After that it was easy.[1]

Howe's invention of the lockstitch sewing machine triggered a complete socioeconomic transformation in the United States and the United Kingdom, with a vertiginous rise in the scale of production of weaving, the massification of the clothing market, the acceleration of exports, and geopolitical expansion. If the short-term consequences were profound for their significance in the methods of textile production, the long-term implications of that dream were more transformational still. It was weaving that first saw the use of binary code to create combinations of different-colored threads, a system that was the precursor to integrated computer circuits.[2]

MORNING MELODIES

Among artists, accounts of creative dreams can be found in abundance. Musicians, for example, often wake up with whole melodies in their heads, which were originally "composed" in their sleeping minds. Anecdotes of these kinds are told about Beethoven, Handel, and many other classical composers. The Italian violinist Giuseppe Tartini claimed to have composed his best-known work, Sonata in G Minor, *The Devil's Trill Sonata,* under the direct influence of a dream:

One night, in the year 1713, I dreamed that I had made a pact with the devil for my soul. Everything went ahead as I had desired: my new servant anticipated all my desires. I thought to give him my violin to see whether it could play me some beautiful melodies. How great was my astonishment to hear a sonata so wondrous and so beautiful, played with such mastery and intelligence, such as I had never conceived of even in my highest flights of fantasy. I felt enraptured, transported, enchanted: my breathing failed me. I was woken up by this violent feeling. I immediately grabbed hold of my violin so as to preserve, at least in part, the impression from my dream. But in vain! The piece of music I composed at that moment is indeed the best that I have written, and I even call it "The Devil's Sonata," but the difference between that and what had so moved me is so great that I would have destroyed my instrument and bade farewell to music for ever if it would have been possible for me to live without the pleasure it afforded me.[3]

The phenomenon is not restricted to just one musical style in particular, of course. The song "Yesterday," composed by British songwriter Paul McCartney, was attributed to a dream:

I woke up with a lovely tune in my head. I thought, "That's great! I wonder what it is?" There was an upright piano next to me, to the right of the bed, by the window. I got out of the bed, sat at the piano, found G, found F# minor 7th, and that leads you through then to B, E minor, and finally back to E. It all moves forward logically. I liked the melody a lot, but because I dreamed it, I couldn't believe I had written it. I thought, "No, I have never written anything like this before," but I had the tune, which was the most magical thing.

Even McCartney himself found it hard to assert his own authorship:

For about a month, I went round to people in the music business and asked them whether they had ever heard it before. . . . It became like handing something in to the police. I thought that if nobody claimed it after a few weeks, then I would have it.[4]

MATERIALS AND METHODS

In the visual arts, oneiric influence is no less present. Albrecht Dürer, the German Renaissance master of engraving and painting, recorded his use of dreams to obtain valuable pictorial images. In his treatise on painting called *Nourishment for Young Painters,* Dürer described the profusion of images and the difficulty in capturing them: "How often do I see great art in my sleep, but on waking cannot recall it; as soon as I wake my memory forgets it."[5]

A decade later, he painted a dream scene of great symbolic power. At the bottom of the watercolor, Dürer described his dream:

> In 1525, during the night between Wednesday and Thursday after Whitsun, I had this vision as I slept, and I saw such torrents of water falling from the sky. The first struck the ground about six kilometers away from me with such terrible force, a huge clamor and splashing that drowned the whole land. I was so shocked by this that I woke before the downpour. And the downpour that followed was enormous. Some of the waters fell at some distance and others more close by. And they came from such a height that they seemed to be falling at an equally slow pace. But the first water that struck the ground fell suddenly at such speed and it was accompanied by such terrifying wind and roaring that when I woke, my whole body trembled and it took me a long time to be able to recover myself. When I got up in the morning, I painted what was above, as I had seen it. May the Lord make all things better.[6]

The picture of an open field without many trees is dominated by the huge column of water coming down from the skies and flooding the earth, while various other smaller columns represent the rain that is about to fall to the ground. It is believed that this dream is a reverberation of the religious uncertainties of the Protestant Reformation, a real torrent that was threatening to flood the world at the start of the sixteenth century. When Dürer painted the watercolor, Luther had already won his fight with the papacy, published the New Testament in German, and begun to organize the new church. Four centuries later, the Franco-Russian painter Marc Chagall produced several pictures inspired by the biblical dream of Jacob, in which the patriarch of the Israelites saw a ladder climbing up to heaven, and he could see and hear God directly and make a pact with Him.

If the relationship between dreams and God was important to Dürer and Chagall, to the Catalan painter Salvador Dalí, the dream production of images moved away from religion and toward technique. Dalí, one of the icons of twentieth-century art, practiced a method of his own devising in order to remain on the threshold of the unconscious for as much time as possible, in order to gather dream images. With a weighty metal key or spoon held in his fingers, the dream hunter would start to nod off until the object fell noisily to the floor, pulling him out of his sleep to bring the profusion of hypnagogic images he'd been immersed in directly onto the canvas. The technique resulted in amazing works whose titles sound like the "Materials and Methods" section of a scientific paper: *Dream Caused by the Flight of a Bee Around a Pomegranate a Second before Waking.*

The focus on the dream phenomenon meant that in the first decades of the twentieth century, psychoanalysis had a striking influence on the Dadaist and Surrealist vanguards, offering inspiration to artists deeply interested in the creative trance, in the flow of consciousness and the free exploration of the unconscious. In the revolutionary *Un Chien Andalou,* the début movie by the Spanish-Mexican filmmaker Luís Buñuel, which was produced in collaboration with Dalí in 1928, what stands out are the Freudian-inspired dream associations, discontinuities, and fragmentations.

DREAMING AND LITERATURE

In literature, it is no different. Since the start of the historical record, countless writers and poets have availed themselves of the inspiration of dreams to begin, develop, or resolve their plots. In addition to this, because they are so varied and unpredictable, dreams became a narrative resource of huge practical usefulness, because they allow the tackling of any subject, however bizarre it might be.

In the classic *Dream of Scipio,* for example, Cicero used a famous dream as a trick for illustrating different points of view. The story begins after the arrival in Africa of the Roman patrician Scipio Aemilianus, who is visited in a dream by the spirit of his adopted grandfather, the famous general Scipio Africanus. Aemilianus sees himself looking at the city of Carthage from "a high place filled with stars, bright and splendid," and sees the tiny Earth in the vastness of space. Then his grandfather predicts that his grandson will reach the position of consul, the highest elected post in Rome, praising his military virtues and promising him a place of honor in the Via Lactea after his death. In a spectacular vision of the universe, Scipio Aemilianus sees that it is made up of nine heavenly spheres, with Earth in the center followed by the moon, Mercury, Venus, the sun, Mars, Jupiter, Saturn, and finally heaven itself, where the stars are fixed. As he contemplates the universe, Aemilianus learns that the spheres emit sounds and is taken by a vision of the planet's climatic belts. This fictional meme from antiquity, preserved for posterity in the work of Macrobius,[7] had a decisive influence on medieval thought, supporting the geocentric model of the planetary system and serving as a philosophical framework for discussions of the soul, virtue, and divinity.[8]

If dreams in churches and monasteries were contested by angels and devils, and therefore a matter of life and death, among poets and troubadours the use of oneiric visions to present revelations became ever more commonplace.[9] Dante Alighieri's *The Divine Comedy* presents prophetic dreams at the end of each of the three nights the narrator spends in Purgatory, with two other dreams recorded in Hell. References to dreams and dreaming appear 211 times in the writings of William Shakespeare, in thirty different works including *A Midsummer Night's Dream*.

Miguel de Cervantes made use of the fact that dreams become more vivid after sleep deprivation as a narrative tool in recounting the adventures and misfortunes of his most memorable character, Don Quixote. The adventures begin when the bankrupt old nobleman, "with too little sleep and too much reading,"[10] allows his wildest fantasies about medieval knights errant to invade his shaky mind. He fits himself up as a knight and sets out on horseback in search of gallant deeds, steeped in a generous, old-fashioned solidarity. There follow a series of episodes of delusional behavior totally at odds with the world, such as in the fight with the windmills, which he takes to be giants. Over the course of Don Quixote's whole heroic, psychotic mental process, his faithful squire Sancho Panza sleeps (and eats) copiously. It is no coincidence that he retains his sanity and his common sense despite his boss's insanities. At the end of the story, Don Quixote falls seriously ill, and is confined to bed. He has an episode of sleep that is "more than six hours long," and when he awakes, his sanity is restored! He offers some final words, and dies.

Romanticism brought great prestige to dreams, not only for use in plots but as a source of artistic creativity. Under the influence of poets like Lord Byron, the English writer Mary Shelley transformed an onei-

ric vision into her celebrated novel *Frankenstein*, a pioneering work of
science fiction, which was published in 1818. The English poet Samuel Taylor Coleridge composed his best-known poem "Kubla Khan"
after having taken opium and fallen asleep reading a book about the
mythical city of Xanadu, the Mongol emperor's summer palace. The
poet described how he composed more than two hundred lines on this
subject while dreaming. When he woke up, he wrote down fifty-four
striking lines that still entrance readers all over the world to this day.
The poem, full of rhythm and color, was left incomplete, because at
one point in his reverie Coleridge was interrupted to deal with prosaic
matters. When he was finally able to free himself and went back to his
writing, he hardly remembered anything. The poem's alternative title
points to the imagistic rapture of the dream, but also to the difficulty
in maintaining the whole intact in memory after an interruption: "A
Vision in a Dream: A Fragment."

REVOLUTION, CATASTROPHE, AND ADAPTATION

This is not a literary resource that has gone out of style. In *Ulysses*,
the masterpiece of narrative fiction by the Irish writer James Joyce,
published in 1922, dreams are mentioned fifty-nine times, as a driver of
situations that make the text progress in parallel with the journey of
the Homeric Ulysses in *The Odyssey*. The masterly Portuguese poet Fernando Pessoa, who wrote in the early twentieth century under many
heteronyms, visited dreams countless times to reflect on memory,
forgetting, and desire. His heteronym Bernardo Soares wrote: "I've
dreamed a lot. I'm tired now from dreaming but not tired of dreaming.
No one tires of dreaming, because to dream is to forget, and forgetting does not weigh on us, it is a dreamless sleep throughout which we
remain awake. In dreams I have achieved everything. I've also woken
up, but what does that matter? How many countless Caesars I have
been!"[11] This is echoed by the heteronym Álvaro de Campos: "I am
nothing. I will always be nothing. I can want only to be nothing. Otherwise, I have in me all the dreams of the world."[12]

   In truth, there is almost no literature that doesn't have dreams playing some part either in the plot or in the creative method. The Angolan
writer José Eduardo Agualusa, for example, attributes an important
role in the creation of his work to dreams:

In most of my novels I dream about the endings of chapters, solutions to plots, about the names of characters, and sometimes whole lines. In *The Book of Chameleons,* the main character, a seller of pasts, appeared to me in a dream. In *Life in the Sky,* a novel for younger readers, I dreamed the title and the whole story developed out of that.[13]

After so much drinking from this source, Agualusa published a novel in 2017 in which dream activity appears as the main strand running through the narrative. The chapters of *The Society of Reluctant Dreamers* each recount the dreams of different characters, who include a Mozambican photographer who makes photos out of her dreams, a Brazilian neuroscientist who decodes them, and a veteran of the Angolan war who has the gift and the misfortune of appearing in other people's dreams. The toppling of the tyrant who believes himself enthroned forever is the collective dream that sustains the narrative right up to its dénouement.

In addition to great courage, trying to change the structure of an oppressive society also requires the ability to take flight with one's thoughts, to imagine alternative futures and manage disappointments. This is made quite clear in *Animal Farm,* the brilliant fable by the Indian-British writer George Orwell about the hopes and failures of the Russian Revolution. The animals' rebellion against the farm owner starts when a prize-winning old pig called Old Major, an alter ego of the revolutionaries Karl Marx and Vladimir Ilyich Lenin, tells the animals of a dream in which he saw the world after human beings had been wiped out. Old Major dies soon afterward, but his dream triggers a revolt that ends up driving all the humans off the farm, leading to a government made up exclusively of animals, based around the slogan "All animals are equal." Nevertheless, the pigs are considered the most intelligent of the animals, and soon two rival leaders appear, the pigs Snowball and Napoleon, representing Leon Trotsky and Joseph Stalin, respectively. Napoleon, who is much more ruthless than the others, ends up taking power, expelling his rival, and finally going back to collaborating with the humans to the detriment of the animals, now under a new slogan: "All animals are equal, but some animals are more equal than others."

In the disturbing *Nineteen Eighty-Four,* likewise by Orwell, a dream is

the starting point for the psychic insurrection of Winston Smith against
Big Brother and his society of telescreens, which control all behavior
and cannot be switched off. Personal nonconformity is transformed
into rebellion when Winston is overtaken by a forbidden passion for
another character, Julia. Hounded and tortured mercilessly by the
state, with the dream of amorous liberation buried in a nightmare of
betrayal, the lovers end up being bitterly parted from each other.

In real life, the most exciting dreams of the great rebels often result
in stories of frustration and failure, plunging them into more troubling
visions, simulacra of reality in which they grope for the truth of what
went wrong. In 1935, exiled with no refuge and tirelessly hunted by
Stalin's agents, Trotsky described a dream in his diary that revealed the
incredible fragility of his position at that moment:

> Last night, or rather early this morning, I dreamed I had a con-
> versation with Lenin. Judging by the surroundings, it was on a
> ship . . . He was questioning me anxiously about my illness. "You
> seem to have accumulated nervous fatigue, you must rest . . ."
> I answered that I had always recovered from fatigue quickly,
> thanks to my native *Schwungkraft*, but that this time the trouble
> seemed to lie in some deeper processes . . . I answered that I had
> already had many consultations and began to tell him about my
> trip to Berlin; but looking at Lenin I recalled that he was dead. I
> immediately tried to drive away this thought, so as to finish the
> conversation. When I had finished telling him about my thera-
> peutic trip to Berlin in 1926, I wanted to add, "This was after your
> death"; but I checked myself and said, "After you fell ill . . ."[14]

The gloomy dream clearly reveals the profound loneliness experienced by Trotsky, the legendary Red Army commander, after his comrade Lenin's death. In 1940, in Mexico, Trotsky was executed in his own
home by a murderer under instructions from Stalin.

When faced with great defeats, it is necessary to reinvent points of
view, in everyday life as well as in politics. In August 1939, a few days
before the outbreak of the Second World War, George Orwell was
struggling to reconcile his revolutionary socialist convictions with the
urgent need to defend Britain against German aggression. The resolution to this conflict came in a dream: one day before the announcement

of the Treaty of Non-Aggression between the Nazis and the Soviets, Orwell dreamed that the war had begun:

It was one of those dreams which, whatever Freudian inner meaning they may have, do sometimes reveal to you the real state of your feelings. It taught me two things, first, that I should be simply relieved when the long-dreaded war started, secondly, that I was patriotic at heart, would not sabotage or act against my own side, would support the war, would fight in it if possible.[15]

Between real life and fiction, three different spheres of dream influence are dynamically, powerfully intertwined: narrative key, artistic inspiration, and political compass.

DREAMING AND SCIENTIFIC CREATIVITY

Creativity involves a radical change in perspective, a recombining of prosaic ideas in order to produce the extraordinary. Oneiric creativity happens even when it is subjected to the quantitative rigor of science, fulfilling a fundamental role in its development. The best-known example is the discovery of the benzene ring by the organic chemist August Kekulé, as published in 1865.[16] A few years earlier, Kekulé had correctly suggested that carbon is tetravalent, that is, that it makes four chemical bonds. He also knew that hydrogen makes only one chemical bond, and that the benzene molecule is made of six atoms of carbon and six of hydrogen. Kekulé was obsessed with discovering the structure of benzene, which could not be some linear combination because the number of carbon atoms was the same as the number of hydrogen atoms. Thinking extensively about the subject as he sat in front of a fire (or on a bus—there is some controversy over this), Kekulé described how he fell asleep and dreamed about a snake eating its own tail, like the alchemical symbol Ouroboros, whose origins go back to the funerary papyri of ancient Egypt.[17] On waking, Kekulé had his answer in the shape of a very clear picture: the structure of benzene is hexagonal.

It may be that this famous case is a fabrication, as Kekulé was later accused of using the dream narrative to legitimize a piece of plagiarism he was committing, stealing ideas from the French chemist Auguste Laurent.[18] The accusation is controversial, and the subject continues to be disputed in the history of chemistry.[19]

Another example of great scientific significance over which no such suspicions are hovering was the experimental demonstration of the chemical transmission of information between the nervous system and the heart, carried out by the German physiologist Otto Loewi. At the time when Loewi became interested in this subject, a controversy was raging about the nature of this communication: was it chemical or electric? Otto Loewi described his experience thus:

> In the night of Easter Saturday, 1921, I awoke, turned on the light, and jotted down a few notes on a tiny slip of paper. Then I fell asleep again. It occurred to me at six o'clock in the morning that during the night I had written down something most important, but I was unable to decipher the scrawl. That Sunday was the most desperate day in my whole scientific life. During the next night, however, I awoke again, at three o'clock, and I remembered what it was. This time I did not take any risk; I got up immediately, went to the laboratory, made the experiment on the frog's heart . . . and at five o'clock the chemical transmission of the nervous impulse was conclusively proved.[20]

The famous experiment first consisted of isolating two frogs' hearts, one of which was connected to the vagus nerve while the other wasn't. Then Loewi applied an electrical stimulus to the nerve, causing brady-cardia (a reduction in heart rate) in the stimulated heart. Finally, Loewi sucked a bit of the liquid around the slower-beating heart and applied it to the other heart. To his delight, the beating of the second heart also slowed—and therefore the transmission was chemical. Loewi called the responsible molecule *Vagusstoff*, the "substance of the vagus," which we know as acetylcholine today. The discovery earned him the 1936 Nobel Prize in Physiology or Medicine.

A good contender for the list of the most transformational ideas of all time, the atomic organization of the elements as expressed in the periodic table, was also the product of a dream. In 1869, the Russian physicist-chemist Dmitri Mendeleyev had been obsessing for months over the search for a natural classification of the chemical elements, an order that was defined intrinsically by their own attributes. He decided to write the names and properties of the elements on cards and started experimenting with different ways of arranging them. Mendeleyev felt that the atomic numbers were relevant, but after several hours he fell asleep on the cards without having managed to understand their pattern. Then he dreamed he could see a table into which each element slotted in its rightful place, perfectly positioned according to its atomic number in groups with similar properties, which get repeated periodically. The understanding that the substances are made up of elements whose relationships obey well-defined mathematical laws concluded the process of transforming alchemy into chemistry.

We know today that the periodic table expresses very well-defined physical interactions among subatomic particles, but Mendeleyev didn't know this. The moment of pure creativity does not depend on understanding all of the theory behind the phenomenon. In a vision, in a revelation, in the epiphanic "Eureka!" moment, in an insight, a flash, in that mental process the Greeks called abduction and which those who study sleep today call the restructuring of memories, the most important thing is capturing the general principles that organize the reality that one wants to be revealed: the gist of the matter. The imagining of a new idea does not need to be precise for it to work. For this reason, abduction does not obey the strict empiricism of induction, nor the logical generalizations of deduction. It is the freest mental process of all, in which the mind is transported to solutions that are not obvious, that are apparently distant and generally surprising.

The oneiric capacity for successfully combining scientific ideas was evident in the story of the nineteenth-century British naturalist Alfred Russel Wallace. On his two-decades-long travels around Brazil and Southeast Asia, in the middle of the nineteenth century, he established that species evolve into other species, and are constantly creating diversity. Wallace believed he had an extensive observational basis for this radical idea, which had been argued about since the days of the French naturalist Jean-Baptiste de Lamarck, almost a hundred years earlier, but was still strongly opposed in academic milieus and unsupported by any mechanisms that could explain the evolution of species. In Wallace's words, "The problem then was, not only how and why do species change, but how and why do they change into new and well-defined species."[21]

In February 1858, on a remote Indonesian island, Wallace experienced intermittent attacks of fever, possibly caused by malaria. During this fever, he had dream visions that related the problem of the evolution of species to the theory that the abundance of surplus resources is limited by the growth of the population, as proposed at the end of the eighteenth century by the English demographer Thomas Malthus. When he woke from his trance, Wallace realized that the reverse was also true: if resources are limited, species evolve in an environment of fierce competition, which tends to select the best-suited individuals in each generation. Everything became suddenly clear: the thing that causes the evolution in species is natural selection. As soon as he had recovered, Wallace communicated his discovery in detail to another English naturalist with whom he came to correspond collaboratively. This was Charles Darwin, who had reached similar conclusions independently, after nearly five years' traveling and researching mostly around South America.

NUMBERS AND INTUITIONS

If dreams have revolutionized chemistry and biology, they have also been crucial in the much more abstract work of mathematicians—but not necessarily for doing calculations. At the age of twenty-three, a restless René Descartes had already studied at Jesuit college, completed his law course, enlisted in the Dutch army, written a book of music theory, and traveled widely around Europe. Trying to escape a storm on the bank of the Danube River, sitting by a stove in search of heat, the traveling polymath had three dreams that revolutionized the way we understand the world.[22]

In the first dream, a nightmare, Descartes was assailed by ghosts and picked up by a whirlwind. He tried to get back to school, but he was unable to keep his body upright and kept stumbling as he walked. Then a person appeared and told him respectfully that a certain Mr N. had a present for him. Descartes thought this must be a fruit from distant lands, and then realized that the people who were gathering round him were all standing up straight, whereas he could barely stay on his feet.

He awoke afraid and prayed to God to keep the nightmare's harm away from him. Soon afterward, he fell asleep again, he dreamed of thunderclaps and again woke in fear, but this time he resorted to reason to confirm that he was in fact awake, quickly opening and shutting his eyes until he had calmed down. Once again he fell asleep, and then he had a dream that was transformative, quite different from the others. In a setting that was calm and contemplative, Descartes found a book called *Dictionary* on a table—and behind this book, a collection of poems. He opened it onto a random page and found a line in Latin from the poet Ausonius: *What path in life shall I follow?* Then suddenly a stranger appeared and showed him a fragment of a line: *Yes and no.* Descartes tried to show him which part of the book the poem might be found in, but the volume disappeared and then mysteriously reappeared. He had the sensation of some piece of knowledge having gotten lost, until he told the man he would show him a better poem starting with the same line. At that point the man, the book, and then the entire dream disappeared. Descartes was deeply struck by this; he prayed and asked the Virgin Mary for protection for him to make a pilgrimage on foot from Italy to France. In his interpretation, the books he had dreamed pointed toward the unification of all science through a single language and a single method.

With the dream-given clue as a starting point, Descartes discovered the path he was to take in life. When he published his *Discourse on the Method of Rightly Conducting One's Reason and of Seeking Truth in the Sciences* eighteen years later, he advocated a new scientific method: the accepting of only what is evident beyond all doubt; dividing each question into smaller questions; constructing thought from the simple to the complex; and confirming conclusions in the light of the broadest possible knowledge. The same publication includes original treatises on optics, meteorology, and geometry, demonstrating the power of the Cartesian method to imagine the rational world described by mathematics.

Descartes created analytic geometry and became one of the most important formulators of algebra. Strangely, in spite of the dream elucidation of his own important intellectual mission, he belatedly expressed real suspicion about the usefulness of dream hallucinations. That is something that did not happen with the seventeenth-century German mathematician Gottfried Leibniz, the co-inventor of integral and differential calculus, who considered dream visions "a formation more elegant than any which we can attain by much thought while awake."[23]

Apart from these examples, there is a noticeable lack of reports of dream discoveries in the lives and works of some of the greatest mathematicians of all time, such as Gauss, Euler, Galois, Cauchy, Jacobi, and Gödel. Though creativity is commended by mathematicians, it would seem that it is during the waking hours that theorems are proved.

The French mathematician Henri Poincaré testified quite explicitly to the importance of relaxation and abduction for his work:

> Most striking at first is this appearance of sudden illumination, a manifest sign of long, unconscious prior work. The role of this unconscious work in mathematical invention seems to me incontestable . . . Often when one works at a hard question, nothing good is accomplished at the first attack. Then, one takes a rest . . . and sits down anew to the work. During the first half-hour, as before, nothing is found, and then all of a sudden the decisive idea presents itself to the mind. It might be said that the conscious work has been more fruitful because it has been interrupted and the rest has given back to the mind its force and freshness.

But Poincaré did not report any theorems that had derived from dreams. The abductive relaxation he used in his work was a waking phenomenon:

> One evening, contrary to my custom, I drank black coffee and could not sleep. Ideas rose in crowds. I felt them collide until pairs interlocked, so to speak, making a stable combination.

This description, written almost a hundred years before the study of the dream reverberation of the game Tetris, emphasizes the great capacity for recombination and spatial articulation between representations in the dream space. The conclusions could not have been dearer to Freud and Jung:

> The subliminal self is in no way inferior to the conscious self; it is not purely automatic; it is capable of discernment; it has tact, delicacy; it knows how to choose, to divine . . . It knows better how to divine than the conscious self, since it succeeds where that has failed.[24]

In 1945, the French mathematician Jacques Hadamard published a seminal book on mathematical creativity, based on questions posed to a number of renowned sages, among them the German physicist Albert Einstein, winner of the 1921 Nobel Prize in Physics, and the U.S. mathematician Norbert Wiener, creator of cybernetics.[25] Hadamard concluded that mathematical creation is made up of four distinct phases: preparation, incubation, illumination, and verification. This well-determined sequence of creative phases echoes many of the dream traditions of antiquity, which prescribed the soliciting and obtaining of oneiric revelations for the resolving of specific problems. However, while he acknowledged the existence of dreams that can offer new mathematical solutions, Hadamard pointed out their rarity among professionals in that field. It is possible that the scarcity is due to the use of mathematical notation, since in dreams it is very unusual to be able to read and write anything reliably. This difficulty probably reflects the recent appearance of reading in our species, a sophisticated behavioral capacity that needed to "hijack" certain areas of the cerebral cortex that had evolved to enable much older skills such as facial recognition.[26] Studies on the ability to perform mathematical calculations while dreaming point to difficulties that are much greater than in waking life, possibly due to the reduction in short-term memory.[27]

One indication that mathematical notation might be an obstacle to dream creativity is the fascinating story of Srinivasa Ramanujan, the Hindu mathematician with no formal education whose fundamental discoveries in number theory and infinite series were only understood many decades later. Nowadays, physicians and mathematicians who are interested in black holes, quantum gravity, and superstring theory pore over the brilliant theorems of this rural-born autodidact. In 1912, at the age of twenty-five, working as an accounting clerk in the city of Chennai, Ramanujan dispatched dozens of unproved theorems to Godfrey Hardy at the University of Cambridge. Many of Hardy's eminent colleagues had ignored similar messages, but after some initial skepticism, the reaction of the renowned British mathematician was one of amazed admiration for the raw talent of his young correspondent. These theories "must be true because, if they were not true, no one would have had the imagination to invent them."[28]

After an intense exchange of letters, Hardy invited Ramanujan to England so that the two men might work together. But a journey across the sea was an offense against the sacred purity of his caste. The Indian man's family, who were worshippers of the goddess Lakshmi Namagiri, a local version of Vishnu's wife, opposed the trip. Ramanujan turned the offer down, but after much insistence by Hardy—and then a dream in which his mother had been with the goddess, who ordered the end of her opposition to the journey—Ramanujan embarked on a ship for chilly England, leaving his wife, his family, and his culture behind him.

The work with his mentor Hardy was intense and incredibly prolific, leading to the publication of twenty-one original articles. Despite having no university diploma, Ramanujan became a professor at Cambridge and was elected a Fellow of the prestigious Royal Society of London.[29] For all the honors he received, he never managed to integrate completely. Far from his family and his goddess, faced with racial discrimination in a society that saw his habits as savage, Ramanujan became depressed and started to exhibit symptoms of tuberculosis. He returned to India in 1919, and died shortly afterward, aged thirty-two, at the height of his mathematical creativity. On his deathbed, he wrote a letter to Hardy recording some mysterious functions he had seen in dreams, enigmas that only at the start of the twenty-first century, almost a century later, began to make sense. Theories formulated by various mathematicians born after Ramanujan's death were based on these functions. Where had they come from?

Through the goddess Lakshmi, Ramanujan described receiving his complex mathematical visions in dreams:

> While asleep, I had an unusual experience. There was a red screen formed by flowing blood, as it were. I was observing it. Suddenly a hand began to write on the screen. I became all attention. That hand wrote a number of elliptical integrals. They stuck to my mind. As soon as I woke up, I committed them to writing.[30]

As a fervent follower of Hinduism, Ramanujan was an enthusiastic expert in the interpretation of dreams. He simply saw no separation between mathematics and spirituality, as his contact with discovery did not pass through reason alone, but rather through revelation; not through the logical demonstration of symbols, but through their beauty. It is likely that Ramanujan's fertile relationship to dream creation, so rare in Western mathematicians, reflects particular aspects of Indian mathematics, which had been characterized by a strong oral tradition, less symbolic restriction in the forming of concepts, and an intimate relationship to the gods.

A DOUBLE UNCERTAINTY

Although the above examples and many others point toward the important role of sleep and dreams in human creativity, to reveal this role scientifically is no easy task. You can expect anything from a dream, or almost everything. If letters, numbers, and books are rare, it is also not true to say that they do not appear in dreams at all. The British mathematician and philosopher Bertrand Russell, winner of the 1950 Nobel Prize in Literature, expressed this truth very simply: "I do not believe that I am now dreaming, but I cannot prove that I am not."[31]

When somebody attributes a discovery to a dream, we encounter a double layer of uncertainty. The question "What might the dream mean?" leads us also to ask, "And did the dreamer really have this dream?" And what got lost? What got added? What is the difference between firsthand experience and its account to other people? This is significant, because the account of a discovery's having emerged from a dream naturalizes, justifies, and above all legitimizes it, concealing other creative processes and also, possibly, any plagiarism. This is why the whole rich collection of anecdotes about dreams and creativity remained no more than conjecture until scientists managed to deal with the question empirically.

CAPTURING THE EUREKA MOMENT

How is one to capture and measure in a laboratory a phenomenon that is as fleeting as the sudden occurrence of a new idea during sleep? The "Eureka!" moment, a restructuring of memories that is capable of changing the world, is an unpredictable, singular event that happens only once in each mind—after this it is already just repetition. The new idea arises with the potential to be propagated through countless other minds, but in the mind that produced it, it's already old, it's irreversibly already-had. In the words of the Brazilian poet Arnaldo Antunes, "what went (was) has been (gone)."[32]

In 2004, the German neuroscientists Jan Born, Ulrich Wagner, and Steffen Gais managed to quantify the relationship between sleep and insight in humans for the first time. The researchers availed themselves of a classic psychological test in which the solution to the problem is encrypted as a palindrome, that is, a sequence of symbols which reads the same backward and forward. The participants were not told of this structure, and so started out by attempting to analyze the sequence as a whole, even though this was not strictly necessary. Of the participants who slept after carrying out the test, 60 percent showed some aware-ness of the hidden information when they were re-tested the following day, while the same happened to just 20 percent of those participants who did not sleep.[33]

This experiment, which was published in the journal *Nature*, was the first quantitative demonstration of the close relationship between sleep and creativity, but it did not allow the researchers to identify which phase of sleep is most closely related to creativity, nor whether there are different types of creativity that might benefit more from a particular type of sleep. In the last two decades, these questions were tackled by a series of experiments carried out by Robert Stickgold, Mat-thew Walker, and Sara Mednick. The evidence showed that the creative solving of a problem—whether the generating of anagrams[34] or the flexibility of word association[35]—is aided by REM sleep that occurs between the presentation of the problem and its resolution.

RESTRUCTURING MEMORIES

What property might REM sleep have that stimulates the restructur-ing of memories? Besides presenting more cortical activity than slow-wave sleep, REM sleep is characterized by reduced levels of synchrony between neurons, with a small degree of repetition of activation sequences. The idea that sleep introduces some kind of informational noise and that this can be useful for learning prompted interesting experiments on the zebra finch. Males of that species start to learn to sing two weeks after they are born, attempting to copy the song of their father. Early exposure to the adult's song is enough to generate a memory that lasts a lifetime. Even brief exposure to a song from the father, from some other adult tutor, or even from a recording played by a wooden bird, is enough to form a robust memory that serves as an internal model for practicing vocal imitation.

Through multiple repetitions over the course of the two following months, the juvenile bird gradually develops a reliable imitation of the model song. The song produced by the juvenile is modified erratically over time until it finally crystallizes into a sequence quite similar to the father's song. The Israeli neuroethologist Ofer Tchernichovski and his team at Hunter College–CUNY carried out an intensive study of this phenomenon, recording all the vocalizations produced by each juvenile from their exposure to the model right up to the crystallization of their song.[36] Their first discovery was that the songs are modified gradually over the course of the day, becoming ever more similar to the model as the repetitions progress. The second discovery was that the songs produced the following morning are less like the model than those from the end of the previous day had been. In other words, in each night of sleep there is a deterioration in the resemblance between the song produced by the juvenile and its model. The diurnal gains exceed the nocturnal losses, so that the juvenile slowly but surely climbs the mountain of resemblance to the model song, two steps forward and one step back, day after day, until the changes are stabilized. The effect happened both in natural sleep and in melatonin-induced sleep.

The third discovery was the most surprising of all: the animals with the greatest degree of deterioration between the start of the night and the following morning, those that had slipped furthest from the model song overnight, were precisely those that were best at copying it at the end of the whole process, a few months later. In other words, those that stumbled most along the way were the ones that had the greatest success in the final learning.

What kind of mechanism could explain this phenomenon? In 2016, the U.S. neuroscientist Timothy Gardner from Boston University and an international team of colleagues published studies of the neuronal activity of the nucleus HVC of zebra finches during the performing of their song and during sleep. The HVC is a region in the brains of song-birds whose activation is essential for starting the electrical propagation that ultimately reaches the syrinx vocal organ and is there transformed into song. The researchers injected a virus into the birds' brains to make certain neurons produce a protein that turned fluorescent when the cells were electrically activated. Using tiny one-inch-long microscopes implanted in the birds' delicate skulls, Gardner was able to visualize the nighttime activation of the groups of neurons that codify song. The result was surprising: while the song itself is stable from one day to the next, the patterns of neuronal activation of the HVC change substantially over consecutive nights.[37]

It is as if the brain, in the search for the best synaptic organization for producing a song to resemble the model, were with every night of sleep wiping out some of the song produced yesterday, in order to continue on the search for the best possible imitation. Sleep would appear to prevent the system from settling into a sub-optimal solution, adding noise to the memory each night. The phenomenon resembles the cycles of heating and cooling used to temper steel alloys, in a process that first hardens the metal and then makes it flexible. The development of the zebra finch's song is like the shroud that Penelope, in *The Odyssey*, wove during the day and unpicked overnight, to buy time to wait for Odysseus to return. To paraphrase the song by the Brazilian band Chico Science & Nação Zumbi, you first need to disorganize before you can organize.

How exactly is this neurophysiological phenomenon, which happens at a cellular and molecular level, reflected in the contents of a dream? We have no way of asking the zebra finch this question, but humans are more cooperative—especially when they are undergraduate students being paid to play a really fun video game. When Bob Stickgold became a global authority in the newly revived field of research into the dream processing of memories, shepherding many well-financed projects and serving on the advisory board of many big companies, he didn't wait to be asked, but allocated his money to a piece of equipment that no one else would have had the nerve to install in a laboratory: a huge interactive video game capable of simulating—in 3D and in lifelike detail—the thrills of Alpine downhill skiing.

The forty-three people who took part in his study played the game enthusiastically in the lab during the day. At night, back home, where they were set up with a piece of equipment to measure the movements of their bodies and eyes, they slept. At intervals varying between fifteen and thirty seconds after their sleep began, the participants were automatically interrupted and their dream accounts recorded. The multisensorial, interactive experience of virtual skiing proved extremely good at finding its way into the participants' dreams. While the game Tetris appeared in about 7 percent of dream accounts, images relating to virtual skiing appeared in 24 percent. Curiously, the same phenomenon happened with almost the same intensity in a control group who did not play the game, but closely watched somebody else playing it. The reverberation of the memories of the game declined clearly with time: the images became ever more abstract, less realistic. More and more old memories began to appear, however, revealing a process of intercalating recent and distant memories that seems to reflect the incorporating of the former into the latter.

Trustworthy accounts of the game, which are typical of the first seconds after sleep has begun—such as "I get like flashes of that . . . game in my head, virtual reality skiing game"—become much freer a few minutes later, while still maintaining a connection to the game. For example, a participant recalled, "I was picturing stacking wood this time . . . I felt like I was doing it at . . . at a ski resort that I had been to before, like five years ago maybe."[38] The increased abstraction in the dream images as the sleep progresses might perhaps be due to the increase in hippocampal activity, which is capable of reactivating old memories that then get mixed up with recently acquired memories and integrate the new facts of life into everything that went before.

ABRUPT REPROGRAMMING

In situations of great maladjustment, of real cognitive difficulty, there is the possibility of resolutional dreams that can seem miraculous. A totally new, highly adaptive behavior can be installed literally from one day to the next, prompting great amazement. One neuroscientist told me that during his master's studies he went to Argentina to take an intensive Spanish course. To his horror, he discovered he was simply unable to communicate with anybody, he could understand almost nothing, let alone say anything comprehensible. After a few days of growing embarrassment, one night he dreamed that he was reading and writing the language fluently. The following day, he was able to do it, taking a real leap in his ability to use new words.

I was told of another impressive case of acquiring motor skills by a man who as a child had been unable to keep his balance on a bicycle and was ashamed of this. When he was an adolescent, knowing that bicycles basically ride themselves once they are impelled forward, he decided to try again. He made little progress in two days of practice and then dreamed that he was riding the bicycle, going around on it with great facility, thinking how easy it was. The next day he succeeded on his first attempt. He had learned to ride a bike.

The ability of dreams to suddenly transport the dreamer to mastering new skills and content is like a flight between distant points, a real abduction. I learned from the Mozambican writer Mia Couto that in some of Mozambique's languages, the words for dreaming, imagining, and flying are the same. Paragliding is a very appropriate description of the huge increase in perspective that dreaming can provide.

One of the most magnificent historical examples of a sudden gain in perspective allowed by dream flight was recounted by the sixteenth-century Italian philosopher Giordano Bruno, a former Dominican friar who became known all across Europe for his intelligence, erudition, polemical ideas, biting style, and astonishing abilities of memorization—which some of his contemporaries put down to magic, though Bruno himself described the elaborate mnemonic models he used in his book *The Art of Memory.*[39]

Among Bruno's many books, one was dedicated entirely to *The Interpretation of Dreams*—the same name that Freud chose to give his own seminal book more than three centuries later. When he was thirty years old, Bruno experienced a dream vision that would become legendary. In those days, the great majority of astronomers still swore by the ancient Ptolemaic system, with Earth at the center of the solar system and a celestial vault made up of fixed stars on a transparent sphere. The heliocentric theory of the sixteenth-century Polish astronomer Nicolaus Copernicus had few adherents, but even in the Copernican model the solar system was still the center of the universe. Bruno, however, had contact with cosmological texts from antiquity that assumed the existence of multiple worlds. It is also possible that he read works by the twelfth-century Iranian philosopher Fakhr al-Din al-Razi and the sixteenth-century English astronomer Thomas Digges, which made some reference to the infinitude of the universe.

It was in this context that Bruno experienced his great dream. According to his account, his spirit left his body and rose up into the sky till it was far away from Earth. The documentary television series *Cosmos: A Spacetime Odyssey*, in the new version presented by the U.S. astrophysicist Neil deGrasse Tyson, described Giordano Bruno's experience thus:

> I spread confident wings to space and soared toward the infinite, leaving far behind me what others strained to see from a distance. Here, there was no up, no down, no edge, no center. I saw that the Sun was just another star. And the stars were other suns,

each escorted by other Earths like our own. The revelation of this immensity was like falling in love.[40]

Whether fact or myth, Giordano Bruno's magical dream account updated the dream of Scipio imagined by Cicero a thousand years earlier, by taking a journey far from Earth, but going further still, crossing the conceptual last frontier of the spheres as concentric ceilings and finally exploding the perspective, reaching out as far as infinity in all directions. Spinning around in space and understanding how tiny we are compared to all that there is, Bruno understood in his own dreamed body that the universe is incredibly vast and that the sun is no more than one of its countless stars, each of them surrounded by its own planets. The sun is not at the center of the universe, nor does a center around which everything orbits even seem to exist.[41]

This profound astronomical truth, rejected during Bruno's lifetime by the German astronomer Johannes Kepler, only began to be confirmed four years after his death, when the Italian astronomer Galileo Galilei became the first to look at a star from the Milky Way through a telescope. Meanwhile the empirical confirmation of the existence of multiple galaxies would only be possible three hundred years later, through the use of spectroscopic methods.

Some of Bruno's ideas, such as the plurality of worlds and life on other planets, were far ahead of their time, despite having ancient roots in Greek and Islamic philosophy.[42] Thanks to his combative style, Bruno made powerful enemies, especially within the Church. His nightmare began in 1592, when he was arrested in Venice and handed over to the Inquisition, which transported him to Rome and tried him for heresy, blasphemy, and immoral behavior. During the trial the philosopher had opportunities to retract, but preferred to remain consistent and inflexible regarding the fundamental aspects of his doctrine.

In 1600, after seven years of dungeons and torture, having categorically refused to renounce his ideas, the brilliant and unbowed Bruno was muzzled, humiliated through the streets of Rome, and burned alive in the public square. Today, in the Campo de' Fiori, where this barbaric crime took place, a solemn statue of Bruno presides over the fruit and flower market on Sunday mornings. On its base, a moving inscription:

For Bruno
From the age that he foresaw
Here where the fire burned.

THE MUTATION AND SELECTION OF IDEAS

It is curious that Kepler, who was as creative and original a thinker as
Bruno, should have written a letter to Galileo expressing his horror at
the idea of multiple suns and planets, just scattered with no hierarchy
across an infinite universe.[43] It was not that Kepler had anything against
dreams, quite the contrary. He was a pioneer of science fiction with his
book *Somnium* (1634), in which an alter ego of his has a dream about a
trip to the moon, and gives a detailed description of what Earth looks
like from that vantage point. While in this case the dream activity was
only a narrative tool, it is impossible not to recognize the use of a
dream to justify a major gain in perspective, provided in this case by
the lunar point of view.

During sleep, the brain lives through the clash between cognitive
flexibility and fidelity, whose mechanisms are the restructuring and the
strengthening of memories. While the fidelity of memory activation is
one of the attributes of slow-wave sleep—a physiological state that is
phylogenetically very ancient, which favors the strict remembering of
preestablished contacts with reality—the reorganization of memories
seems to be an attribute of REM sleep, a more recent physiological
state that facilitates the resolving of new problems. Reordering memories is a very adaptive ability in a challenging environment, especially
one that is constantly changing in unpredictable ways. But an excess
of oneiric creativity can lead to dangerous ideas in the real world, and
it is safer to subject them first to the scrutiny of a faithful simulation
of reality.

The alternation between slow-wave sleep and REM sleep over the
course of the night gives the brain the opportunity to pass through
several cycles of selecting and mutating ideas. In the first half of the
night, memories are reverberated at the peaks of the slow waves that
overtake most of the brain, with the most important new memories
being reinforced and the rest eliminated. In the second half of the night,
there are episodes of REM sleep that get longer and longer, with higher
and higher levels of cortisol (the stress hormone), in a simulation of
waking levels of alertness. The deactivation of the frontal lobe of the

cerebral cortex and the shutdown of overall noradrenaline release during REM sleep reduces the accuracy of decision-making and of the ordered carrying-out of plans, creating a discontinuity in the logical makeup of the dream narrative. This gives rise to the dislocations, condensations, fragmentations, and associations between the elements of the dream, combining memories in unexpected ways. The heightened, noisy cortical activity observed during REM sleep, as it generates "processing errors" that loosen the neuronal synchrony, does indeed create new pathways for the electrical propagation.

As we have seen, REM sleep promotes the expression of genes that are necessary for long-term potentiation, which leads to the synaptic strengthening of the memories that are restructured as one falls asleep. Working alongside two talented computational neuroscientists, the Cuban Wilfredo Blanco and the Brazilian César Rennó-Costa, I used simulations of neural circuits right across the sleep / wake cycle to show that the long-term potentiation prompted by REM sleep, in addition to strengthening memories, also causes their reorganization. This means that the mere potentiation of some connections is enough to cause a redistribution of the synaptic forces, directly or indirectly altering huge portions of the neural network. An analogy should help to clarify the phenomenon: a balloon filled with air is a closed system, so when our hand squeezes it, the other side of it is distorted.

GYPSY MEMORIES

But how is it possible that certain memories remain stable for so many years, if even the cells that they use as their framework are replaced over time? Old memories tend to be much more resistant to forgetting than recent memories. Just as gypsies migrate constantly with no fixed territory and no return to their origins, memories seem never to stop migrating out to the wide borders of the cortical network, burrowing further and further in as life goes on, ever more extensive and resistant to disturbances. Richly detailed memories can last more than a century in the mind of somebody very old, but it is not possible to say that they have always been the same since childhood. On the contrary, experimental evidence indicates a constant transformation and intercerebral migration of memories over the course of a life, with a particular role reserved for sleep as the state propitiating these changes.

This part of the story begins in 1942, when Donald Hebb concluded his extensive analysis of neurological patients with lesions in the hippocampus.[44] As these patients presented serious amnesiac conditions, with an inability to form new declarative memories, it started to become clear that the hippocampus is directly involved in the acquisition of new memories. In the following decade, a case of surgical damage became famous and clarified the question. The patient, Henry Gustav Molaison, since known by the initials H.M., presented with a serious condition of convulsive attacks triggered bilaterally in the left and right hippocampi. After the complete removal of these epileptic focuses, the patient was cured of his convulsions, but he started to display amnesia for declarative memories, that is, memories that can be stated verbally about events involving people, things, and places.[45]

H.M. was able to learn the name of a new person for a few minutes, but forgot it again soon afterward. This amnesia was total in anterograde terms, that is, for all memories from the period of his life after the surgery. However, his retrograde amnesia, relating to those facts that preceded the surgery, was only partial. While recent facts had been completely forgotten, the old memories, especially those from childhood, remained well preserved. H.M. was studied intensively throughout his post-surgical life, and his clinical condition did not subsequently change. The very detailed study of this patient made it clear that the hippocampus is essential for the formation of new declarative memories. It is there that the relationships between the different perceptual attributes of each memory are codified—images, sounds, textures, smells, and tastes that are characteristic of each object represented, all of which are separately codified in the cerebral cortex but initially integrated via the hippocampus. This explains why this region has such a crucial role in the mapping of the surrounding environment, which allows spatial navigation, and in the codifying of complex events, like the movement and action of multiple objects and actors in a succession of scenes.

In order for declarative memories to last, it is vital that there be an intact hippocampus both at the moment of acquisition and in the hours following. However, as time passes, the memories become progressively less represented in the hippocampus and more in the cerebral cortex, to the point where they survive without any major upset in neurological patients who, whether through accident or from surgery, have undergone a complete bilateral removal of the hippocampus. This progressively increasing engagement of the cerebral cortex in the codifying of declarative memories is called corticalization, a phenomenon we have been aware of since the 1950s but whose mechanisms remained unknown until recently.

BRAINS IN THE SNOW

In 1999, Claudio Mello, Constantine Pavlides, and I formulated the hypothesis that it is sleep that induces the corticalization of memories. Taking advantage of the final months at the end of my doctorate, in November and December of the turn of the millennium, I carried out experiments to measure the expression of immediate early genes in the sleep of rats that had previously been subjected to electrical stimulation of the hippocampus. The aim was to implant in the hippocampus an artificial memory, the "long-term potentiation" discovered by Lømo and Bliss—and follow its tracks through the brain over the course of the sleeping and waking time that followed. A pilot experiment suggested that the initial expression in the hippocampus would decline rapidly and increase in the cerebral cortex after a few hours. We then designed a comprehensive experiment to surround the preliminary findings with good controls and an adequate number of animals for each group. We concluded that it was necessary to carry out daily experiments for eight consecutive weeks. The snow was falling incessantly on those short days and the Twin Towers of the World Trade Center were still in place. I carried out the task patiently. Each day, at the end of the experiment, every rat was killed and its brain frozen at $-123°F$ for subsequent use.

When I had completed the experiments and finished collecting all the brains, it still remained for me to section them into very thin slices and treat them chemically to reveal the levels of gene expression. But before I could do that, I needed to write my thesis, graduate, publish papers, and spend some time with my family and friends in Brazil. So a full year passed, and it was not until January 1, 2001, that I finally moved to the city of Durham, North Carolina, to start my postdoc at Duke University. I decided to take the frozen brains with me to process at Duke itself, in the laboratory of my colleague and collaborator Erich Jarvis, a U.S. neurobiologist who was then a professor at the same department and who had kindly offered his help. I put the dozens of brains into a large cooler full of dry ice, sealed them up with adhesive tape, and took a cab to LaGuardia Airport in a ferocious blizzard. The airport was total chaos, with a number of canceled flights and people sleeping on the floor. My own flight was canceled, but a transfer was arranged to another airline. Unable to take the cooler with me, I dispatched all the brains to North Carolina, took my boarding pass, and hoped everything would work out.

It didn't. I waited in vain by the baggage carousel until the last bag was collected. Many other passengers, like me, were complaining about their belongings going astray in that frozen New Year disruption. I learned at the airline company desk that the cooler had been transferred to another airline because of the canceled flights. They promised that my luggage would be located and that it would arrive on the airline's second and last flight on that route, twelve hours later. I went to Durham, I started to settle myself in, and after the length of time stipulated I returned to the airport, but the cooler did not appear. They promised it would come the following day and I spent the night unable to sleep, thinking about the dry ice in its slow but inexorable sublimation. The next day, I was the first person to arrive at the airline desk—but no sign of the cooler.

This tragic scene was repeated for three long days, with two daily trips to the distant airport and three nights with little sleep, dominated by ever more pessimistic imaginings of the state of the dry ice and the brains in the cooler. For the first time I understood why in the Greco-Roman world it was said that a dream about a fixed idea is insomnia. Not even all the Olympian gods combined could help me now that disaster was inevitable. I prayed to my Orishas—the Yoruba divinities present in Afro-Brazilian religions—and handed the fate of the brains over to chance.

On the fourth morning, when I ran into the airport hall and saw the cooler in the distance, covered in tags from the various airports it had transited through, I feared the worst. I frantically unrolled the pieces of adhesive tape and almost fell over when I saw what was inside: brains, perfectly preserved, beneath a good-sized layer of dry ice, still intact. I gave effusive thanks to the living and the dead for the opportunity to continue researching those brains.

MEMORIES MIGRATE DURING SLEEP

The results of that research revealed in some detail a process of migration of memories out of the hippocampus during REM sleep.[46] We were able to document a sequence of three distinct waves of gene regulation following the stimulation of the hippocampus. The first wave begins in the hippocampus itself half an hour after the stimulation, it reaches cortical areas close to the place of stimulation after three hours of waking, and it ends during the first episode of slow-wave sleep. A second wave begins during REM sleep in the cortical areas close to the stimulus, it is propagated to remote cerebral regions during the subsequent waking, and is brought to a close in the new phase of slow-wave sleep. Finally, a third wave of gene regulation begins during the next episode of REM sleep in various cortical regions—and we don't know where it ends, because that was where the experiment stopped.

The hippocampus presented a gradual reduction in gene expression from the first wave to the third. The most remote cortical regions, several synapses away from the place of the initial electrical stimulation, showed the opposite profile, with a gradual increase in the gene expression as the waves progressed. These results supplied us with the first experimental evidence that REM sleep may participate in the transfer of memories from the hippocampus to the cerebral cortex via waves of molecular plasticity that get deeper with each sleep cycle. Subsequent studies using the exploration of novel objects instead of long-term potentiation confirmed that the effects persist in the cortex but not in the hippocampus.[47] While the synaptic changes are renewed and propagated cortically, in the hippocampus the changes stop and the memories quickly fall away.

It is important to keep in mind that the hippocampus is much smaller in size than the cerebral cortex, with much less capacity for codifying memories. During post-learning sleep, the hippocampus undergoes a transient activation of mechanisms of molecular plasticity that are enduring in the cerebral cortex. This is why the hippocampus gives up its role in each lately acquired memory, bit by bit, becoming less and less relevant as the memory matures. As a trade-off for this "forgetting," the hippocampus nightly renews its capacity to learn again, freeing codifying space for the new memories from the following day.

Memories are not in fact reliable. They lose pieces, they acquire new associations, they combine with each other, they are stripped of some details and acquire others, they pass through filters of desire and censorship, and most of all they change the biological structure that supports them, coming to be represented in different neuronal circuits, generating new ideas, yet still maintaining the appearance of stability. A masterpiece of permanence in the midst of the constant transformation, a wonder of flexibility with their identity still fully intact.

13

REM Sleep Isn't Dreaming

By considering the functioning of those mechanisms described in the previous chapters, we can understand why it is that sleep is cognitively so important, but that doesn't help us to decipher the intimate and potentially instructive meanings of dreaming. Ions, genes, and proteins do have a very active story to report over the course of the night, but we don't need to know of their existence for them to act upon us. Nor does knowing about them explain the content of dreams. The events of dreams do not happen only at a molecular or synaptic level, or at the level of isolated cells, but principally within exceptionally complicated patterns of electrical activity, propagated through vast meshes of neurons that represent the world's objects according to very particular rules.

When two neurons are activated in synchrony to the point where electrical firings are generated in a third downstream neuron, an association at a cellular level occurs. When words are associated through semantic, syntactical, or phonetic congruence, a different order of association happens, a psychological one, which in turn is implemented via a plethora of underlying cellular associations.

The space of mental representations is not to be confused with the space of the neuronal mesh, because one emerges from the other, just as the synchronized movement of a large shoal is the result of the interaction among all the fish but cannot be explained by what happens within each fish individually. The mind operates according to its own symbolic laws—association, dislocation, condensation, repression, and transference—which are microscopically anchored in the mechanisms of synaptic plasticity presented in the previous chapter, but certainly cannot be reduced to them.

WHEN SCIENCE DENIED DREAMING

Today there is no longer any doubt that, beyond the role of sleep in the processing of memories, dreams have specific meanings to their dreamers. This truth, so obvious to anyone who has ever paid attention to their own dreams, has been denied, in multiple different ways, by a range of anti-Freudian philosophers and scientists who held up REM sleep as conclusive evidence of dreams' irrelevance. Why waste time investigating subjective accounts of nighttime hallucinations, when there is a measurable physiological state within reach of any serious, minimally well-equipped researcher?

For the whole second half of the twentieth century, this sophism was used to drain the enthusiasm for dream research, which came increasingly to be considered unscientific. This hollowing out of dreaming happened in favor of strictly neurophysiological investigations into the properties of REM sleep. As if in a reverse bit of sleight-of-hand, the whole ancient mystery of dreams stopped being a problem worthy of study. Dreams were a matter for charlatans, fortune-tellers, priests, psychoanalysts, and other professionals in the metaphysics business. The secondary gain of this opting for ignorance was calming the general public about the bizarre and often embarrassing nature of the dream narrative. Dreams were mere meaningless epiphenomena of REM sleep, purely random by-products of a strictly physiological underlying reality, and therefore of no psychological significance.

What happened to the status of the relationship between REM sleep and dreaming is just one specific example of a more general phenomenon in science. In their rush to deal with a difficult question, scientists often lapse into the mistake of declaring its nonexistence. This happens to this day with the problem of consciousness, which many psychologists and philosophers find easy to solve by reducing the subjectivity of consciousness to a group of objective neural operations. The same thing happened with the geneticist Barbara McClintock, who discovered the transposition of genes while studying the enormous variety of color patterns of maize. McClintock produced detailed documentation of the existence of mysterious genetic leaps within the maize genome, with the insertion, deletion, and translocation of genes between chromosomes. Yet she was still totally discredited, and in 1953 stopped publishing her results. In time, research came together to validate the existence of gene transposition in animals, plants, fungi, and bacteria, turning her work into a compulsory subject in any genetics textbook. In 1983, McClintock was recognized with the first Nobel Prize in Physiology or Medicine ever to be awarded solely to a woman.

But let us return to the distinction between dreaming and REM sleep. The argument of the irrelevance of one to the other, while naïve, thrived in the biomedical field and spread widely through the lay public via the media, isolating any dissenting voices that claimed to be dissatisfied with the impoverishing of the discussion. The reductionist position became hegemonic, and remained so until the end of the millennium, when it had to face its first empirical challenge.

The long wait of almost a century after Freud's *The Interpretation of Dreams* was worth it, however, as the new evidence couldn't have been more enlightening. The difficult task of rescuing dreams as an independent psychological phenomenon, an individual expression of adaptive processes worthy of scientific interest, fell to the South African neurologist and psychoanalyst Mark Solms. Born in Namibia, a postgraduate at the University of the Witwatersrand in Johannesburg, who trained at the Institute of Psychoanalysis in London and had extensive experience researching at University College London and at the Royal London Hospital, Solms spent many years developing a really good question: are there people incapable of dreaming even during REM sleep?

Intrigued and troubled by the ideological bias of the debate, Solms decided to test out the hypothesis that REM sleep and dreams are distinct phenomena and must therefore correspond to different cerebral mechanisms. To do this, he examined neurological cases in search of cerebral lesions which, by sheer chance, might happen to dissociate REM sleep from dreaming. As it happens, however, typical neurological lesions are not like the experimental surgical lesions made in controlled laboratory conditions. They are unique, complex, individual scars, produced by accidents that are as idiosyncratic as the marks they leave on their survivors. Searching for a profile of lesions that might eliminate dreams without affecting REM sleep must have seemed as hard to Solms as looking for a needle in a haystack. The multiplicity of particularities of each case must at first have suggested the nonexistence of any regularity or pattern. In order to understand where Solms found the patience and the breadth of vision to gradually produce an answer to his question, we need a better understanding of his character.

In addition to developing academic and clinical interests, Solms is an active dreamer in the field of social engineering. Following the democ-ratization of South Africa, he returned to the country and decided to transform a farm that had belonged to his family for more than three hundred years into a winery. Understanding that the farm was inhabited by impoverished workers who were descended from genera-tions of slaves, he encouraged excavations to reveal the place's past and shared his 50 percent of the property with the farm's staff to establish a winery that was socially innovative and award-winning. With this kind of strength of purpose and imagination, it is no surprise that Solms assembled and compared a vast collection of neurological cases marked by disruptions to the capacity to dream—some of them familiar clas-sics, others real rarities.

Solms observed that different kinds of cerebral lesion are capable of altering aspects of sleep or of dreams.[1] Deep lesions in the pons, which can reduce or even eliminate REM sleep, when they don't actually kill the patient, only rarely do away with the ability to dream.[2] Lesions in temporal limbic areas cause epileptic discharges, which in turn provoke recurring stereotypical nightmares. Lesions in frontal limbic areas lead to an unusual syndrome: the patients not only retain their ability to dream, but start to dream excessively, even all night long. However, they lose the ability to distinguish reality from dreams. A clinical inter-view illustrates this condition:

> [PATIENT:] I wasn't actually dreaming at night, but sort of think-ing in pictures. It's as if my thinking would become real—as I would think about something, so I would see it actually hap-pening before my eyes and then I would also be very confused and I wouldn't know sometimes what had really happened and what I was just thinking.
> [EXAMINER:] Were you awake when you had these thoughts?
> [PATIENT:] It's hard to say. It's as if I didn't sleep at all, because so much was happening to me. But of course it wasn't really happening, I was just dreaming these things; but they weren't

like normal dreams either, it's as if these things were really happening to me . . .

[AN EXAMPLE:] I had a vision of my [deceased] husband; he came into my room and gave me medicine, and spoke some kind things to me, and the next morning I asked my daughter: "Tell me the truth, is he really dead?" and she said "Yes Mama." So it must have been a dream . . .

[ANOTHER EXAMPLE:] I was lying in my bed thinking, and then it sort of just happened that my husband was there talking to me. And then I went and bathed the children, and then all of a sudden I opened my eyes and "Where am I?"—and I'm alone!

[EXAMINER:] Had you fallen asleep?

[PATIENT:] I don't think so, it's as if my thoughts had turned into reality.[3]

After several years of research, Solms compiled 110 cases of patients with their REM sleep intact from a physiological point of view, but who were unable to report dreams.* These included cases of Charcot-Wilbrand syndrome, characterized by a difficulty in recognizing visual scenes and objects (visual agnosia), and by a loss in the ability to imagine or dream in visual images. This syndrome, observed in patients with thromboses and captured in pioneering descriptions by Jean-Martin Charcot in 1883 and Hermann Wilbrand in 1887, is associated with temporal-occipital damage and with the preservation of REM sleep.[4] These patients are not capable of reporting thoughts and images, even when they are woken in the middle of an episode of REM sleep. For them, dreams are replaced by a state of deep unconsciousness.[5]

Neuropathological examinations using tomography or histology revealed a spectacular discovery. The patients' brain lesions, in their great diversity, were divided into two main types. The first were related to the junction region between the parietal, temporal, and occipital cortices, known for its involvement in visual, auditory, tactile, and semantic processing.[6] The other type involved the axons or cell bodies of dopamine-

---

* In several cases, the ability to dream returned in time, probably through mechanisms of neuroplasticity.

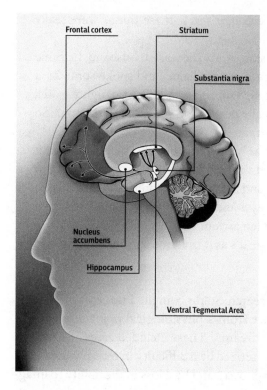

Figure 14. The VTA is a small nucleus of cells that projects dopaminergic axons around huge parts of the frontal and medial cortices, as well as to subcortical structures such as the nucleus accumbens. Damage to the VTA or its projections eliminates dreaming without getting rid of REM sleep.

producing neurons, located in a small area deep in the brain, the ventral tegmental area (VTA; see fig. 14). The dopaminergic neurons in this area distribute their axons across vast portions of the brain and are principally responsible for the neurochemical signaling that allows animals to avoid pain and seek pleasure.[7] Recent studies of rodents suggest that the acquisition, processing, and recovery of memories that are important to the animal's survival depend on the interaction between the VTA and the hippocampus and prefrontal cortex.[8]

Damage to the VTA or its axonal projections does away with dreaming entirely without affecting REM sleep. These lesions are also followed by a loss of motivation, a lack of pleasure, and a reduction of intentionality during waking life. This happens because the VTA is a vital part of the brain's system of punishment and reward, a cerebral structure that enables us to pursue aims, avoid aversive stimuli, satisfy our libido, and learn from positive and negative experiences. This system effectively allows us to have, to satisfy, and to frustrate expectations, and is crucial for the expression of the instinct that makes us fight with all our strength to survive, even in situations that are hopeless.

Memory formation is a selective process in which reward contingency—the rule by which behavior leads to positive reinforcement—determines which memories will be retained and which forgotten. Sleep plays a fundamental role in the long-term maintaining of information, specifically benefiting those memories associated with reward. The key to the consolidation of memory during sleep is the reactivation of recently codified representations, which seems to include dopaminergic neurons.[9]

In order to investigate the cognitive effects of the activation of the dopamine receptor during sleep, Jan Born's team first trained research volunteers to associate different visual images with large or small rewards. During the subsequent sleep, the researchers administered a substance to the volunteers that was capable of activating the dopamine receptor. Twenty-four hours later, the volunteers were tested to check the retrieval of previously viewed scenes when mixed with new scenes, a task that requires the hippocampus to be intact if it is to be carried out successfully. When they were given a placebo, the participants turned out to be much better at learning those images associated with large rewards. However, when the drug that activated the dopamine receptor was administered to them, there were no apparent differences between those images associated with large or small rewards, creating a learning deficit. The results supported the idea that the preferential consolidation of those memories associated with large rewards involves a selective dopaminergic activation of the hippocampus.[10]

And so Mark Solms's unlikely question ended up being answered. Having reached the bottom of this haystack, having found the needle, it proved so sharp that it punctured the overinflated balloon of anti-Freudian theories that had equated dreams and REM sleep. The discovery of the autonomy of dreams in relation to REM sleep—despite the role of dopamine in the origins of both—completed the long path of a hypothesis that had been intuited at the end of the nineteenth century, but which could not possibly be solved at that time, in which the chemical and anatomical mechanisms were still totally unknown.

The Freudian proposition that desire is the motor of dreams is much more factual than its critics would acknowledge. The surgical precision of the hypothesis was probably in part masked by its poetic appearance, since it would take a hundred years of accumulated knowledge about the neural mechanisms of motivation before the line made biological sense. What Freud's shrewd clinical observation revealed, just through the analysis of the behaviors and memories externalized by his patients, was the likely existence of the mechanisms that Solms, a century later, would finally be able to identify. Dreaming "is" desire because both "are" dopamine. This conclusion is intimately related to the fact that dopamine is essential for the very occurrence of REM sleep, as we have seen above. The involvement of the dopaminergic reward system is a serious refutation of Karl Popper's attack on Freud: psychoanalytic theory is testable, definitively.

Faced with Solms's empirical findings, a number of anti-Freudian arguments that had been repeated over the course of the twentieth century like catchphrases with a scientific gloss lost steam. It is no longer possible, for example, to trivialize the rich and intriguing significance of dreams as a useless by-product of REM sleep. Nor is it possible to go on accepting that dreams represent a simply random chain of images. Evidence points to a succession of images organized by the dopaminergic system of reward and punishment, a process capable of trying, evaluating, and selecting adaptive behaviors without however subjecting the body to any risks, as everything is simulated in the safe environment of one's own mind.

A SIMULACRUM OF LIFE

This theory allows us to reach a better understanding of where the subjective quality of dreams comes from, as they are always concerned with people and things interacting in complex scenarios and never with the constituent pieces of these representations. Nobody dreams about color without shape, angles in the abstract, contrast and nothing else. Our dreams are way more intricate than that. It follows that the subjective experience of the dream cannot be explained merely by the activation of the primary visual cortex, which is the first part of the cerebral cortex to receive visual stimuli and which processes very basic attributes of the image, such as spatial position, contrast in brightness, object orientation, and angles.

The visual component, which is dominant for most people, can create incredibly beautiful subjective experiences, with colors and movements that are mesmerizing. In spite of the hegemony of vision, however, dreams can involve mental experiences relating to all of the senses, combined in various ways according to rules that are as yet little understood. Auditory, gustatory, tactile, and olfactory dreams exist, and they are memorable, as if the astonishment of dreaming were greater when it invades sensorial territory beyond light. There are also dreams that are characterized by a strong feeling of movement, associated with motor representation and the vestibular system, which is responsible for the body's balance. Its potential to reflect all of the dimensions of waking experience makes dreaming a disconcerting simulacrum of reality.

Nor does a typical dream usually present parts of the body with autonomy and a lead role of their own: few people dream about foreheads, noses, lips, or elbows independently. We almost always dream about entire objects, be they people, animals, or things, even if they are sometimes combined into chimeras with bits on loan from some other complete representation of an object, or split by some dramatic event within the narrative itself. If the strong electrical reverberation during REM sleep is the main cause of the vividness of dreaming, its occurrence in several different cortical areas involved in the complex representation of visual objects explains the equally complex quality of dream images. A dream does not show all the possible levels of sensorial representation, but only the most elaborate ones. The headquarters of dreams are those most remote areas of the brain, those that are furthest from the sensorial and motor periphery and therefore better able to associate and integrate pieces of information coming from the senses. These areas comprise vast multisensorial portions of the cerebral cortex, as well as an intricate subcortical circuitry that includes the hippocampus and the amygdala,[11] which are involved, respectively, in the acquisition of declarative memories and the evaluating of these memories as reward or punishment.

DEFAULT MODE NETWORK

So many areas are activated during REM sleep that it is more useful to think of them not as a collection of separate parts but as a large, complex cerebral circuit. Intriguingly, this circuit overlaps with what is known as the default mode network (DMN).[12] People who suffer damage to the medial prefrontal cortex, which is a very important region of the DMN, experience severe damage to their ability to dream.[13] This network was discovered in 2001 by the team of U.S. neurologist Marcus Raichle, from Washington University in St. Louis, Missouri, and was originally described as the collection of areas that reduce their activity during the carrying out of tasks geared toward attaining goals but are activated when the brain is "resting,"[14] like the engine of a car in neutral. During waking, the DMN network is activated when the person's mind is wandering, "doing nothing." In sleep, the effects depend on the state in question, as the activity of the DMN network reduces during slow-wave sleep and increases during REM sleep.[15] The DMN's activity during REM sleep alternates with the activity in the cortical areas that are anatomically closest to the sensory organs.[16] In addition, similar patterns of DMN activity, albeit attenuated and partial, are observed during the reveries of the waking mind.[17] These advances over the past decade give a strangely refreshing flavor to chapter 2, verse 58, of the *Bhagavad Gita*, the masterpiece of Vedic literature, dated between the fifth and the second centuries B.C.E., in the heart of the Axial Age: "When this man, like a tortoise retracting its limbs, entirely withdraws his senses from the objects of sense, his mentality is stabilized."[18] When we withdraw our senses to dream, the part of our brain that is activated is the DMN.

So would one then expect that pharmacological inducers of dreamlike states, such as ayahuasca or LSD, would increase DMN activity? This question prompted the neuroscientist Fernanda Palhano, then a doctoral student in Dráulio de Araújo's laboratory, to investigate signs of functional magnetic resonance recorded in people under the influence of ayahuasca during a period of rest. The data published in 2015 in the journal *PLOS One* were quite clear: in a state of calm while subjects were awake under the effects of ayahuasca, DMN activity was *reduced*, as was the strength of the functional connections between the areas that constitute it.[19] As one of the researchers who took part in this study, I was pretty surprised at this result, but four months later an almost identical result using psilocybin was published in the journal *PNAS* by the British team of Robin Carhart-Harris, David Nutt, and the great pioneer of psychedelic neuroscience, Lady Amanda Feilding.[20] About a year later, the same researchers showed a similar phenomenon using LSD.[21] Curiously, the weakening of the DMN was correlated positively with the reduction in the "time traveling" carried out by the mind.[22]

Araújo and Palhano believed that the key to understanding this apparent paradox is the comparison with meditative states, which also see a reduction in DMN activity.[23] Psychedelics and meditation share many psychological characteristics, such as the increase in introspection and of self-perception in the present moment.[24] In the words of the Dalai Lama, "one type of meditation" is to "let out whatever thoughts come and go. Whatever comes, let it come and let it go. Never make an attempt to hold on to them. Through that practice, thought automatically becomes weaker, weaker, weaker. Then eventually, there is some kind of cessation of thought."[25] The reduction in DMN activity during meditation has been associated with a lessening of daydreaming.[26] This is not the case with psychedelics, since seasoned users present more daydreaming, not less.[27] On the other hand, the awareness of the daydreaming is altered in both states. DMN activity increases during periods of daydreaming, but decreases with the awareness that the mind daydreamed,[28] as is the case in psychedelic experiences. The dream experience seems to be more closely related to traveling in autobiographical time than psychedelic or meditative experiences. It is as if these experiences of contemplation led to an altered point of view, changing the perspective from actor to alert spectator—which differs from the dream experience.

The fact that the production of dreams depends on the system of punishment and reward reinforces the theory that dreams are simulations of situations that are significant to the dreamer. To dream about conquering the object of desire is an important aspect of dream life from earliest childhood, as clearly illustrated by the frustrated dream of being given a tricycle I recounted in Chapter 5, or the simple, happy dream of the boy Dean in the bathtub in the same chapter. Both are well-defined examples of the Freudian concept of wish fulfillment, in which the fabric of the dream narrative represents the obtaining of some reward. Most of the dreams we have, however, are characterized by the frustrated search for the fulfillment of our wishes, in which the simulation of the search for assorted goals occurs through attempts that are incomplete, imperfect, and, most important of all, unsuccessful. The occurrence of frustrated desires in dream narratives is noteworthy, such as in dreams about attacking the fridge in people who are on a diet, or about consuming addictive drugs in people in withdrawal, or about freedom in people who are incarcerated.[29]

DREAMING AND IMAGINING ARE SIMILAR CEREBRAL PROCESSES

Imagining oneself as a character in a possible future event makes it possible to plan to act effectively upon what is yet to happen. Experiments carried out by various psychology researchers, such as Daniel Schacter and his team at Harvard University, showed that the capacity to imagine the future is closely related to the capacity to recall the past. This discovery goes back to the start of the 1980s, when Schacter, who had completed his doctorate at the University of Toronto under the supervision of Estonian-Canadian psychologist Endel Tulving, carried out research into episodic memories in amnesiac patients with brain damage.

One day a completely amnesiac patient, who is known only by the initials K.C., showed up to do the psychological exam. The patient presented a huge lesion in the temporal and frontal lobes and was totally without episodic memory, unable to recount any event that occurred at specific times and places. To Tulving and Schacter's astonishment, K.C. also proved incapable of imagining the future:

> E.T.: "Let's try the question again about the future. What will you
>   be doing tomorrow?"
> (There is a 15-second pause.)
> K.C. smiles faintly, then says: "I don't know."
> E.T.: "Do you remember the question?"
> K.C.: "About what I'll be doing tomorrow?"
> E.T.: "Yes. How would you describe your state of mind when you
>   try to think about it?"
> (A 5-second pause.)
> K.C.: "Blank, I guess."[30]

Patient K.C.'s neurological condition was the first in a series of similar cases that were somewhat surprising, contradicting as they did the widely held sense that the past and the future are opposites. There was to be an even greater surprise in 2007, when Dan Schacter and Donna Addis published the first study of brain imaging that compared tasks of envisioning the future and recollecting the past. It became clear that the brain regions used for the two processes are practically the same: hippocampus, precuneus, retrosplenial cortex, lateral temporal cortex, lateral parietal cortex, and medial prefrontal cortex. This is why patients with lesions in these areas present deficits both in episodic memory and in the imagining of future situations.

UNCONSCIOUS REPROGRAMMING OF MEMORIES

In principle, this process of simulation would not need to be conscious in order to generate adaptive changes in behavior. At some point in the evolution of mammals, maybe 200 million years ago, dreaming began to be positively selected as an unconscious reprogrammer of memories, a biological mechanism capable of reactivating, reinforcing, and editing memories in order then to test them out in quite trustworthy simulations of reality. Much later, now in the lineage of our speaking hominid ancestors, the capacity to dream came to be even more evolutionarily favored for the repercussions of its conscious verbal recall on the actions of waking life, not only of the dreamer, but of the whole family group who are exposed to the constantly renewed morning narrative. In the relative monotony of the Paleolithic, with the routine of producing chipped stones and the protracted flow of migratory hunts, dream accounts must have been one of the most stimulating and anticipated moments of day-to-day life, filled with hope but also with fear. There are countless cultures that seek out and propitiate revelatory or healing dreams, creating expectations in the whole community surrounding the dreamer. The collective imperative of a society that believes in the utility of dreams certainly facilitates their being remembered and interpreted. In the words of the Brazilian indigenous leader and writer Ailton Krenak, "The dream is an institution. And it admits dreamers."[31]

The great delay in urban and technological civilization in recognizing that dreams are adaptive, that is, that they facilitate the adaptation of the individual, results from our forgetting the art of dreaming, and from science's delay in examining the subject seriously. It was not until 2010 that Stickgold and his group provided a quantitative demonstration that dreaming about a new task correlates to a better subsequent performance at this task. The participants were exploring a virtual maze, and the time they took to complete the journey was measured. Next, half of the participants went to sleep, and the other half remained awake. Each of these groups was divided up depending on the occurrence or not, during this period, of mental images related to the maze. After five hours, each participant navigated though the maze once again, and the times it took them to complete the task were compared to the previous measurements.

Among those participants who remained awake, there was little improvement in performance, independent of the image content of their daydreams—that is, whether or not images related to the game had appeared spontaneously was irrelevant to the results. However, for those participants who slept, the image content made a great deal of difference. Those who reported dream images relating to the navigation of the maze completed their navigation very much faster than they had before sleeping. In contrast, participants who did not report any dreams related to the task showed no improvement in performance. This experiment was the first demonstration that the actual content of dreams, and not just the time spent in REM sleep, encourages an adaptation to the environment.[32] Have shamans and psychoanalysts known this all along?

In contemporary urban reality, remembering a dream upon waking requires much more than a mere wish to remember. During REM sleep, the cerebral levels of the neurotransmitter noradrenaline are practically zero. Noradrenaline strengthens the voluntary evocation of memories, which is why it is not surprising that we wake from REM sleep with a great difficulty in remembering our dreams. Since we live in a society that asks nothing from dreams but gives them nothing either, we get out of bed already feeling the need to satisfy desires, whether it's to pee or to have a coffee. We abandon the thread of our recollections of the dreams we've had and immediately think ahead in time; we begin our projection toward the future, which consists of mentally examining what we need to do in that new day. Those memories used to simulate our plans of action are strengthened by the release of noradrenaline, which is directly involved in the process of paying attention to the sensorial stimuli to which we are constantly exposed while awake. And so, in the space between the bed and the bathroom, the possibility of rescuing the dream vanishes. By the time we reach the toothpaste, a few minutes later, the chance to recall the final dream of the morning has already been completely squandered.

THE ART OF DREAMING

Dreaming is a physiological construct, a specific trajectory of mnemonic activations firmly directed by the compass of desire, but not always capable of producing a chain of images that is vigorous, moving, or beautiful. Each dream is itself a test, a possibility of representations that might fail at the first image, stumble at the first scene, or continue in a dynamic process of production until it has formed a cathedral of meanings, with a huge freedom of variations that run from a dream with imperfect and dull images, a shadow dance coming apart into painful associations that can cause shock, sadness, or regret, to the weaving of plots that resonate deeply with the dreamer's most vital emotions, filled with details that slot together movingly to create a composition that is authorial and truthful about their interior life.

Sometimes the dreamer interrupts the dream, whether to deal with their baby or go to the bathroom, and then resumes the same oneiric narrative they had been dreaming before as though returning to a novel, a long, complex, interconnected sequence of scenes marked by the actions of characters who are interrelated, with their own missions and purposes. In these cases it becomes obvious that there is a coherence and organization between the disparate parts in the dream time, as well as an internal memory of the experience, in such a way that the end of the dream can refer to intentions that were defined at the start. Far from being a strange phenomenon, the resumption of the dream's thread indicates that it is very much not random and that it possesses driving emotions that don't change from one moment to the next. A well-formed dream is a fair simulation of the successful search for an initially desired reward, or of the successful flight from an initially feared punishment, without any forgetting halfway, without a loss of control of the will, without any dissolving of the wish or of the fear at any point in the trajectory.

CRAZY HORSE DREAMS HIS DESTINY

The heroic and tragic story of the Lakota people movingly exemplifies the importance of experiencing well-formed dreams. The beginnings of Lakota history go back to the ninth century, when builders of earthen mounds for funerary and residential purposes occupied the valleys of the Mississippi and Ohio Rivers. Between the sixteenth and seventeenth centuries, these people moved toward the great prairies between the Missouri River and the Rocky Mountains,[33] a huge corridor full of buffalo and stretching from Canada to Mexico. In the south, the Apache, the Navajo, and the Comanche became dominant; in the north, the Sioux, Cheyenne, Arapaho, Crow, Cree, Kiowa, Pawnee, and various other groups fiercely divided up the territory. All of them formed alliances and fought among themselves and most of all against the white invaders—the French, the Spanish, the English, and finally the citizens of the United States of America—in a chaotic process of cultural attrition that gradually destroyed most of the powerful Amerindian populations north of the Rio Grande.

For a long while, the notable exceptions to this were those Native American peoples who learned to master horses, and were furnished with mounted warriors comparable to the Huns or the Mongols: the Apache, the Comanche, and the Sioux. The last was a derogatory term that meant "little snake" or "enemy," used by the whites and by other Native American groups to refer to the Lakota and their Dakota and Nakota cousins. In the first half of the nineteenth century, the Lakota conquered a great part of the central Plains corridor. Their culture of combat and honor, marked by blitz attacks to steal horses and take scalps, was led by warriors who were frequently members of secret societies and practitioners of religious beliefs with painful sacrifices for obtaining oneiric visions.

The war with the United States government began in 1854 in Fort Laramie, just three years after the signing of a treaty for the cession of lands with eight different indigenous peoples. On that occasion, a minority of older chiefs, among them the venerable Conquering Bear, were used by the government to try to legitimize a demarcation that was detrimental to the Lakota and the Cheyenne. These peoples had never recognized the territories ceded to the Crow under that treaty, a vast stretch of lands that included the valley of the Little Bighorn River. The repressed tension between the Lakota and the whites exploded after a Native American man killed a cow belonging to a settler. A lieutenant called John Grattan at the head of twenty-nine soldiers invaded a camp with thousands of Lakota to demand aggressively that the person responsible for the cow's death be handed over. Conquering Bear tried to pacify the enraged soldiers, but he was one of the first to be shot.

The violence spilled over, and within minutes the whole platoon lay butchered.

The Grattan Massacre went down in history as the first time the Lakota were in open conflict with the United States Army. It was very probably also the first time that the powerful warrior Red Cloud killed a white man. It was also a horrific baptism of blood for a shy boy called In-the-Wilderness, who witnessed the whole conflict with eyes wide open and from that episode onward was fated to play a crucial role in this war. In the weeks that followed, the troops retaliated barbarically, with an ever more traumatic effect upon the fair-skinned, curly-haired little boy, until he finally made a very personal choice: the road of vengeance. In-the-Wilderness's father then took him to a sacred river to undertake a solitary four-day spiritual journey, fasting atop a rocky peak in search of a vision of his own destiny. He dreamed of a warrior on horseback emerging from a lake, as if he were floating. The warrior wore simple clothes, he had no paint on his face, and for ornaments wore only a feather in his hair and a brown pebble behind his ear. He passed through a shower of bullets and arrows without getting hit, but then he was engulfed by a storm and the people raised their arms to hold on to him. At the end of the dream, the warrior extricated himself, and then the lightning struck him, marking his body with hail and his face with a lightning bolt. Black Elk recounted that his cousin In-the-Wilderness

dreamed and went into the world where there is nothing but the spirits of all things. That is the real world that is behind this one, and everything we see here is something like a shadow from that world. He was on his horse in that world, and the horse and himself on it and the trees and the grass and the stones and everything were made of spirit, and nothing was hard, and everything seemed to float. His horse was standing still there, and yet it danced around like a horse made only of shadow, and that is how he got his name, which does not mean that his horse was crazy or wild, but that in his vision it danced around in that queer way.

It was this vision that gave him his great power, for when he went into a fight, he had only to think of that world to be in it again, so that he could go through anything and not be hurt.[34]

In-the-Wilderness's father interpreted the boy's incredible vision as evidence that one day the child would grow to be a great warrior untouched by arrows and bullets, so long as he avoided ornamentation, sought simplicity, and took nothing from his people, never coveting the rewards of military glory. When he returned to be once again among his people, the boy took the name Crazy Horse. In the years that followed, he would grow and become stronger until he was one of the stoutest pillars of indigenous resistance on the northern prairies. When he was going to fight, he would cover his body in white dots like hail, and paint a lightning bolt on his cheek. As a sign of humility and devotion, he never used a headdress but rather a single feather. In time, Crazy Horse became Red Cloud's right-hand man, and came to have a central role in the main battles between the Lakota and the avalanche of civilian and military invaders. Driven by violent emotions, Crazy Horse became the white men's worst nightmare.

PERSONAL MEANINGS

The narrative that we have developed hitherto provides the reader with a number of different points of view to inform the interpretation of dreams. While it is neither possible nor desirable simply to reduce dreams to biological mechanisms like electrical reverberation, when deciphering dreamed symbols it is important to keep in mind that they are generated by the raised levels of electrical activity in REM sleep, but governed by the dreamer's expectations and desires as sensorial and motor representations are reactivated. We must also remember that dream narratives are inscribed in the memory bank via the gene expression triggered by REM sleep. Keeping all these levels in perspective, independent in their own terms but causally connected to one another, it becomes easier to see why understanding the motivation for a dream requires understanding the dreamer's subjective context at the present time. Only in this context is dream interpretation possible. Symbols usually have very personal meanings, which are provided by the networks of association that bring meanings together according to conceptual or phonetic similarities, through individual polysemous signs that don't lend themselves to the use of general keys shared by different people or cultures. A dream is a very private mental object.

One ambiguous historical dream provides a good illustration of the traps of oneiric interpretation, as a poorly deciphered oracle can lead to incorrect clues and catastrophic results. In the first century B.C.E., Pompey the Great was a powerful general and Roman consul who built an imposing new theater for the people and was compared by the Roman biographer Plutarch to Alexander the Great.[35] In 59 B.C.E., Pompey forged an alliance with the promising politician and soldier Julius Caesar, and married his daughter, Julia. At first, father-in-law and son-in-law helped each other, but as the years passed, Julius Caesar's power grew and Pompey's declined, and the two leaders drifted apart. Julia's unexpected death undid what little bond yet remained between the generals. Pompey allied himself with the conservative senators against Julius Caesar's populism, and civil war began to brew. When Caesar crossed the Rubicon and marched toward Rome, Pompey fled with his troops to Macedonia. However, a year later, Julius Caesar crossed the Adriatic Sea in pursuit of the fugitives, catching up with them in central Greece, not far from Mount Olympus.

From the top of a broad hill, Pompey's forty-five thousand soldiers, who were well rested, well equipped, and supplied with provisions, looked down at the plain below to Julius Caesar's twenty-two thousand soldiers, tired and hungry from their journey. All the same, Pompey was considering avoiding open warfare and simply starving the enemy to death. The night before the decisive battle, the old general had a powerful vision in a dream that made him hesitate. He dreamed that he was inside the theater that he'd had built in Rome, offering spoils of war to Venus Victrix, the goddess of victory, to the intense applause of the crowds. While the dream apparently foretold a resounding victory, Pompey did not sleep easily. After all, was it not from Venus herself that Julius Caesar's family claimed to descend? Might the spoils of war be a representation not of what the dreamer would conquer in the fighting but of what he was about to lose forever?

When day broke, Pompey did not know whether the dream was a divine augury or the satisfying of a hopeless wish, and he hesitated to give the order to commence the fatal struggle. The battle almost didn't happen because Julius Caesar, recognizing his opponent's numerical advantage, started to withdraw his troops. However, the die was cast, at least in the hearts and minds of Pompey's followers, who were already busying themselves delightedly with the sharing out of senior positions in the Republic. Eager for the spoils of war and intoxicated by blind confidence in the numbers, they pushed Pompey into the conflict.

Abruptly his men left their strategic positions on the hill and launched the attack, but despite having double the number of infantry soldiers and seven times more cavalry, they were roundly defeated by Caesar's hardened veterans. In a panic, Pompey abandoned his own men on the battlefield and fled in disguise on board a ship. When he disembarked in Alexandria, he was stabbed to death by a Roman centurion and agents of the Egyptian king Ptolemy XIII, who was eager to please the winning side. When Caesar arrived in Alexandria, he received his ex-son-in-law's head in a sack. Contrary to what the Egyptian ruler had expected, the Roman refused to open it and ordered those responsible for the crime to be executed; then he deposed Ptolemy XIII and had a son with Cleopatra, the deposed king's sister. He ordered Pompey's head to be buried beneath the temple of the goddess Nemesis, responsible for the punishing of arrogance, and returned to Rome to assume absolute power as dictator. It was the beginning of the end for the Roman Republic.

14

Desires, Emotions, and Nightmares

The symbolic richness of historical dreams often camouflages simple, visceral plots about the struggle for survival. In order better to understand a report of a specific dream from a particular person, it is essential to begin by imagining the dream repertoires of other mammals. This makes it possible to distinguish which elements in the dream reflect components that are atavistic and ancestral—of huge importance to the individual, connected to the ecology of life under the merciless reign of Mother Nature—as opposed to those aspects pertaining to human culture, in all its opulence, complexity, and, I might as well say it, futility.

But how is one to infer the dream repertoire of other animals without resorting to fantastical speculation? We can assume that the typical mammalian dreams reflect those mammals' most immediate and important habitual problems, which are renewed daily throughout their lives: the inescapable needs to feed, to avoid predation, and to find sexual partners so as to leave fertile offspring. These are the essential problem of any sexed being, the Darwinian imperatives of evolution. Although the comforts of contemporary life might allow middle-class and upper-class individuals to experience lower relative levels of anxiety as concerns food and predation, the same cannot be said for the eternal struggle for true love. Dreams of hunger and assassination are not commonplace on psychoanalysts' couches, but the anticipation, the satisfaction, and the incompleteness of romantic love still make their mark very clearly on today's dreams. On the other hand, the huge wretched mass of the homeless and refugees across the planet, who couldn't be farther from a couch, still have desperate dreams about not having anything to eat or about being killed by death squads.[1] These are dreams that are intimately connected to simple raw survival, not all that far from what one would expect from other mammals roaming freely in nature, as the struggle to remain alive is a daily one that happens very close to the threshold of death.

While we can't ask the capybaras on the Brazilian Pantanal wetlands whether they dream about jaguars every night, we can ask what sort of dreams are experienced by people repeatedly subjected to imminent danger, such as soldiers in a war zone. The answer is that they often dream about the possibility of being attacked, or about attacks themselves and their consequences. These terrifying dreams reawaken particularly violent events, but they also simulate possible future catastrophes, mixing past and future in a spiral of memories that feed on fear and take on lives of their own. With each electric reactivation, gene expression is mobilized, creating waves of plasticity that sculpt the mind through the stubborn recurrence of the same stark choice: kill or die, kill or die, kill or die . . . As sleep is repeated nightly, these memories become so strong that they cause recurring nightmares that can last many years after the danger has passed. These nightmares are one of the most characteristic symptoms of post-traumatic stress disorder.

What must the first dream of all have been like, presumably dreamed by the ancestor shared by all of today's existing mammals, some 200 million years ago? This was an animal the size of a mouse, probably nocturnal or crepuscular and dependent on underground hiding-places to survive the terrible conditions imposed by the unquestionable masters of the planet at the time, the dinosaurs.[2] The fearful occupation of this so very narrow ecological niche suggests that the first dream of all would probably have been a nightmare.

The work of Finnish psychologist and philosopher Antti Revonsuo, a professor at the universities of Skövde and Turku, supports the connection between intense stress in the daytime struggle to survive and the recurrence of nightmares. Based on a comparison of children's dreams collected in countries that were culturally similar but very different in their degrees of violence (such as the Gaza Strip and Galilee), Revonsuo and his colleague Katja Valli were able to confirm a higher prevalence of nightmares in violent societies. This led them to propose the theory that the prototypical dream, the father of all dreams at the origins of the typically mammalian consciousness, was indeed the bad dream.[3] A nightmare, being able to simulate possible dangers to be avoided in real life, can prepare the dreamer to face the dangers of the following day, practicing scripts for action or simply increasing their alertness.

THREAT SIMULATION THEORY

The main prediction that comes out of the threat simulation theory is that in situations of the greatest stress, at the frontier between life and death, the dream narrative will relate directly to the threats of real life. Of all the symptoms that can appear in sleep disorders, the recurring nightmares of post-traumatic stress disorder might be the most disturbing from a psychic point of view. This syndrome, which is systematically recognized in war combatants and survivors of genocide, can occur in anybody who experiences any acute stress that is powerful enough to leave an emotional scar. A longitudinal study of Vietnam War veterans, published in 2015 in the journal *JAMA Psychiatry*, concluded that post-traumatic stress disorder, even forty years after the end of the conflict, continues to affect around 270,000 ex-combatants.[4]

Very violent experiences, such as life-or-death struggles, serious accidents, and sexual abuse, can be followed by behavioral disorders that resemble but are not to be confused with a panic attack. The syndrome includes flashbacks in which the trauma is relived with symptoms such as tachycardia, heavy perspiration, frightening intrusive thoughts, an aversion to places, events, objects, thoughts, or feelings related to the traumatic event, being easily frightened, permanent tension, a difficulty sleeping, bursts of temper, a difficulty remembering the main characteristics of the traumatic event, a prevalence of negative thoughts about oneself or about the world, feelings of guilt, a lack of interest in pleasurable activities, and—of course—disturbances to REM sleep.[5]

Besides all these symptoms, one of the characteristics of trauma is the repeating of nightmares related to the event that caused it or to the circumstances associated with it.[6] In the late Middle Ages, there is a record of a French nobleman called Pierre de Béarn who began to suffer from terribly disturbed sleep after a trauma in single combat with a huge Pyrenean bear. The man would become agitated in his sleep, roaring and brandishing his sword threateningly, to the point where he was abandoned by his family.[7] Scientific studies today show that war veterans can dream about traumatic events for decades on end, with a wealth of repeated details.[8] People subjected to persecution, abuse, and torture also present with recurring nightmares.[9]

DUMUZID'S DESPAIR

It is no accident, therefore, that the first recorded dream in history should have been the nightmare of a mythical man pursued by pitiless killers. This was Dumuzid, the Shepherd, the fifth predynastic Sumerian king, who was said to have reigned in the legendary time before the flood, around five thousand years ago. According to tradition, Dumuzid was the husband of the goddess Inanna, with whom he lived in an erotic idyll followed by a tragic outcome, as recorded anonymously in cuneiform characters on ancient clay tablets. At the start of the poem *The Dream of Dumuzid*,[10] he calls, weeping and desperate, for his wise sister Gestinanna so that she might interpret the fearsome vision he has just had:

> A dream, my sister! A dream! In my dream, rushes were rising up for me, rushes kept growing for me, a single reed was shaking its head at me; twin reeds—one was being separated from me. Tall trees in the forest were rising up together over me. Water was poured over my holy coals for me, the cover of my holy churn was being removed, my holy drinking cup was torn down from the peg where it hung, my shepherd's stick disappeared from me. An owl took a lamb from the sheep house, a falcon caught a sparrow on the reed fence, my male goats were dragging their dark beards in the dust for me, my rams were scratching the earth with their thick legs for me. The churns were lying on their side, no milk was poured, the drinking cups were lying on their side, Dumuzid was dead, the sheepfold was haunted.

In the repetitive style of humanity's earliest texts, Gestinanna interprets the dream as a clear premonition of death:

> My brother, your dream is not favourable, don't tell me any more of it! Dumuzid, your dream is not favourable, don't tell me any more of it! The rushes rising up for you, which kept growing for you, are bandits rising against you from their ambush. The

single reed shaking its head at you is your mother who bore you, shaking her head for you. The twin reeds of which one was being separated from you is you and I—one will be separated from you. The tall trees in the forest rising up together over you are the evil men catching you within the walls. That water was poured over your holy coals means the sheepfold will become a house of silence.

Gestinanna goes on to specify the terrifying meaning of each element of the dream, until she sees the attack is imminent. What follows is the purest expression of the panic experienced by somebody being hunted: "My brother, your demons are coming for you! Duck down your head in the grass!" He begs: "My sister, I will duck down my head in the grass! Don't reveal my whereabouts to them! I will duck down my head in the short grass! Don't reveal my whereabouts to them!" Gestinanna replies: "If I reveal your whereabouts to them, may your dog devour me! The black dog, your shepherd dog, the noble dog, your lordly dog, may your dog devour me!"

The mere description of Dumuzid's enemies inspires the ancient terror of being preyed on by strangers with absolutely no possibility of negotiation or compassion: "Those who came for the king . . . who know not food, who know not drink, who eat no sprinkled flour, who drink no poured water, who accept no pleasant gifts, who do not enjoy a wife's embraces, who never kiss dear little children . . ." Ten men from five different cities surround the house, shouting: "Man run after man." In reality, these are demons that have come to carry Dumuzid off to the underworld of the dead. The demons try to bribe Gestinanna to reveal Dumuzid's hiding place, but she refuses to help them. Then they try to corrupt one of the fugitive's friends, who ends up betraying him and revealing his whereabouts. Captured, tied up, and wounded, Dumuzid weeps and begs for clemency from his brother-in-law Utu, the sun god, Inanna's brother, that he might transform his hands and feet into gazelle legs so that he can run away from his captors. Utu accepts the tears as an offering and grants the request. Dumuzid flees to another city, but once again the demons find him. This misfortune is repeated three times until Dumuzid hides in his sister Gestinanna's holy sheepfold, where each part of the prophecy then comes true and Dumuzid meets his sorry end. When the final demon enters the story, "the drinking cups lay on their side, Dumuzid was dead, the sheepfold was haunted."

WOUNDS OF LIFE

It is not hard to understand where a traumatized person's nightmares come from. The memory of the violent event, having been so powerfully codified, is too intense, it possesses very strong synaptic connections, which makes them capture and monopolize the electrical activity produced during sleep. But not all nightmares are motivated by some specific trauma. Dreams often possess a negative tenor, from the most terrifying nightmares to dreams of frustration and anxiety, which are experienced weekly by between 4 and 10 percent of the urban population.

Traditional cultures seem to be no different. Among rural Mexicans in the town of Tzintzuntzan, in Michoacán, Mexico, it is believed that poor nutrition in childhood leaves a person prone to nightmares. Around a third of the dreams presented by this population are of an openly disagreeable, frightening, or even threatening tenor, from fatal arguments between neighbors to sudden floods that are capable of tearing a man out of his own bed. Sexual impotence and loneliness also appear in around 10 percent of accounts.

Despite being prevalent, nightmares can be skirted around by dreamers who are experienced and well trained, as in this dream reported by the shaman Davi Kopenawa, an important leader among the proud Yanomami people on the Brazilian-Venezuelan border:

> I was also often frightened by an enormous jaguar in my dreams. It followed my tracks in the forest and got closer and closer. I ran from it with all my strength, without ever succeeding to put it off the track. Eventually I would trip in the tangled forest and fall before the fierce jaguar. Then it leapt on me, but just as it was about to devour me, I would suddenly come to my senses, crying. Other times I tried to escape it by climbing up a tree. But it chased me, scaling the tree trunk with its sharp claws. Horrified, I hurried to the tree's highest branches. I did not know where to run anymore. The only escape was to throw myself into the void from the top of the tree where I had sought refuge. I started to flap my arms in desperation, as if they were wings,

and suddenly I could fly. I glided in circles, high above the forest like a vulture. In the end I would find myself standing in another forest, on another shore, and the jaguar could no longer reach me.[11]

As dreaming simulates the satisfying of desires and fears, emotions of yearning, fulfillment, and being frustrated are frequently reactivated during the dream experience. This psychological observation has backing from studies of functional imaging during REM sleep. These experiments confirmed a powerful activation of the amygdala, a subcortical region directly involved in the emotional gauging of interaction with the world.[12] This reinforces the idea that a dream is a simulation of behaviors that can provoke reward or punishment. A world that is like a tutorial, virtual and imaginary, in which the mammal can test out strategies that are essential for its survival without incurring any real risks. Insofar as they are applied to the undetermined future, we are talking about a probabilistic oracle.

The idea that dreams are painted in the colors of the emotion prompted by the satisfying or non-satisfying of a desire has backing both from psychoanalysis and from neurology. On the other hand, the oneiric content reported by healthy subjects often shows a surprising degree of affective neutrality even when faced with elements that are frightening, grotesque, or bizarre. It is likely that this effect results from the deactivation during REM sleep of the regions of the prefrontal cerebral cortex that are active in the making of decisions and in the ordered carrying-out of plans. This reduction causes a deficit in working memory, that short-lived and easily disposable recollection of actionable information, which might explain simultaneously the discontinuities in the logical composition of dreams* and the numbing of the alarm system that warns against incongruities—which in real life produces a fight-or-flight response. The loosening of the chain of memories—which leads to estrangement from the situations simulated

---

* While it is believed that the oneiric chain of memories is intrinsically less coherent than that which occurs during waking, it is possible that the course of a dream is logical and coherent but that the memory deficits of REM sleep make it impossible to report it. The methods of neural decodification described in Chapter 17 have the potential to clarify this mystery.

in dreams—is also a loosening of criticism and censorship. If anything is possible in a dream, anything can also be acceptable.

This reasoning is supported by the neuropsychological results obtained by Matthew Walker, of the University of California–Berkeley, and Jessica Payne, of the University of Notre Dame. These U.S. researchers independently showed that REM sleep plays a key role in the processing of emotional memories and the attenuating of the impact of aversive experiences after a night's sleep. The evidence shows that REM sleep recalibrates the connections between different regions of the nervous system that are involved in emotional processing, such as the anterior cingulate cortex, the amygdala, the hippocampus, and the autonomic system. With a lack of REM sleep, the excess of activation in these regions can lead to irritability and a reduction in memory. For most people, spending a whole night without sleep makes it very hard adequately to regulate their emotions, especially negative ones.

But what are nightmares actually for? The simulation of behaviors and images conjured up in dreams gives us the possibility of experiencing—without risk, in a sustained way, without waking—situations that could potentially be harmful in real life. As an exploration of the space of probabilities, it is an invaluable tool for inhibiting impulsive behaviors that are dominated by emotion. Consider the following simple, revealing example.

A doctoral student woke early and headed for the laboratory, intending to collect the communal-use car, which he had reserved, in order to carry out an experiment at the field station two hours from the university. He was frustrated not to find the keys to the car in the office. He went down to the garage and his fears were confirmed: the car wasn't there. It only took one phone call to discover that a junior colleague had taken the car the previous day and hadn't yet brought it back. The day's work was wasted, and he was very angry. That night, the researcher dreamed he was with his colleague in the morning and that he blew up at him, conveying his annoyance with shouting and expletives. His colleague, who was well over six feet tall, started to hurt him, with punches and kicks. When he woke the next morning, the doctoral student felt as afraid as he was angry. He realized right away that he needed to be diplomatic. When he met his giant colleague, he was nice and accepted his sincere apologies.

A nightmare will often have an immediate protective meaning, warning against an imminent danger. The function of accident prevention is quite clear even when the premonition is partially realized. Returning home from a party, exactly a year after the death of a close friend in a car accident, a woman fell asleep at the wheel and dreamed images of the car crash in slow motion, going up onto the curb and colliding with a wall. It was only the following day that she realized that the start of the dream really had happened, as she had indeed climbed the curb, though she had not crashed the car into the wall.

In other cases, the almost-happening of a serious accident triggers a persistent reverberation, motivated by a fear of what could have happened. Two couples who were friends, each of them with their small child, were staying in a tropical hotel, spending a weekend relaxing by the seaside. One evening, playing on a riverbank, a current suddenly carried the children some distance away toward the mouth of the river, almost sweeping them into the ocean. The parents swam out furiously and managed to save the children, to everybody's relief. The night after the incident, one of the parents had a number of nightmares reiterating the danger, genuine replicas of the situation.

HUNTED ONE DAY, HUNTER THE NEXT

The fear of losing one's children is a particularly dreadful aspect of the fear of predation, but the dividing line between predator and prey is a tenuous one, since victory is volatile. We need only examine our species' interminable war record to understand the concrete role that fear and nightmares play in a conflict situation. In 1865, the United States government began the construction—without permission and with no notice—of three new forts right in indigenous territory.[13] The fortifications profaned the traditional hunting grounds between the Bighorn Mountains and the sacred mountains of Paha Sapa, or the Black Hills. To the whites, these rugged giants, up to 7,242 feet tall, which nowadays display huge sculptures of U.S. presidents on Mount Rushmore, were merely an obstacle on their way to the gold mines in the West. In stark contrast, to the seven Lakota tribes and other Native American peoples of the central Plains, the Paha Sapa were the center of an entire cultural universe, the mythical "heart of all there is."[14]

When government officials finally met with Lakota and Cheyenne chiefs to negotiate, more than a thousand soldiers were already marching on the region under the command of Colonel Henry Carrington.

Red Cloud was furious and left the meeting, and began to organize attacks on the whites, demanding their unconditional withdrawal from the forts. It was at this time that the young Crazy Horse became close to the Lakota leader, who was almost twenty years his senior.

From the time that Fort Phil Kearny was constructed, the attacks were a deadly routine.[15] Neither soldiers nor civilians were prepared for the terrible conditions in that remote fort, more than sixty miles away from the nearest military support. The claustrophobic wooden constructions were always too hot or too cold—and the sickening stench was inescapable. While the flow of settlers heading westward remained constant, the number of soldiers diminished month after month owing to casualties in combat, escort missions, and desertions toward the gold-filled hills beyond the western horizon.

Carrington requested replacement troops, which were promised but hardly came at all. Since the battalion only had access to outdated weapons, scarce munitions, and just a handful of good horses, Carrington was hesitant to attack the Native Americans. However, even hidden behind the fort's high walls, the soldiers felt the wrath of Red Cloud. Between June and December there were fifty attacks, with seventy deaths. The cemetery filled with crosses began to unsettle the 360 men of the Second Battalion of the Eighteenth Infantry Regiment. Though the Native American losses were also high, the autumn advance sowed an unstoppable fear among the isolated troops.

But there were exceptions. In November 1866, Captain Will Judd Fetterman, a Civil War veteran, arrived at Fort Phil Kearny. Fetterman saw the growing frustration with Carrington's command, and spotted a chance to stand out. Within days he had shown himself openly critical of his superior over rounds of whiskey, scornful of the commander who refused to fight the enemy. Eager to prove his supremacy in battle, Fetterman spoke the line that would become famous: "With eighty men I could ride through the entire Sioux nation."[16] He was supported in his bravado by Lieutenant George Grummond, another young Civil War veteran. Both men saw Red Cloud's revolt as a unique chance for rapid fame and everlasting glory. Nevertheless, for the families of the soldiers and civilians, for all those who did not want to win medals, and for Carrington himself, the escalation in violence was cause for extreme concern. The constant attacks and the piling up of the wounded led

*Dream Stele between the
front legs of the Great
Sphinx of Giza*

Joseph Interpreting Pharaoh's Dream *(1894), by Reginald Arthur*

Vishnu dreaming the universe, Dashavatara Temple, Deogarh, India

TO ΟΡΑΜΑ ΤΟ ΑΠΟCΟΛΥ ΠΑΥΛΥ

ΔΙΑΒΑC ΕΙC ΜΑΚΕΔΟΝΙΑΝ
ΒΟΗΘΗCΟΝ ΗΜΙΝ

Ο ΑΠΟCΤΛΟC
ΠΑΥΛΟC

The Apostle Paul's Macedonian vision. Mosaic in the Greek city of Veria, to the left of the altar erected on the spot where Paul is believed to have preached to the crowds in the year 51.

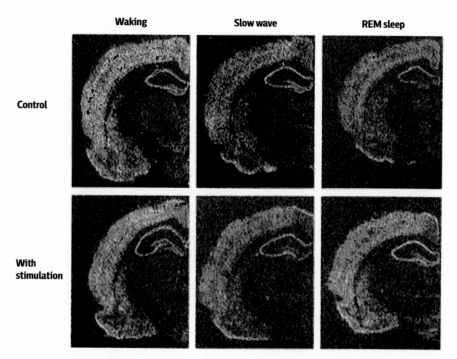

Impressions of the day: molecular day residues. In control animals, which were not experiencing new stimuli, the expression of the immediate gene Zif-268 declines during sleep. In animals subjected to new stimuli, the expression of the gene is re-induced during REM sleep. The images represent the frontal cortex of a single cerebral hemisphere. The color scale goes from red to blue, corresponding to raised and lowered levels of gene expression, respectively.

Dürer's torrential dream transformed into a watercolor (1525). Among the visual arts, this is one of the oldest-known examples of explicit dream inspiration.

Chagall's Jacob's Dream, *1966*

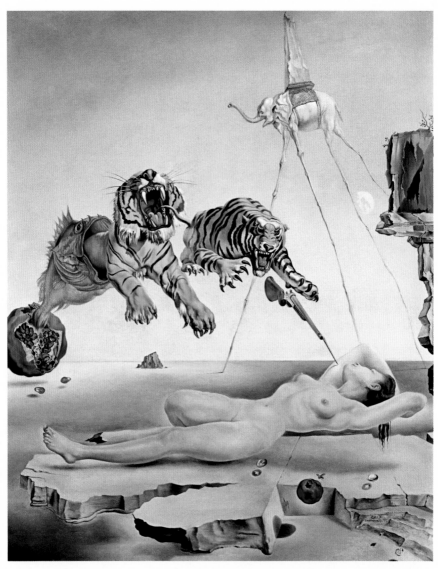

Dalí's Dream Caused by the Flight of a Bee Around a Pomegranate a Second Before Waking *(1944). The painter pursued the multiplicity of meanings through his paranoiac-critical method, inspired by Freudian theories.*

The Battle of Little Bighorn depicted by the Lakota. (Above) Painting by Kicking Bear (1898). To the left, the figure dressed in buckskin is Custer. The figures just sketched out in a ghostlike manner in the top left, behind the figures of the dead soldiers, are the spirits of those killed in the battle. In the center are the figures of Sitting Bull, Rain-in-the-Face, Crazy Horse, and Kicking Bear. (Below) Painting by Amos Bad Heart Bull (c. 1890), depicting Crazy Horse in the center of the battle, with spots painted all over his body.

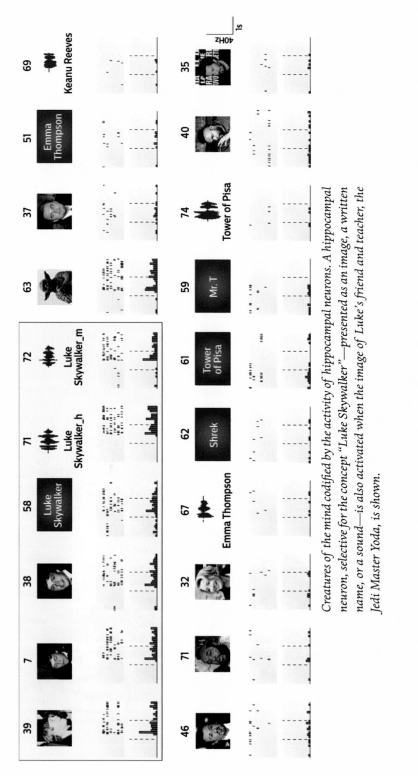

Creatures of the mind codified by the activity of hippocampal neurons. A hippocampal neuron, selective for the concept "Luke Skywalker"—presented as an image, a written name, or a sound—is also activated when the image of Luke's friend and teacher, the Jedi Master Yoda, is shown.

the group to decide that if the worst were to happen, they would kill each other rather than fall into the enemy's hands.

The ordeal of Frances Grummond, George Grummond's twenty-one-year-old wife who had arrived at the fort pregnant, is well recorded for posterity. There were months of torment on terrible, dangerous roads until at last, in September 1866, the young woman was relieved to see the stockade. When the wagon that was carrying her family was already approaching the gates, it had to give way to an ambulance that was conveying a man's torso separately from his scalped head. Besides the deep hatchet cut on his back, the body bore other marks of savagery: the corpse had been disemboweled and a fire kindled inside the abdominal cavity. That night Frances had a panic attack and could not sleep.

The day dawned over a thick layer of snow. Carrington ordered two columns to leave the fort to rescue the cart of firewood whose drivers were being harassed by one of Red Cloud's warrior bands. As they pursued the fleeing Native Americans, the soldiers were amazed to see a curly-haired brave dismounting from his mustang to examine its hooves, as if it were injured. This man was Crazy Horse.

Calmly, he let the soldiers approach. When they were very close, he remounted and galloped away at full tilt. The American soldiers took the bait, abandoned their defensive formation, and raced furiously up the escarpment in his pursuit. When reinforcements finally arrived, the disaster was confirmed: a lieutenant impaled on a tree stump, a sergeant with his skull split, and five other soldiers injured.

Terror spread through the fort when the columns returned with the news. Frances recorded the nightmares she had that night:

A sense of apprehension that I seemed to have been conscious of ever since my arrival at the Post, deepened from that hour. No sleep came to my weary eyes, except fitfully, for many nights, and even then in my dreams I could see [my husband] riding madly from me with the Indians in pursuit.[17]

A SIMULATION OF REALITY

A few days earlier, with the arrival of the first snowflakes, scouts from the Crow people had sighted a large camp sixty miles away. It was no less than Red Cloud himself at the head of a formidable coalition of two thousand Lakota, Arapaho, and Northern Cheyenne warriors. Even after being warned by their scouts of the approaching danger, Carrington decided to carry on with his plans to raise the fort's flag, on a marvel of a flagpole more than a hundred feet high, the final missing piece to complete the construction. On the lawn in front of the palisade, soldiers in formation listened to the cornets of the battalion band, speeches, and a salvo of cannon fire that shattered the icy silence around them.

Carrington was not aware that Red Cloud's camp was moving over to the other side of the mountains in front of the fort, just a mile or two away. Crazy Horse, who had previously been reserved, was now the most eloquent voice in favor of one final blow. On the afternoon of December 20, 1866, a fortune-teller was summoned by Red Cloud to predict the future. How many soldiers would they kill in the coming battle? After performing a number of ritual gallops with a blanket over his head in search of a vision, the fortune-teller declared that he could see in his fists more than a hundred dead soldiers. That was all Red Cloud needed to know.

The twenty-first dawned to bright sunlight and dry air. First thing in the morning, a big wagon filled with soldiers came out of the fort to fetch firewood and, as usual, they were attacked. Under pressure from Fetterman and Grummond, Carrington decided to send a detachment of eighty men to teach the Native American warriors a lesson. As he was about to burst through the gates of the fort at the head of the troop, Fetterman received explicit orders from his commander not to go farther than the crest of the mountains, in order to maintain constant visual contact with the fort. The detachment galloped away in search of a fight.

At noon the soldiers could be seen exactly on the crest, watching a group of warriors who were coming and going at the edges of the guns' range, provoking and harrying the U.S. troops. The latter had paused in their advance, torn between obeying Carrington's prohibition and yielding to the impulse to punish the "savages'" taunting. At that moment, the young Oglala warrior emerged, with his curly hair and a falcon feather, on the back of his bay horse with its white muzzle and feet. Crazy Horse was shouting insults in English; then he dismounted to look at his horse's hooves, ignoring the shots that whistled very close to his feet. When the soldiers approached, he mounted and moved away again, stopping again soon after. He even started a fire, acting as if he was intending to hand himself over to the soldiers as a sacrifice. Even so, in spite of all the provocations, the soldiers didn't dare advance to the other side of the mountain.

And it was then that the intrepid Lakota warrior tried his final trick: he dropped his pants and showed his buttocks to the stunned soldiers. The insult worked perfectly. Disobeying Carrington's orders, Fetterman raised his saber and ordered a cavalry charge, leading the furious gallop of the whole detachment beyond the crest of the mountain. Fetterman and his troop went into the valley ready to massacre their enemies, only to discover they were totally surrounded, by hundreds of hidden Lakota and Cheyenne warriors who had been lying in wait for vengeance for the invasion of the Paha Sapa. Fetterman and his eighty men—all he needed to "ride through the entire Sioux nation"—were annihilated by Red Cloud's clenched fist of teeming braves. After the battle, Carrington handed the widowed Frances Grummond a sealed envelope with a lock of George's hair. Her worst nightmare had become a reality.

Six days after the massacre, *The New York Times* reported with great fanfare that the dead represented 8 percent of all the United States' military losses in indigenous conflicts up to that moment. It was the greatest defeat without survivors that the American army had hitherto experienced. Even though the first three days had been days of bitter mourning in the camp of the Lakota and Cheyennes, on the fourth day the Native Americans had a euphoric celebration of their victory at the battle of the "hundred in the hand." As a sign of respect for his outstanding leadership in the fight, Crazy Horse was invited to sit close to the fire, with the older chiefs.

Amazed at the might of the warriors they had judged so lightly, the generals in Washington could see that they had lost this war. In August, the forts were vacated and Red Cloud went personally to burn them, accompanied by his warriors. It would be another year before the great Lakota chief would finally agree to sign a peace treaty with the whites.[18] On November 6, 1868, the United States for the first time signed an agreement on the Native Americans' terms, committing to withdraw all their troops from the "Great Sioux Reservation," a vast territory spanning from the Bighorn Mountains in the west to the Missouri River in the east, from the 46th parallel in the north to the boundaries of the state of Nebraska and the Dakota Territory in the south. These were lands that the United States government believed to be without value, just stops on the way to the gold mines in the Rocky Mountains. But to the tribes of the northern prairies, the most precious lands in the world, all around the sacred Paha Sapa mountains, would remain that way. Red Cloud had triumphed in his war, and the dream of Native American self-determination remained alive.

ROMAN NOSE DISBELIEVES

If the story of Crazy Horse suggests that believing is being, Roman Nose's trajectory shows the curse of not believing. The leading warrior of the Northern Cheyenne, who died in combat in the prime of his life, used as his protection against enemy bullets a sacred headdress, a unique artifact ritually prepared for him by the shaman White Bull.[19] The headdress was made based on a dream that Roman Nose had when he fasted for four days in search of visions, on an island in a lake in Montana. He reported a dream in which he found a snake with a single horn on its head. As a result, his headdress did not have two horns on its sides as was the custom among the Cheyenne, but just one imposing central horn, with a long tail that reached almost down to the ground when Roman Nose rode a horse.

During the making of the headdress, the shaman had avoided contact with any object originating from the world of the whites. When he offered the headdress to Roman Nose, he established a series of alimentary and social restrictions that the chief needed to uphold to preserve the magic, forbidding him from shaking hands with anyone or ingesting any food "contaminated" by metal, under penalty of dying in the next battle. In order to be able to access the artifact's magical powers, Roman Nose also needed to observe strict rituals about using and keeping it, raising it repeatedly in the four cardinal directions. The use of the headdress was accompanied by sacred painting for the battle: red on the nose, yellow on the top of the head, black on the jaw. Roman Nose never received a single serious injury until his final battle. To the terror of his opponents and in apparent contradiction of White Bull's restrictions, he wore a blue cavalry jacket with gold epaulets, a symbol of the military subjugation of the enemy. Painted and adorned with his powerful headdress, the imposing chief, well over six feet tall and muscular, with broad shoulders and an aquiline nose, made a powerful impression on the warriors he led. Beside him, they, too, felt invincible.

For his whole lifetime, Roman Nose was vigorously opposed to the signing of peace treaties with the whites, attacking caravans, military posts, railroad stations, and telegraph lines. The Cheyenne warrior was a match for Crazy Horse in bravery and in dislike for the world of the whites. Both dreamed of the total expulsion of their enemy and a return to the pure life of the ancestors. In September 1865, in a Cheyenne and Lakota attack that Roman Nose asked to lead, the two warriors engaged in an incredible test of courage in the face of hundreds of enemy soldiers. Crazy Horse caused a commotion by galloping along the ranks of soldiers to run down their ammunition and their enthusiasm. Then Roman Nose repeated the same maneuver several times, whirling around and rearing up his horse as if he were immune to the intense gunfire, giving war cries and leaving his warriors totally astounded. He continued in these maneuvers under a hail of bullets fired by ever more alarmed soldiers, until his horse was hit and he left the battlefield unscathed. The feat was widely acknowledged as evidence of his invulnerability and, therefore, of the effectiveness of his dream headdress.

But Roman Nose's dream collapsed on September 17, 1868, when he was fatally struck in the dry bed of the Arikaree River, in the state of Colorado. A few days earlier, the Lakota had offered a feast to pay tribute to the most important Cheyenne fighters. Roman Nose forgot to stress his dietary restrictions. When he did remember to ask the cooks if they had used any metal utensils, the answer was affirmative. Before he was able to carry out all the purification rituals that the situation demanded, scouts from the United States Army were sighted by Lakota lookouts, and an expedition was sent out to destroy them. A number of chiefs called Roman Nose to join the fight, but he asked them to wait, believing that he would die if he did not purify himself. For most of the day, the fighting proceeded in his absence. The scouts were equipped with repeating rifles, a technological novelty the Native Americans knew little about and that led to their losing many men without managing to break through the enemy line.

When Roman Nose finally appeared on the battlefield, on a white horse, with his magnificent headdress trembling in the wind, he was still hesitant about fighting, as he had not concluded the necessary rituals. Near sunset, repeatedly urged and even provoked by the Native American chiefs, the Cheyenne giant yielded to fatalism and commanded his last attack on the whites. He took a shot in the hip and retired, still on horseback, but with a bullet having hit his spine. He died at sunset, aged about thirty, certain that he had lost his faith. The dream-inspired sacred headdress had lost its magical powers.

THE WORST PAIN

Among the best-defined categories of dream, with a long trajectory going back to prehistory, are dreams of mourning. Death, whether one's own or that of one's loved ones, is the most absolute limitation that anybody can encounter. The loss of somebody loved means the disappearance of the external object, but not of the internal object, that is, of the mental representation of that person. With a person's decease, it is possible for that internal object not just to die but to be preserved and transformed. Accepting a sudden death is the human being's greatest emotional challenge, especially when the death is violent. When a loved one dies, or has an accident, or receives news of a terminal illness, the need to simulate the not-death can verge on the absurd, revealing an apparently inexhaustible exploration of alternative hypotheses for the disappearance, in an evident attempt to satisfy the wish for life.

This becomes very clear in the testimony of Crimeia Alice Schmidt de Almeida, a communist militant and ex–political prisoner who was tortured barbarically in her seventh month of pregnancy during the Brazilian military regime (1964–1985). A member of the Commission of the Relatives of the Political Dead and Disappeared, Crimeia saw her husband, André Grabois, for the last time in 1972, when she left a guerrilla band operating up the Araguaia River in the Amazon rainforest to seek medical supervision in São Paulo, owing to problems with the pregnancy. Like so many others, André Grabois was never found. The lack of a body keeps the flame of hope painfully alight, which is revealed in dream fantasies. In Crimeia's words, "It's irrational, but it happens, in dreams: what if he hasn't died yet? What if he's still being tortured? What if he's lost his memory? They are questions that torture me."[20]

Even when relationships are more distant, death evokes dream feelings of strangeness and inadequacy. A sixty-year-old lady read a story in a newspaper about a stabbing and recognized a work colleague and that woman's husband among the victims. She wasn't able to attend the customary thirtieth-day Mass, and soon afterward dreamed that she was going to the wake and found instead a joyful gathering with pastries. The colleague's husband had indeed died but she herself was still alive, going around smiling and elegant. Gradually the dreamer began to realize that nobody else at the wake was able to see the resuscitated victim, only her. She asked her colleague directly what was going on, and the woman changed the subject, saying she'd retired.

Dreams simulate those possibilities that are present in the context of a dominant desire. Faced with irreversible facts like death, desire acts upon the dream as a driving force of the electrical reverberation of memories, even reversing reality to find satisfaction in the simulation of a here and now that is impossible. When it completely changes the direction of the facts, reviving dead people or restoring ended relationships, a dream will produce a reaction of great disappointment at the moment of waking, as the death or the separation are real and the resuscitation in the dream makes it necessary to accept the death anew. This probably works as a punishment of the neural network that was activated during the "good dream," reducing the likelihood of its occurring again in the future. This is why dreams of satisfying desires whose object is the actual lost beloved are typical only in the initial stages of mourning.

THE END OF MOURNING

It is common for those closest to somebody who has died not to be able to dream about them for days, months, or even years. The disappearance of a loved person causes huge disorder in the mental world of their closest relatives and friends, often leading to a conscious suppression or unconscious repression of the memories associated with the person lost. But what was repressed ends up returning. As things progress, dreams about the loss of a loved one, whether real or symbolic, give those widowed or separated the chance of a goodbye or a settling of accounts. In the 2007 documentary *Jogo de cena (Playing)*, by the Brazilian filmmaker Eduardo Coutinho, a mother describes having ended a five-year period of mourning for her murdered son because of a dream in which he appeared and told her to be happy, "because he's an angel now." The end of the mother's mourning is explained by the pleasure she felt on waking, which must have positively reinforced the neural circuit that connected the symbol of the angel to her dead son. The dream with its extremely positive content did not collide with reality at the moment of waking, since it did not deny the death but sublimated it into an irrefutable fantasy. The new neural circuit, strengthened by the good dream, was installed as the strongest representation and came to dominate the stream of consciousness. What had once been nothing but an intrusive, recurring rumination of horrific thoughts relating to her son's murder, now had a chance of leading to the benign conclusion that her son was still happy, immortal, and infinitely good. A conversion from a dead person to something divine, not all that different from the deification of people who have died that took place throughout antiquity and that still happens in hunter-gatherer societies today.

But the redemptive interpretation does not always dominate the dream narrative. Negative memories function as attractors of electrical activity, reiterating the trauma, delving deeper with each nightmare. The fear of having nightmares causes nightmares. In such cases, psychotherapy is essential to interrupt the vicious circle, revisiting the trauma many times over in a context that is mild and harmless, until it is resignified by the association with other symbols, by the conscious processing of negative memories and by the evoking of concurrent positive content.

The following series of dreams exemplify the successive dynamic of first reiterating and then resignifying the trauma. A young woman was kidnapped and spent twelve hours in the hands of her armed captors, to then be set free in a remote area. In the months that followed this trauma, the woman had countless nightmares that almost perfectly replicated the situation she experienced during the kidnapping. As time passed, the nightmares started to become more abstract, with ever-greater variations. Then another phase began in which other dreams took hold and developed, but they were suddenly interrupted to be replaced by the kidnapping dream again. In this phase the dreamer was literally kidnapped from her dream and into the nightmare. After a while the dreamer could no longer experience the nightmare without remembering that none of that was real. Gradually the dream lost its meaning, rendering that prediction of a bad future less and less probable, since the prediction was based on the increasingly implausible repetition of a past negative event. The overcoming of the series of traumatic nightmares only occurred fully when Thanatos gave way to Eros. After seeing an erotic bestseller for women in an airport bookstore, the woman had her first lustful and pleasurable dream since the day of the kidnapping. She bought all the books in the series and for a whole week was sated with erotic dreams. And then, as quickly as they had come, these delightful dreams also disappeared. She was freed from her nightmares.

DARWINIAN IMPERATIVES AND CULTURE

The threat simulation theory is supported by many biological and psychological facts, but it still cannot claim to account for the dream narrative in its entirety. The diversity of dreaming requires us to expand the scope of our theories to include the positive side of motivation, too, made up of appetite, rewards, and pleasure. In strictly logical terms, the predator's nightmare is the prey's pleasurable dream—and vice versa. On the savanna, lionesses and zebras must dream almost identical narratives, desperate headlong races across the grasslands with jumps and back-kicks, with blood and sweat, broken teeth, slashed throats, and more blood and flesh and fat and bones. Dreams with identical content, but with affects reversed and objects swapped around. Large predators usually have a success rate of less than 20 percent.[21] When a zebra manages to escape, which happens frequently, the exhausted, hungry lionesses must have nightmares about hooves, stripes, and starvation. Out in the wild, faced with prey and predators, the typical dream narrative of humans corresponds to the same repertoire of concerns as any other animal in the same conditions: kill to eat, survive, and procreate.

But the dreams of the human species in civilization reflect much more than those Darwinian imperatives. It is safe to say that our dream repertoire diversified as improvements to language, tools, and knowledge increased the everyday distance between us and death. Violent narratives and fierce appetites probably dominated the dreams of our ancestors until the development of livestock and agriculture, which for the first time enabled the stable abundance of food. The last ten thousand years have seen the creation of those technologies that have allowed many people to know food security. Nightmares about hunger became less frequent, but they never stopped happening entirely, since malnutrition persists to this day among the poorest people. Besides, the whole development of civilization has been marked by war and persecution.

Dumuzid's dream expresses an ancestral fear in civilization, the rising up of wicked and hungry men capable of implacably hunting somebody down, forcing him to hide like a wild animal before coldly executing him. This is unlikely to be much different from the fear that drives the indigenous Munduruku people in their resistance struggle against the building of dams in the Tapajós River deep in the Amazon, or the Juruna people—formerly the most populous in the area of the Xingu River, and almost completely destroyed by rubber tappers in the nineteenth century—who denounced the impact of the Belo Monte hydroelectric mega-dam whose hasty construction devastated a huge area of the Amazon rainforest and displaced more than twenty thousand people. A fear all the greater because they believe that dreams are an involuntary insertion into a dream world that is as concrete and possibly even more dangerous than the world experienced during waking. The intimate Amerindian connection between dreams, metamorphosis, and death is related to this belief.

THE MULTIPLICITY OF SMALL DESIRES

Even if the archaic dream of a desperate prey confronted by a deadly predator describes the everyday life of so many minorities and marginal populations across the globe, especially those who are war refugees,[22] it is undeniable that the urbanization that followed the invention of agriculture, which gave our ancestors the security of having shelters in which to spend the night, with walled homes guarded by armed watchmen, enabled a reduction in violence.[23] As the daily fear of death cooled, it increased the mental space and emotional availability for dreaming and creating. Dream narratives became more complex just as culture did.

We no longer commonly dream about lionesses chasing us, but the challenges of real life that become pressing and relevant appear very clearly on the dream landscape. As we have seen, one quite common dream, a distant descendant of the dreams of Paleolithic hunts, pertains to the taking of academic tests in the near future. These dreams often just reflect the fear of something going wrong and of missing or failing the test. We dream about a pen blowing up, about arriving late to the place where the test is being held, about not having any clothes for the exam, or even totally forgetting the subject at the exact moment the test is taking place. But there are also dreams that are specifically related to the actual taking of the test. Students all over the world have dreams about Pythagoras's theorem, about Mendel's genetic inheritance, and Mendeleyev's periodic table.

The scheduling of highly stressful tests some time in advance can reveal a curious aspect of dreams: the previous and unconscious programming for a specific challenge. A doctoral student had scheduled her thesis defense quite a long way ahead, but subsequently rescheduled it for some months later. The night after the date originally scheduled for the defense, she had a long, intense nightmare in which she was presenting her research but feeling completely unprepared. It was as if the dream were expressing the program that had been set in motion some time ago, staging the future challenge with amazing—albeit ill-informed—precision.

While it is true that some attempts to explain dreams are lacking in introspection, such as Crick and Mitchison's theory about the randomness of dreams, other attempts often get lost in perspectives that are unreasonably anthropocentric or ethnocentric. The U.S. philosopher Owen Flanagan, for example, gained some notoriety for writing that dreams cannot have any adaptive function as he, Flanagan, had never had a dream that helped him to solve a problem in real life. It is quite likely that the life of a full professor at Duke University, living in privilege and free of the greatest stressors, is not the best candidate to reveal dreams' primitive functions. Based on the monotony of his own dreams, Flanagan the anti-Freudian concluded that dreams are devoid of any meaning or function: "Dreams are the spandrels of sleep."[24]

On the other hand, the Freudian tradition was criticized and even ridiculed for insisting that dreams are an attempt to fulfill wishes, and for considering the censoring of obscene thoughts a universal function of dreams. We know today that this heightened censorship was a specific cultural hallmark of the conservative Viennese society in which Freud lived and produced his work.[25] In any case, it is clear that the scientific and industrial revolution of the sixteenth to nineteenth centuries greatly soothed the main human problems, especially for the middle and upper classes. In the twentieth century, the appearance of radio, cinema, and television caused a combinatorial boost in possible dream narratives. The Pantanal jaguar might dream of a thousand different ways of slaughtering a capybara, but they are still all hunting dreams, very similar to one another.

Not us, though. The multiple needs of the human species have created the conditions for dreams to become disordered collections of images, patchwork quilts of wishes. The typical dream of our times is a blend of meanings, a kaleidoscope of wants, fragmented by the multiplicity of desires of our age.

15

The Probabilistic Oracle

Any reader who has followed our journey attentively to this point will be well placed to understand why dreams were considered divinatory in so many different ancient civilizations and contemporary cultures. Ever since our ancestors began to make a written record of their thoughts, around 4,500 years ago, there has been ample documentation of dream narratives about what is yet to happen, as well as how to go about intervening in the future through behaviors represented in the dream.

When a dream reverberates memories from the past, it is reflecting the dreamer's expectations for the future. Most of all, it is reflecting their chances of success or failure in the small or large private epics triggered by desire. These expectations do not only incorporate what the dreamer is contemplating consciously, but also—and in fact principally—their unconscious perception of the whole context in which they find themselves, with its meanderings, promises, and chasms. It is the broad, diffuse sum total of impressions, collected both above and below the threshold of consciousness, which forms the basis for intuition and which gives dreams their life.

In the words of Jonathan Winson, "the dream expresses what is happening to you right now," but it's a now that is determined by the situations that have already been experienced and infused with possibilities for the future. The practical keys to interpreting any dream are its past and future elements, as the brain's present is impregnated with memories and simulations. Two dreams of great historical importance allow us to illustrate this point directly.

"IN THIS SIGN YOU WILL CONQUER"

Dreams played a fundamental role in all the phases of ancient Rome, including the latter one. In the third century, the gigantic Roman Empire sank into a serious military anarchy that almost broke it apart once and for all. The crisis was cooled with the rise of the Emperor Diocletian, who managed to run the huge territory through a tetrarchy shared with three other Augustuses or Caesars under his authority. For many years, Diocletian governed Asia Minor, while his right-hand man Maximian governed Italy, Constantius reigned in England, and Galerius waged wars in the east. When Constantius died, his son Constantine was proclaimed Augustus by the western troops. In Rome, however, Maximian's son Maxentius was crowned emperor. The conflict between the two remained dormant until Constantine invaded Italy and besieged Rome.

Fortified within the capital with his large numbers of troops, Maxentius was preparing to break the siege of Constantine's army at dawn, fired up by the oracular prediction that an enemy of Rome was to die that day. Earlier, however, as he marched with his troops, Constantine had had a striking vision, a solar halo shaped like a cross with an inscription in Greek: "In this sign you will conquer." That night, Constantine dreamed that Jesus Christ was instructing him to mark his soldiers' shields with the Greek letters *chi* and *rho*, the initial letters of his holy name. At the dawn of October 28, 312, beneath military standards topped with *chi* and *rho*, Constantine's army destroyed Maxentius's troops on Milvian Bridge, with Maxentius himself drowning in the Tiber River. The war ended, Constantine publicly embraced his new faith, and the beliefs of the oppressed Christians became the official ideology of the Roman state. An imperial dream changed history.

DREAMING ABOUT THINGS THAT HAVE NEVER BEEN DONE

Premonitory dreams were also at the center of the decisive conflicts in the intense drama of the war between Native Americans and whites in the United States. Red Cloud's victory was followed by a fragile peace that lasted only a year. To the northeast, the Lakota Hunkpapa chief Sitting Bull made it clear he wouldn't sign any agreement with the whites, which earned him Crazy Horse's approval. Soldiers continued to press for the Native Americans to live to the east and only hunt in the west—where the gold was. After the suspension of commercial exchanges and the restricting of food rations, Red Cloud decided to go to complain directly to the Great White Father, President Ulysses S. Grant. On the long train journey, Red Cloud witnessed the flow of thousands of settlers, the huge industrialized cities, and the massive military power that his hosts made a point of displaying. The acute perception of death on an industrial scale destroyed his urge to fight. He retired to his reservation, having resolved never again to pick up arms against the whites.

Crazy Horse's disappointment in the great Oglala chief could not have been more bitter. Convinced that any contact with the whites was harmful, he moved farther and farther away from the reservations that the whites had imposed, resolving to strengthen his people's traditions and freely occupy the lands of their ancestors. In 1874, however, large seams of gold were found in the Black Hills, the sacred hunting grounds of Paha Sapa, which up until then the United States government had considered valueless. Pressure on the lands grew, but emissaries who tried to buy them were driven away by Sitting Bull and Crazy Horse: "One does not sell the earth upon which the people walk."[1] It was then that the U.S. Secretary of the Interior gave an ultimatum: all the Lakota must return to their reservations by January 1876, on pain of being considered enemies.

Winter came and went, and the Lakota did not give in. Inexorably, the brutal war machinery of white civilization began to move. Columns with thousands of soldiers were mobilized to surround the Native Americans, who were armed almost exclusively with bows and arrows. Under pressure from conflicts in several different points of their own territory, thousands of Lakota, Northern Cheyenne, and Arapaho met in the territories yielded by the whites to their Crow enemies and pitched a camp in the valley of the Little Bighorn River, in Montana. When summer finally came, the Lakota were trapped. Faced with the repeating rifles, machine guns, mortars, and cannon, they needed to learn new forms of fighting, which were more effective and deadly than all the weapons of the whites. They urgently needed something miraculous to happen. They needed to kill and not die.

A week before the famous Battle of Little Bighorn, Crazy Horse led a series of disturbing attacks on a column of a thousand soldiers under the command of General George Crook, a veteran of the Civil War. Crazy Horse used a number of guerrilla tactics that day that he claimed to have first experienced in dreams. In the words of the U.S. writer Dee Brown, "On this day, June 17, 1876, Crazy Horse dreamed himself into the real world, and he showed the Sioux how to do many things they had never done before."[2] The fighting stretched on until nightfall. When the sun came up, General Crook had beaten a retreat and Lakota sovereignty had lived to fight another day.

"THESE HAVE NO EARS"

The conflict reached its climax on June 25, 1876. One week earlier the camp had doubled in size, reaching around a thousand teepees and a population of almost seven thousand people, among them about two thousand warriors. One of the main promoters of the gathering of so many different groups was Sitting Bull. He was a member of the Buffalo and Thunderbird societies, two secret organizations of dreamers distinguished by the totemic spirits that had appeared to them in visions. After the tragic death of Roman Nose, the stunned Northern Cheyenne found in Sitting Bull a replacement who was equal to the role, both in his contempt for the whites and in his strict observance of the religious sacrifices necessary to protect the fighters. For the same reasons, Crazy Horse and his Lakota Oglala warriors came to consider Sitting Bull a leader.

Days before Lieutenant Colonel George Armstrong Custer's attack, Sitting Bull took part in the Sun Dance, a purification ritual carried out on the solstice after the last buffalo hunt, in order to propitiate visions and divine protection. He fasted, he danced, he sacrificed pieces of flesh cut from his arms, he danced, he suffered, and he danced some more until he dreamed. He saw a shower of soldiers tumbling from the sky like locusts onto the green grasses, falling headfirst and losing their hats while a thundering voice said: "These have no ears!"[3] The interpretation of the dream was obvious. How often had they warned the white men that they would not tolerate the invasion of their hunting grounds? The white men absolutely had not heard them: they "had no ears" and this would be the end of them. The Great Spirit had spoken.

Inspired by Sitting Bull's vision, the chiefs withdrew their warriors from the enormous camp, positioning them secretly in a ravine behind a nearby hill. Having been informed by their Crow scouts about the huge concentration on the banks of the Little Bighorn River, Custer led his seven hundred men across almost entirely unfamiliar terrain, determined to inflict an unforgettable defeat on the indomitable Sioux and their allies. When they approached, the scouts confirmed that there were not many warriors in the camp: it was summer and the adult men must all be out hunting bison. Faced with the empty camp, Custer ordered a devastating attack, expecting to encounter only old people, women, and children.

And then the prophecy was realized. When the fierce band with unsheathed swords invaded the crowd of conical tents to the sound of shrill bugles and wild cries, the bluecoats did not find what they had expected at all. While the women and children withdrew, the warriors rose over the hill like a swarm of enraged bees. The regiment that had planned to carry out a massacre with no resistance found itself quickly surrounded and staunchly attacked by avalanches of warrior braves who didn't stop coming. Finally the soldiers despaired and broke ranks, fleeing in panic across open ground. From that moment, everything happened fast. Within minutes, the disordered core of the Seventh Cavalry Regiment was surrounded and slaughtered. Two hundred and sixty-eight soldiers died, including Custer, two of his brothers, a nephew, and a brother-in-law.

The event was reported with horror by the same newspapers that in previous months had glorified the bloody exterminations Custer had led against the Plains tribes. Drunk on the frenzy of the press and the general public, the ambitious and vain long-haired commander had died at the height of his fame, falling victim to the dream of a wise "savage." Perhaps Custer would have been luckier if he had given up on the cowardly attack after having his own nightmare about Crazy Horse.

PROSPECTING THE UNCONSCIOUS

To mammals living free in nature, and to those groups of humans that are closest to it, dreaming continues to be an essential biological function for warning against dangers, mapping out possible outcomes for the problems that are prevalent in the dreamer's life, selecting adaptive strategies, and integrating successive learnings into a coherent whole. A dream is a privileged moment for prospecting the unconscious, adding clues about the risks and opportunities of the environment, many of them subliminal but nonetheless able to be integrated into a general impression of what might come to pass. The brain takes yesterday as the basis for simulating what tomorrow might be like. So a dream can be seen as a way of testing out a theory in a simulated environment, with cycles of selective strengthening of memories during slow-wave sleep due to electrical reverberation, genomic storage triggered at the start of REM sleep, and a restructuring of memories during long episodes of REM sleep. Carrying out several consecutive cycles of mutating and selecting memories every night, the sleeping brain consolidates the best strategies it is able to devise in dreams.

The evidence is converging toward the idea that mammals' dreams are probabilistic simulations of past events and future expectations. The main function of these simulations would be to test out specific innovative behaviors against a replica of the world from memory, instead of the real world, leading to learning that is risk-free. This conjecture is a generalization of Revonsuo and Valli's theory of dream simulation of threats, according to which dreams can simulate actions that lead to undesirable consequences and that should therefore be avoided in the real world (suffering predation, for example). It is necessary to extend the reasoning to those actions that lead to a desirable result and that, therefore, should be implemented in the real world (finding food or fertile sexual partners, for example). An examination of mental content during REM sleep revealed that more than 70 percent of accounts included emotions, with an equal balance between positive emotions and negative ones. The idea that nightmares evolved as a way of negatively modulating the simulations of dangerous behaviors, while pleasurable dreams correspond to a rewarding association of pleasure with the simulating of particularly adaptive behaviors, is analogous to the concepts of Thanatos and Eros proposed by Freud as death and life instincts.

Faced with the huge quantity of uncontrolled variables, dream simulation frequently gets its "predictions" wrong. Occasionally, however, the simulation happens to coincide with reality, and then the dreamer confirms that the oracle can indeed, under certain conditions, make correct predictions. Dreaming works, therefore, as a probabilistic oracle, not very different from what was believed in antiquity in terms of its consequences for the dreamer, but quite different in its nature: in the place of a certainty motivated by hypothetical external mechanisms for the generating of the dream, which would be divine or spiritual, the uncertainty that is inherent in its biological nature. Dream images do not, then, reveal the dreamer's destiny tomorrow, but only the apparent course that they are on today.

THE BEST BETS

As a perceptual and motor reverberation, a dream comes up with intentions, actions, and consequences, in a simulacrum of ecologically relevant situations that are staged as imaginary video clips. As an associative narrative, a dream expresses through its explicit or implicit symbols not merely what the dreamer desires, but their own assessments of risk. Using this psychobiological lens, how are we to understand what might have happened to Constantine or Sitting Bull?

It would be tautologous to say that their dreams predicted the future, since we probably only know of Constantine's and Sitting Bull's dreams because the future smiled upon them. Probabilistic oracles operate *a posteriori*, and they evidently tend to be more remembered when they happen to prove "correct." Rather than anticipating a victory thanks to divine intervention, Constantine's dream proposed a proof of faith in the new religion, through the use of the emblem of the holy son of its single powerful God. To understand the military benefits of this conversion, it is important to consider that Christianity was already quite influential among the soldiers and officers of the Roman army when Constantine converted. Adhering to the religion of his own troops, on the verge of a decisive battle against a more numerous enemy, was a quite adaptive response for the emperor being challenged hard in long and costly civil wars.

The elucidating of any dream requires the identification of the dreamer's dominant wish. Constantine was fervently craving control of Rome in order to begin unifying the warring empire. At the gates of the capital of the world, he needed his legions' mystical fury more than ever. The dream expressed, then, the wagering on a path that was risky and yet more likely for victory: an oracle based not on the certainty of success but on its best bet.

The same can be said about Sitting Bull: his dream pointed toward the high likelihood of an event that in other circumstances would be improbable. Custer's rapid invasion of almost unknown enemy territory, followed by a surprise attack on a large Lakota and Cheyenne camp, looked like a suicidal tactic. However, this same tactic was implemented with great success by a range of American military commanders in the wars against the Native American peoples, such as Colonel John Chivington at the Sand Creek Massacre in 1864, General Ranald Mackenzie in the attack on Palo Duro Canyon in 1874, and

Custer himself in the Battle of Washita River in 1868. The Arikara called Custer "Creeping Panther Who Comes in the Night." Among the Crow, the general was the "Son of the Morning Star Who Attacks at Dawn."

The valley of the Little Bighorn River was located in highly contested territory. To the Lakota and the Cheyenne, the Black Hills region was Paha Sapa, the revered mountains belonging to their people for generations. But the Crow claimed those lands, too, based on the Treaty of Fort Laramie of 1851, which was systematically disrespected by the white colonizers and miners attracted by the discovery of gold in the mountains to the west, as well as by the Lakota and Cheyenne who simply did not recognize it. With so many territorial disputes, right in the middle of the gold rush, and with Custer's announcement in 1874 of new deposits of the precious metal in the sacred mountains, it was not hard to predict the excessively aggressive behavior of the American soldiers, who were violent, impulsive, hungry for riches, and unable to listen. Under these conditions, predicting a rash attack was reasonable and even logical. Sitting Bull was fervently hoping to unify the heterogeneous and fragmented groups of warriors gathered at Little Bighorn to defend the camp. Dreaming of a total victory was possible. Sitting Bull's dream was an expression of that likelihood.

Determining how successful Sitting Bull's premonitory dream was depends on the time frame you are considering. In the summer of 1876, the dream seemed like the expression of a manifest destiny of the people of the prairies, a clear signal of the Great Spirit's protection against the warlike invaders. Sitting Bull and Crazy Horse experienced the pleasure of managing what so many other Native American people had failed to achieve: blocking the invasion. The masterful capacity for mounted combat allowed the Lakota to dream of a victory against the invaders in a way that neither the Incas nor the Aztecs, who were much more numerous, came close to achieving. With their skittish mustangs, their accurate arrows, their few firearms and mystical courage, the Lakota fought against the whites with the shrewdness of the fox, the boldness of the bear, and the wisdom of the badger. But a few months after the Battle of the Little Bighorn, in the harsh winter, it was the heavy hand of the Great White Father in Washington that made itself painfully felt. If Red Cloud won the first of the Lakota's wars against the whites, Sitting Bull won the last—because what followed was nothing but terrible misfortune for his people.

The U.S. Congress responded to Custer's humiliating defeat with the "sell or starve" rider to the Indian Appropriations Act, which cut off all food rations until the end of hostilities and until the Black Hills were definitively surrendered. Paha Sapa was invaded, with large military contingents deployed to defeat the insurrection. Frozen, beaten, and hungry, many Native American people perished over that very tough winter. Less than a year after the Battle of the Little Bighorn, in the spring of 1877, the main Lakota and Northern Cheyenne chiefs had already given up. In May, Crazy Horse turned himself in, and Sitting Bull fled to Canada with hundreds of his followers. In September, in custody, Crazy Horse was murdered by a soldier.[4]

Sitting Bull's dream had absolute premonitory validity on the day of the battle, with apparently auspicious consequences for the Native American people for a few weeks, followed by a real nightmare. After years of hunger on the icy Canadian prairies, without any bison thanks to the indiscriminate killing carried out by the professional hunters with their long-range rifles, Sitting Bull and his people returned to the United States and gave themselves up, agreeing to live on a reservation.[5]

Even old and defeated, Sitting Bull continued to be a thorn in the authorities' side, traveling the United States as a star of the "Buffalo Bill's Wild West" shows, always ready to publicly declare his assessment of the deplorable white civilization. He was shocked at the number of people without homes living on the streets of the big cities, and he was seen giving alms to hungry beggars. In 1890, at the age of fifty-nine, he was arrested and killed in custody, struck by gunshots fired by Native American policemen.[6] In retrospect, Sitting Bull's dream had no validity beyond that fateful day in the valley of the Little Bighorn River. Looking at the Amerindian narrative in perspective, from the arrival of Columbus to today, the dreams of the Lakota did not meet with any different fate than the dreams of the Aztec, Comanche, Maya, Inca, Mapuche, Munduruku, Guarani, Krenak . . . it's a long list.

Constantine's premonitory dream, meanwhile, did in fact endure. The Roman Empire remained Christian for almost all its latter existence, Christianity spread across the globe, and today 2.2 billion people, more than 30 percent of the total population of the world, consider themselves Christians. If Pope Francis is successful in his modernizing of the church, with the ordination of women and the welcoming of homosexuals, it might last another thousand years. Clearly, none of this was in Constantine's plans. After all, it's impossible to predict the distant future very far beyond one's own historical context. Most likely, all the emperor wanted was merely to enthuse his troops and defeat the enemy of the day. The probabilistic oracle evolved in the context of day-to-day survival.

But whose survival? Did Constantine really dream about that Christian symbol, or was the dream constructed by him—or his biographers—for military, religious, or political purposes? This question relates both to the imperfection of the historical record and to the intrinsic untrustworthiness of dream accounts, which can lend themselves to all kinds of secondary uses. History is full of examples of accounts of dreams being used for political ends.

Publius Cornelius Scipio Africanus, one of the greatest generals of all time, winner of the Second Punic War against Carthage, rose to power while still young thanks to the political manipulation of dream accounts. In the elections for aedile in 213 B.C.E., Publius's brother was a candidate. Since he did not seem to have much popular support, Publius told his mother of two prophetic dreams in which both brothers were elected. Their mother enthusiastically embraced this supposed revelation and supported Publius's candidacy by making sacrifices to the gods and preparing her son a white toga. Publius was acclaimed in the Forum alongside his brother, and both were elected. Publius continued to spread the story that the gods spoke directly to him in his dreams, manipulating this belief at decisive moments in his story.

The Greek historian Polybius left a record of the calculated use that Publius made of religious beliefs:

[We should not suppose that] Scipio won such an empire for his country by following the suggestion of dreams and omens. But since [he] saw that most men neither readily accept anything unfamiliar to them, nor venture on great risks without the hope of divine help, . . . [he] made the men under his command more sanguine and more ready to face perilous enterprises by instilling into them the belief that his projects were divinely inspired.[7]

If Scipio Africanus manipulated the belief in dreams to climb the steps of the administration of Rome, Julius Caesar's dreams seem in retrospect to have been appropriate. Plutarch reported one striking dream that Julius had shortly before crossing the Rubicon River and entering Italy with a single legion, defying the express orders of the Senate not to approach with the troops that had been victorious in the Gallic campaign. This invasion of his own territory was the beginning of an irresistible taking of power, which Julius exercised successively through the roles of tribune, dictator, and finally consul.

According to Plutarch, Julius Caesar had dreamed of having sex with his own mother on the eve of his crossing of the Rubicon, the first act in a long process that would lead to the destruction of the Republic and the creation of the Empire.[8] While Julius's own initial reaction to his dream was one of embarrassment, soothsayers soon produced an interpretation that was extremely auspicious: the great man was literally preparing himself for possessing his "mother" land. It just so happens that Suetonius[9] dated the same dream eighteen years earlier, when Julius was thirty-three and a quaestor in Spain. The dream had occurred after Julius visited the temple of Hercules and lamented before a statue of Alexander the Great—who conquered the world before dying at thirty-three—that he had not yet managed to achieve anything comparable himself.

The discrepancy between Suetonius's and Plutarch's accounts suggests the shameless political manipulation of dream narratives for the construction of a biography. Both writers used and abused dreams as supposed causes of important historical events. In the case of Julius Caesar's maternal copulation, it is more likely that it was Plutarch who did the manipulating, attributing the dream to the historical moment when it would have the greatest impact. To what end would he have carried out this manipulation? To favor Julius Caesar through evidence of a destiny foretold? Or to show him as a man without scruples, capable of anything? Or even, simply, to spice up the plot of an already tasty narrative? Plutarch was in the habit of attributing multiple meanings to dreams, which allowed him to outline his biographical subjects' characteristic features more freely. Perhaps a different question would be more relevant: what is so special about dreams that they can have any belief at all attributed to them? How did this oracle evolve to be blind and yet, even so, to be so direct and accurate at times?

THE CULTURAL ORIGINS OF THE ORACLE

Let's do a quick recap. Hundreds of millions of years ago, nervous systems became capable of remembering what happened to the organism as a whole. This allowed them to evolve toward being able to simulate, in waking life and in real time, the most likely future as regards the individual's fundamental needs. The capacity to predict the immediate future can be clearly seen in a frog that takes a mosquito mid-flight by anticipating its movements. But the frog probably is not conscious of this, in the sense of having a constantly active representation of the self which is able to comment continuously on its own successes and failures and thereby create a narrative of its own life, open to editing by vanity, pride, fear, irony, compassion, or phlegmatic objectivity.

In spite of the existence of REM sleep in reptiles and birds, everything suggests that it was only in mammals that the mental state of dreaming expanded to be an active "working space" for several minutes in the sleeping animal, capable of simulating the actions of the dreaming self without waking the body. To the extent that the dominant desire is or isn't realized, the oneiric simulation makes it possible to reinforce or to inhibit behaviors based on their likely effects on the environment. By simulating objects of desire and aversion, the dream occasionally came to represent what would in fact happen. This "biological oracle," blind to the future and insightful toward the past, but still able to simulate possible futures, is more accurate the fewer the variables involved, and the greater the relevance of the prediction. In other words, the oracle works best when the number of alternative futures is limited, yet the importance of the possible different outcomes is great.

Mammals that experience a lot of REM sleep—primates, felines, canines—typically occupy high positions in the food chain, whether by having a high potential for predation (tigers), or a cooperative social organization (chimpanzees), or both (wolves). Animals that occupy low positions in the food chain sleep less and have less REM sleep than predators. It is hard to devote much time to sleep when you are being hunted.[10]

In addition to long periods of REM sleep, primates, felines, and canines are characterized by the occurrence—and especially when young—of games with objects and other animals. These games, which we immediately associate with the human behavior of play, are heightened simulations of reality, interactive representations of something absent as though it were present. If in a dream the make-believe is the totality of the experience, in waking play—which is loved both by children and by tiger cubs—the imagining of reality is only partial. The great capacity to play and the immaturity of the nervous system at birth combine to allow mammals safely to practice many specific skills that are dangerous in real life. The tiger cub does not learn how to hunt buffalo by hunting buffalo, but by playing at hunting with its litter companions. The imagination is a mental space that is protected, which is particularly useful for learning high-risk skills. The offspring of the most intelligent and creative mammals are also those that spend the longest time programming their brains before exposing themselves to the risks of adult life.

The capacity to imagine gave us a decisive evolutionary advantage, and it is at the origin of human consciousness. One cortical area essential to the imagination is the BA10, in the frontal lobe. This is the largest histologically well-defined area in the human cerebral cortex, which underwent accelerated evolution in the history of our species, being much larger in humans than in other simians.[11] The BA10 area is necessary for carrying out several tasks at the same time, keeping on standby some imaginary acts that might later become real.[12]

Among other things, the capacity to imagine allowed us to expand and deepen the reliable simulation of the mental states of many other individuals, a capacity that is well developed among primates generally, but taken to extremes of sophistication in hominids. Successfully imagining what others think and feel depends on having a mental model of each particular person, a dynamic representation of that individual's typical actions and stances, with likelihoods of occurrence of specific behaviors defined by past experiences with that person. This ability gave the bands of biped primates unprecedented efficacy in group behavior, which is as important while hunting as it is when taking flight.

Carrying this evolutionary supposition to its logical conclusion, the genesis of the oracle of night would have happened in three distinct stages. A first moment saw the evolution of molecular and neurophysiological mechanisms capable of encouraging the reverberation of memories and their long-term storage, through slow-wave sleep and REM sleep, respectively. The promotion of the restructuring of memories depends on the interaction of these mechanisms, and must date from the same period. Considering what we know about the animals that exist today, it is likely that this happened right at the start of the evolution of land vertebrates, around 340 million years ago. As a result of the operation of these mechanisms, when an animal awoke from its sleep, it was better adapted to its environment, in a way that was unconscious but effective.

A second stage, possibly at the start of the evolution of mammals 220 million years ago, brought the evolution of a longer REM sleep that could stretch to many minutes, reaching a duration in some species three hundred times longer than that found in birds and reptiles. This created the conditions for the electrical activation of long sequences of memories, the biological substrate of dream narratives. The oneiric oracle began to take shape, since the reverberation of memories during REM sleep reflects not only those experiences that have already been lived but also those that are desired. In this second stage, shared by all mammals to a greater or lesser degree, the oracle was still unconscious, but its impact on waking life became potentially large, owing to the recollections of dream reality carried into waking. The dreams of mammals, unlike those of birds and reptiles, became a mental space for the fusion, fission, and evolution of memes, a cauldron of symbolic representations that could truly simulate possible futures. In the words of the neuroscientist Jonathan Winson, "Dreams were never designed to be remembered, but they are the keys to who we are."

This stage of mental functioning corresponds to the concept of primary consciousness, defined by the U.S. biologist Gerald Edelman, who won the 1972 Nobel Prize in Physiology or Medicine for his fundamental discoveries about the chemical structure of antibodies before transforming himself into an influential neuroscientist in the second half of his career. The primary consciousness is the mental representation of the *now*, with its fleeting emotions, sensations, and perceptions, fully alert in the present, but with only diffuse access to the past or future. It is the means of mental functioning that is prevalent among mammals, which are structurally and behaviorally very diverse but all equipped with neural circuits for sensorial perception, motor action, and the processing of short-term memories.[13] These circuits also include the DMN, the activation of which is crucial for the experience of dreaming.

Edelman proposed that the brain is the dynamic product of a constant competition between groups of neurons and their synapses, which are positively or negatively selected according to the interaction with the environment. Edelman's theory was given the name "neural Darwinism," being clearly inspired by mechanisms analogous to those that operate in the immune system and in ecological interactions that shape the evolution of the species.[14] To Edelman, the brain more closely resembled a jungle than a computer. One important aspect of this conception of the nervous system is that the neurons compete among themselves for access to the neural activity and for those substances that are necessary for metabolism. This creates the basis for seeing the development and maturation of the nervous system as a product of the competition between distinct neural populations. From there to the idea that thoughts also compete with one another is just a short hop.

According to Edelman's conception, other animals are lacking in the secondary consciousness that characterizes us, a means of mental functioning based on the interaction between representations of oneself and others to generate counterfactual simulations of possible or probable alternative futures.[15] This skill allows us to go well beyond the present, as we are able not only to go through an experience but also plan it and continuously assess it. If the imagination is a dream that is directed by the conscious will but of low intensity, a waking dream kept faint by the bombardment of perceptions, then an actual dream can be much more intense, even if it is not directed by conscious desire. But what, after all, is consciousness?

In order to understand the mechanisms that produce conscious experience, the French neuroscientists Stanislas Dehaene, Lionel Naccache, and Jean Pierre Changeux carried out a series of extremely revealing experiments that have become classics. They showed that when a person is stimulated with images that are very faint, on the border between perception and non-perception, what determines whether a specific image is seen consciously is the spread of the neuronal activity to quite remote areas of the cerebral cortex, very distant from the regions that receive the sensory inputs from the retina and other sense organs.[16] During the first two hundred milliseconds following the stimulus, neuronal processing happens in really specific, spatially restricted processing networks dedicated to the sensory modality of the stimulus (sight, hearing, etc.). In the next interval of time, nearly a second after the stimulus, the activation can be reduced until it disappears or, on the contrary, it can spread. When it disappears, the image never gets to be consciously perceived, and we say then that the stimulus was subliminal. However, when there is a spread in activity to almost the whole cerebral cortex, the image comes to be perceived consciously. Curiously, in schizophrenic patients, the subliminal process is preserved, but conscious access is reduced.[17]

Among the various theories that attempt to explain consciousness, it is the Global Neuronal Workspace theory, formulated by the Dutch neurobiologist Bernard Baars[18] and extended by Dehaene, Naccache, and Changeux,[19] that explains most experimental findings.[20] According to this theory, conscious experience corresponds to the "ignition" of a vast circuit of neurons distributed all across the cortex, in a transition from multiple isolated parallel processes to a single global process, in which all parts have access to information from the whole. The concept mimics the grid computing that was developed from the 1990s, in which connected machines can share information and carry out cooperative processing, recruiting other machines depending on their availability. In the brain, this work is carried out by neurons from the more superficial layers of the cortex, which possess extremely long axons that are capable of disseminating activation quickly. When the threshold of cortical activity spreading is crossed and consciousness is established, it becomes possible to stabilize any mental object for as long as is necessary, through the feeding back of neuronal activity that selectively amplifies the relevant information.

If the difference between conscious and unconscious thoughts is the greater or lesser cortical spread of electrical activity, how are we to interpret the fact that during REM sleep there is a huge spread of this activity in the cerebral cortex,[21] much more than was believed until recently? This discovery supports the hypothesis that REM sleep played a key role in the move from primary to secondary consciousness. The journey was evolutionarily a long one, since the lifestyles of octopuses and leopards have more in common with each other than they have with us. Even if we are much closer to our fellow mammals than we are to mollusks, our mental software differs from them all thanks to the presence of secondary consciousness.

SPEAKING AND HEARING

Edelman's definitions for primary and secondary consciousness are essentially the same as those put forward by Freud between 1900 and 1917[22] in proposing the concepts of id and ego, respectively. The influence of psychoanalysis was neither accidental nor unconscious, despite the disparagements suffered by Freud in the biomedical field. The dedication to *Bright Air, Brilliant Fire*, Edelman's important 1993 book on consciousness, testifies to this: "To the memory of two intellectual pioneers, Charles Darwin and Sigmund Freud. In much wisdom, much sadness." Darwin clarified our evolutionary continuity with the other animals,[23] including in our emotions.[24] Freud observed that the move from primary to secondary consciousness happens most of all through the acquisition of speech, that is, as the representation of things moves toward the representation of the names of things: from the imagetic to the semantic.

The Gospel according to John states that in the beginning was the Word . . . but where did words come from, after all? Even if vocal communication is spread widely among land vertebrates, only very specific groups of animals are able to learn the signs that are used in those interactions. Chimpanzees living in the wild produce complex mixtures of sounds and gestures that science is gradually beginning to unlock.[25] In captivity, our closest cousins learn to use arbitrary signs to refer to dozens of different objects and actions,[26] enormously expanding their ability to communicate with humans. Nevertheless, some skeptics have argued that this does not represent real symbolic communication, but rather a functional communication based on learning the specific rules of the experimental setting.[27]

Classic field studies of the spontaneous communication of green monkeys (*Cercopithecus aethiops*), our distant cousins on the African savannas, were the first to show that there is no reason to doubt the presence of symbols beyond the human species. Green monkeys naturally present three types of alarm call, corresponding to the presence of land predators, air predators, or predators that slither along the ground. On hearing the alarm calls made by an adult, other adults react speedily to protect themselves, hiding themselves up trees in the case of land predators such as lionesses, below trees in the case of air predators such as eagles, or leaping out of the way to scan the ground around them, in the case of snakes. Adolescent green monkeys are capable of vocalizing the same sounds, but they do this out of the appropriate context, not producing any flight response from the adults. Field experiments have shown that the green monkey alarm system fulfills the criteria for symbols in the strictly semiotic sense, as conceived more than a century ago by the U.S. philosopher and mathematician Charles Sanders Peirce.

In Peirce's semiotics, the interpreter of a sign is informed about the corresponding object according to three and only three possibilities of representation: icon, index, or symbol. Icons convey information through similarity to the object, indexes convey information through spatial-temporal contiguity to the object, symbols convey information through social convention.[28] In order to refer to the object "lion" using only an icon, it would be necessary to show a photo, movie, or drawing of a lion, or play the sound of its roar, or disseminate its smell. To use only an index, you would need to point at a lion. To use only symbols, we could say or write *ngonyama, libaax, simba, león, leão,* or *lion,* in the Xhosa, Somali, Swahili, Spanish, Portuguese, and English languages, respectively. While icons and indexes of lion can be understood generally and possess something intrinsically leonine about them, symbols are totally arbitrary and function only among those people who share the code for deciphering them.

The African green monkeys' system of vocal communication is quite a clear example of the use of symbols in animals that are not human. Among the young monkeys, we see a gradual learning of the context of an appropriate use of the vocalizations, through multiple repetitions of the pairing between visual/olfactory stimuli from the predator and auditory stimuli from the alarm calls made by the watchful adult individuals—followed by the group taking flight. The alarms paired with specific predators initially function as indexes of their presence, but with time, and through multiple repetitions, the youngsters gradually internalize their elders' social convention to interpret these alarms.

And then the move into the symbolic happens: the animal no longer needs sight or scent of the predator for it to seek shelter: a vocal alert is enough. This was demonstrated in classic field studies carried out four decades ago by the U.S. ethologists Dorothy Cheney and Robert Seyfarth. Using loudspeakers to reproduce alarm calls in the middle of the African savanna, Cheney and Seyfarth documented adult green monkeys reacting correctly depending on the particular type of vocalization they were presented with, even without the presence of any predator. This shows the symbolic nature of this communication, since the meaning is transmitted in the absence of the object.[29]

Since the initial discovery of symbolic calls in green monkeys to warn against predators, published in 1980, similar alarm systems have been found in other African primates, such as Diana monkeys, Campbell's monkeys, and chimpanzees, as well as a large variety of non-primate species, including dwarf mongooses, prairie dogs, squirrels, chickens, and meerkats. In addition, bottlenose dolphins are able to learn and interpret human gestures as symbols for the parts of their own bodies.[30]

Computer simulations of artificial creatures representing the interactions between vocalizing prey and three types of predator—running on land, slithering along the ground, and flying in the air—suggest that the code that assigns specific meaning to each type of call arises spontaneously in populations that are equipped for multiple vocalizations, through random variations in the stimulus-vocalization pairing that end up being established and sustained long-term.[31] However, this only happens when the proportion of prey to predators is large enough for the population of prey to survive sufficiently long to pass the referential code around.

ARGUMENTS, NARRATIVES, AND CONSCIOUSNESS

The use of symbols is not, therefore, exclusively human. Referential communication in various non-human species corresponds in Peircean semiotic terms to the concept of dicent symbol, which functions as an index because "its object is a general interpreted as an existent."[32] Through the repetition of the index in the physical presence of the predator ("an existent"), a memory of the association between vocalization and predator is eventually formed, which makes it possible to evoke it symbolically even in its absence ("a general"). In the area of semiotics, what distinguishes human language from the communications systems of other species is our incredible capacity to string symbols together with other symbols, creating potentially infinite chains of representations of representations of representations, corresponding to a composite symbol that Peirce called an "argument."

While countless animal species do use sequential vocalizations to communicate, there is little evidence of meaning being assigned to the order of the vocalizations. The capacity to produce complex arguments through the combination of simpler vocalizations does seem to be extremely rare and possibly exclusively human, if we don't count the examples of suffixes and other sequential modifications found in African animals such as hwamei birds and some primates, including chimpanzees.[33]

The gradual developing of our repertoire of icon vocalizations (onomatopoeia), index vocalizations (demonstrative pronouns), and symbolic vocalizations (nouns, verbs) took hundreds of thousands of years of slow evolution, until we had become the most fearsome predators on the planet. It was not superior claws and teeth that granted us this position, but our effective communicating, our social organization, and our weapons. Hunting in packs using spears and arrows required excellent coordination at a distance, which our ancestors carried out through vocalizations and gestures.

The role of language in human evolution is undeniable, but it is obvious that there are still many pieces of this puzzle missing before we are able to understand the accelerated process by which symbolic repertoires with very limited meanings produced the explosion of referential richness of today's languages. From *lion* and *zebra* to proper names like *Enheduanna,* from simple verbs like *walk* to words like *why, soul, zero,* and *Internet,* a vast number of mental processes went on, compressed into a fairly short period of time relative to the species' anatomical evolution. The shift from the world of icons and indexes to the use of arbitrary symbols and their sophisticated arguments corresponded to an ever-increasing assigning of weight to somebody else's opinion. You don't need to see the lion, it's enough to hear the vocalization produced by somebody who has seen it. The meaning of the signs became ever more dependent on social consensus, creating a human overvaluing of collective beliefs, rooted in the expansion of the ability to simulate and predict the mental states of other individuals—what in neuroscientific jargon is called "theory of mind."[34]

The cognitive leap toward symbolic-argument language changed our interaction with the world forever, radically altering our relationship to dreaming. At some point in the Paleolithic, reports began to appear of experiences that people had had, both in their waking lives and in their dreams. What had previously been a strictly private experience, capable of influencing the dreamer's emotions and actions without anybody else being able to know about it or understand, gradually came to constitute a collective one. The gathering of the clan around the fire to share their experiences from waking and dreams gradually boosted the expanding of vocabulary, the growth in empathy, and the start of memorializing the history of the clan, through tales of the deeds of the ancestors. Memes became ever longer and more complex, forming huge collections of memories that included increasingly sophisticated representations of past and future events, of landmarks, of new words, and of people who had already died. This was a fundamental condition for the emergence of the concept of family lineage, the affective basis of the timeline that recalls the origin of the clan.

Which brings us finally to the third critical moment for the emergence of our consciousness: the birth of a new mental universe that is engaged not only with the present but also with the past and the future, inhabited by ancestors and by the spirits of animals that were dangerous but also very tasty, creatures that were desired and feared, killed or for killing, capable of firing up the attention of our ancestors to the point that they were obsessively depicted in cave paintings.

BREAKING STONES

Every animal has, as the horizon of its future, its next meal, the next attack from a predator, the next mating. But hominids took a leap when they began to work with thoughts about thoughts, using mental objects as tools to operate on other mental objects and thereby simulate not only reality but their actions upon it. Anticipating the movements of the large migratory herbivores over the course of the seasons of the year, Paleolithic hunts often involved the cornering of the animals or driving them recklessly to their deaths off the edges of cliffs.

The ability to imagine the future and combine mental objects was also what allowed for the development of the chipped stone technology that made it possible to confront, slaughter, clean, and butcher their prey. The exhausting work of constructing weapons of stone requires at least four kinds of imagined thing: the desired shape of the stone, the movement of the body required to attain this shape, the movement of the body required to kill with this weapon, and the final effect of the whole thing: the feeding of the group. The activity of gathering plants, mollusks, and insects also requires imagining how to find them and extract them from their shelters or burrows. Capuchin monkeys use stones to break coconuts and shells. Twigs are used as tools by all simians, as well as by crows. Dolphins use sponges. The novelty that appeared in the human lineage was the successive coupling of tools: the poly machine. This process was very slow at first, the cultural accumulation from one generation to the next almost imperceptible.

It is hard to appreciate the extraordinary span of time that this process requires, given that the whole history of the species fits into a footnote of prehistory. From the rudimentary chipped stones of Oldowan technology, which began around 2.6 million years ago, to the biface hand-axes that characterized Acheulean technology, starting around 1.7 million years ago, there was space for a vastness of hours in which the cultural accumulation each generation passed to the next was almost nil. Between then and the Mousterian technology, characterized by the elaborate production of sharpened points and multiple cutting surfaces, starting around 160,000 years ago, another near eternity was spent on the hard toil of chipping stones for obtaining tools. And yet, despite the gigantic scale of this cultural inertia, the progress did happen. Ways of thinking that were ever more complex did slowly evolve, transforming human life forever. The discovery that cave painting occurred in acoustically different spaces depending on the kind of animal represented, with an attenuating of sounds in the case of predators and amplification of sounds in the case of hoofed prey,[35] suggests a sophisticated combination of art, technique, and magic, in order, through the manipulation of echoes, to motivate our Paleolithic ancestors to carry out their dangerous hunts.

From the first rough stones of three million years ago up to the sharp points found shortly before the advent of metallurgy, around forty thousand years ago, there was a long process of acquiring specific manual movements capable of producing surfaces for cutting, piercing, and concussing. Countless times, the culture of individual groups was lost to the darkness of predator defeats, food shortages, floods, and droughts. Technical perseverance and improvement occurred in the comings and goings of the cultural transmission of the Paleolithic, at the beginnings of the human cultural ratchet.

The optimization of chipped stone technology took around three million years to be coupled with a rod, transforming it into a spear. It must have been immensely hard to stabilize the stone at the end, making it firm enough to be able to pierce the resistant skin of an auroch, the huge ancestor of bovines that became extinct less than five hundred years ago. The invention of the spear and the development of a rich and flexible verbal communication, able to organize hunts in real time but also to plan them, using the landscape as part of the traps made up of shouting, movement, and fire, led our ancestors to the top of the food chain. Humans became so deadly that few of the megafauna of the Pleistocene have survived today.

After the invention of the spear, it took our ancestors another four hundred thousand years to get to another revolutionary tool, in which at least three elements need to function together: the wooden bow, the stretched string, and the true arrow. Who could have come up with this idea for the first time? The oldest evidence goes back at least ten thousand years. Was it a nighttime dream, or a waking daydream? We will never know, but the fact is that the idea spread quickly across almost every continent.

In summary: the trajectory of human development is characterized by the complexifying of tools and of the internal mental states that devised them. Over the course of this long journey we developed a rich vocal language based on the generation of new signs for the combination and juxtaposition of elements. The human self does much more transforming of the reality around it than the self of other mammals. Even if the capacity to dream did create the bases for some awareness of the self in a variety of species, it was the capacity to describe one's own experiences, coming both from waking and from dreams, to oneself and to others, that led to the narrative of group cohesion, with its foundation myths, its repertoire of exemplary stories, and its day-to-day gossip.

The fully fledged use of dreams as a probabilistic oracle dates from this third moment, maybe tens, maybe hundreds of thousands of years ago, in which our hominid ancestors found themselves equipped with enormous collections of memories transmitted from generation to generation—memes. Experience came to be culturally inherited in the form of representations of people and the knowledge associated with them, through oral accounts, songs, tombstone arrangement, and paintings, statues and other icons. It was the reverberation of these representations during sleep and then during waking that led to divinatory dreams, whose effects on reality came about both from intrusive, spontaneous reverberation and from voluntary, effortful remembering.

A MENTAL STATE CAPABLE OF SIMULATING LIFE

The capacity to dream, in parallel with waking life, enabled imagetic simulations occurring at varied timescales and, what is more important, uncoupled from the musculoskeletal system. It made possible a hidden, internal space for mental work, capable of simulating the conquest of objectives, situations, and probabilities of outcome, safely and with no interference in real behavior, with no limits to the complexifying of the natural or social relations involved, with no limits to the horizon of the future under consideration. What we call intentional or voluntary actions are behaviors guided at each moment by anticipatory simulations that allow us to make decisions based on their expected results. Dorsal and ventral areas in the cerebral cortex harbor the incessant flow of activity that makes up these simulations and sustains them. When this process is working well, it creates behaviors that are better adapted and so more likely to be transmitted from one generation to the next.

How important imagining the germinating of a seed must have been to make people start sowing deliberately! How important imagining the coming seasons and the phases of the moon must have been for people to choose the time to plant and harvest! The abundance of ideas and material goods launched the reign of multiple small necessities, much beyond the Darwinian imperative of kill, survive, and procreate. Dreams became symbolically richer, but the dream oracle began to find it harder to guess at the immediate future, precisely because of the explosion of combinational possibilities. On the other hand, conscious oracles began to flourish based on accounts of dreams that were shared and interpreted in the light of the existing cultural hoard. The increased awareness of the dream contents allowed our ancestors to construct models of the visible and invisible world, in the hope of trying to reduce errors in predicting the future.

It is important to remember that electrical reverberation is inherently noisy, and neural circuits operate on associations of different kinds, including symbolic ones. As a result, a dream's manifest content is rarely the same as its latent content. Thus dreams that are direct and unambiguous to interpret are rare, and those that are indirect and ambiguous are frequent. As culture developed, with the growth of vocabulary and the development of increasingly rich and diversified memes, the scope of life expanded, and the dream oracle needed to consider an ever growing number of variables. Moreover, the content that was unconscious at the time of the dream could quickly be brought into the consciousness and shared for collective elaboration by the members of the group, since it could be spoken, summarized, reviewed, painted, drawn, and—as of 4,500 years ago—recorded in written words.

It was in this third stage that dreamers began to be conscious of the oracle. They were now able to name it, and to ask for revelations through it. It was only in this third stage that the dream became a central object not just of human attention, but also of human communication. It was through narratives about the past and future that we accumulated and disseminated human culture, that monstrous evolving marvel, an immense force of knowing that took us out of the caves in just a few thousand years and threatens to take us to Mars before we have even learned to inhabit our own planet in peace. And among all these narratives, the most valuable, the most longed-for and respected were the divinatory dreams of chiefs, shamans, and priests about ancestors, totemic animals, and divinities.

CREATURES OF THE MIND IN THE HIPPOCAMPUS

In what part of the brain are these supernatural beings represented? The hippocampus receives information from multiple senses and plays a defining role in the codification of complex representations. While the existence of representations of space and time have already been demonstrated in the hippocampus of rodents, as well as specific responses to objects and other individuals of the same species, in human beings the question is much harder to tackle, mainly because of the practical and bureaucratic obstacles to obtaining neuronal records in humans. The question remained mysterious until 2005, when the Argentine neuroscientist Rodrigo Quian Quiroga of the University of Leicester made a fundamental discovery in epileptic patients. It is common for these patients to be admitted for several days to monitor their brain activity, in order to carry out detailed mapping of the epileptic focuses so that they might be surgically removed with as little neural damage as possible. Taking advantage of this window of opportunity, Quiroga and his team examined the activity of the neurons of the temporal lobe—which includes the hippocampus—in patients who were stimulated with photographs of people, animals, objects, and buildings. The researchers discovered that some of the neurons recorded would be vigorously activated when the patients were stimulated with images of objects or especially of specific people or characters, such as Bill Clinton, Halle Berry, Luke Skywalker, or even Bart Simpson.[36]

This phenomenon occurred despite the large variety of stances and garments in the pictures, as well as the multiplicity of complementary elements. In addition, it was also possible to evoke the desired responses through the names of the characters, whether written or spoken. The cells discovered by Quiroga showed themselves able to learn, becoming sensitive to new stimuli by association with another favorite stimulus. This seems a plausible mechanism for explaining the associativity of the flow of thoughts in which one image leads to another and so on in succession, via quite idiosyncratic pathways.

Quiroga's research was the first proof that neurons in the human temporal lobe can have their activity linked to specific people, whether real or fictional. The results suggest the existence of a sophisticated mechanism for representing individuals and objects in a broad, flexible way. The fact that these representations do not change despite so many contextual differences suggests that they possess a high degree of autonomy and internal consistency, representing real "creatures." The "inside" also has its own internally represented "outside."[37]

The different cortical areas activated during imagination participate in the codification of the different qualities of the objects imagined, as well as the intention to evoke them. When combined, these hippocampo-cortical circuits can recombine memories in a flexible way, to imagine both alternative pasts and possibilities for the future. Some of these same regions, notably the hippocampus and the medial prefrontal cortex, are also activated during REM sleep. The dream inhabits the interface between yesterday and tomorrow, with the potential to have a powerful impact on the dreamer each time they wake. It is therefore plausible that human consciousness in particular, with its immense capacity for recounting the past and imagining the future, stems from an invasion of waking life by dreams. The first mental space for simulating ideas must have been dreams, long before our ancestors learned to do this while awake.

The gradual expansion of the ability to tell stories and travel mentally in time was the fuel for the human cultural explosion of these last millennia. Unlike other simians, who have a limited sense of the temporal dimension, our ancestors became progressively more able to predict the best time for going on a hunt, the best day for gathering fruit, the best month for planting or harvesting. The fact is, at some moment in our recent history, we started to be able to formulate brief narratives of the future based on the past. The capacity to remember and recount longer and longer chains of thoughts, coupled with the active imagination that finds it easy to symbolize, made it possible to develop more and more complex plans with the simulation of more and more variables, further and further into the future. The narration of human existence expanded people's capacity for memorizing, increasingly rich meme repertoires were constructed, and culture was formed and expanded through the accounts of people's lives and deaths.

NECROPHILIA AND CIVILIZATION

In hindsight, the journey from monkey to human was marked by increasing necrophilia. Even if the social norms of human mourning have varied widely over time and space, the lamentation and wonder when faced with death are widely prevalent behaviors in our species.[38] Their origins might go back to the common ancestors of *Homo sapiens* and other primates, and maybe even long before that, as there are even descriptions of the phenomenon in elephants and dolphins.[39] Nevertheless, it is chimpanzees and gorillas that most clearly show the reluctance and sadness involved in the act of separating oneself from the dead bodies of one's relatives. The naturally mummified bodies of chimpanzee babies and children can continue to be tended to by their mothers for days or even weeks after the death, getting carried around and cared for as though they were alive. Mothers share their space in the nest with their dead children and show obvious distress when separated from the bodies. Violent deaths of adults tend to cause frenzies, while the natural wasting away of elderly individuals can be accompanied by pre-death care, regular inspections of their bodies, searches for signs of life, aggression, or cleaning the corpse, protracted remaining of the offspring close to the body, or, to the contrary, avoiding the place where the death occurred.[40] Similar though simpler behavior appears in primates that are more distant from us, such as the gelada,[41] a sturdy

Ethiopian monkey akin to the baboon. These behaviors significantly resemble human beings' responses when they encounter the death of a loved one, indicating phylogenetic continuity in primate mourning.

However, unlike what we find in other animals, it is common for humans to keep the dead close to the living for years or decades, buried or kept in homes or their surroundings, on altars and in shrines, inside villages or on their outskirts as well as in special geographical features, those sacred trees, rocks, caves, waterfalls, and mountains that house imaginary beings. The capacity to imagine what others feel and think was projected onto animals, plants, and lifeless things alike, making up a theory of mind that is free to attribute intentionality to any object, animate or inanimate. Surrounded by dangerous predators and by prey that were necessary for survival, our ancestors started to wake the human consciousness through cosmogonic narratives that frequently mixed men and animals to explain events.

The myths about the origin of the world, which are very recent in the evolution of our species, derive from the unprecedented expansion of our capacity to mentally represent real and imaginary beings, both humans and wild animals, that were syncretized with our ancestors. The neurophysiological facility for recombining memes in our dreams must have contributed to this zoomorphism—the mixing of people and beasts—which has been observed in our culture since that time. Indeed, the mixture with other creatures, plants, and geographical features was almost inevitable, since in a dream there is nothing to prevent representations from blending. Naturally, a fantastical mental fauna presented itself on countless mornings to the waking consciousness of our dumbfounded ancestors. The result was the widespread prevalence of zoomorphism in human culture, animals mixed with people like the Lord of Beasts from the Paleolithic, the powerful Egyptian god Anubis, the Great Sphinx of Giza, the Cretan Minotaur, the Hindu god Ganesh, or Sagittarius in the zodiac (the "circle of animals"). But this is not just a primitive trait with no contemporary correspondences, as zoomorphism also rules among soccer team mascots and Walt Disney characters. We have been wild animals for as long as we've been people.

SUBJECTIVITY IS BORN OUT OF NOSTALGIA

According to the Brazilian anthropologist Eduardo Viveiros de Castro, from the National Museum of the Federal University of Rio de

Janeiro, "Amazonian ideas about 'spirits' do not point to a class or type of beings, but to a disjunctive synthesis between the human and the non-human."[42] The first gods were probably combinations of ancestors and animals, producing the animism, totemism, and genealogical myths of so many traditional cultures.[43] Given the scarcity of objective data about this phase of human mental evolution, we can only look at it in current hunter-gatherer populations, whose self-denominations are almost always synonymous with "real people." This way of life was predominant from the oldest hominid bipeds seven million years ago, until very recently, between eleven thousand and seven thousand years ago, when the gathering of wild grain developed into agriculture. Today's hunter-gatherers, who are nomadic or semi-nomadic, many of them seasonal or occasional practitioners of agriculture, hold essential keys to our understanding of the emergence of human consciousness. Their way of life, which is older than the oldest unit of time, runs all the way through our transition from wild animals to people.

In Amerindian and Siberian cultures, there is a belief that shamans are capable of shape-shifting, taking on the body of a panther, a wolf, or a bird. Among the Huaorani of the Ecuadorian Amazon, for example, the shaman adopts a jaguar-spirit and meets with it perilously in dreams—while asleep or having consumed ayahuasca—to receive guidance about the hunt.[44] These encounters occur in the realm of what anthropology calls perspectivism, according to which the world is inhabited by a huge variety of human and non-human subjects with very different and reciprocal points of view.[45] All animals are not simply and equally endowed with souls, as in the original conception of animism,[46] nor does each people's humanity end at its borders, as intended by the most radical ethnocentrism. Each species would be a center of consciousness with its own perspective, in such a way that the same criteria used by the members of a population to distinguish themselves from another indigenous group would be applied by animals to humans and other animals.[47] Thus, just as an indigenous person would consider themselves human or jaguar while they are hunting wild pigs, a jaguar would conversely see itself as jaguar or even human[48] when hunting an indigenous person—who would be a wild pig to them.

In the words of Viveiros de Castro, in several different Amerindian cultures, "Whatever possesses a soul is a subject and whoever has a soul is capable of having a point of view."[49] He continues:

Animals are people, or see themselves as person. Such a notion is virtually always associated with the idea that the manifest form of each species is a mere envelope (a "clothing") which conceals an internal human form, usually only visible to the eyes of the particular species or to certain trans-specific beings such as shamans. This internal form is the "soul" or "spirit" of the animal: an intentionality or subjectivity formally identical to human consciousness, materializable . . . in a human body schema concealed behind an animal mask . . . This notion of "clothing" is one of the privileged expressions of metamorphosis—spirits, the dead and shamans who assume animal form, beasts that turn into other beasts, humans that are inadvertently turned into animals . . . This perspectivism and cosmological transformism . . . can also be found . . . in the far north of North America and Asia, as well as amongst hunter populations of other parts of the world.[50]

In those pre- or semi-agricultural cultures, predation appears as the main key for constructing the self and its social relationships, through physical or symbolic appropriations for personal gain. However, as the world is dominated by predatory relationships, there is the possibility of a reversal of perspective, that is, the hunter can always become the hunted. Since life is conceived of as the constant struggle to impose one's own point of view on beings endowed with spirits and with points of view of their own, it is believed that the link between predator and prey continues even after the violent event, with consequences for both. It is common for the hunter to carry out rituals to appease the spirit of the slaughtered prey, avoiding its vengeance. If there is not exactly guilt, there is a fear of retaliation, the panic of turning from predator into prey through the actions of that dead creature still living in the imagination. As the shaman Davi Kopenawa stated:

Animals are also human beings. This is why they turn away from us when we mistreat them. In the time of dream, I sometimes hear their unhappy and angry talk when they want to refuse themselves to the hunters. If you are really hungry for meat, you have to arrow the game with care and it must die on the spot. If it happens like this, the animals are satisfied to have been rightly killed. Otherwise they flee far away, wounded and furious at humans.[51]

Just as in hunting and in war, one frequently finds in dreams a risk of another's perspective being imposed. For a Juruna Indian, a dream about slaughtered pigs means that their soul has been successful in the hunt and so the waking hunter will meet with the same success. On the other hand, a dream about pigs running freely through the forests means that enemies are pursuing their soul and so they will appear in the hunter's path. This would lead them to keep themselves safe for a few days, telling nobody their dream.[52] Among the Juruna of the Xingu Indigenous Park, a dream about vultures close to a person is a sign that they will die even if they appeared alive in the dream, because "vultures only eat carrion." Dreams are particularly open to the imposition of different perspectives, in this case that of the vultures.[53]

It is very likely that the belief in gods and spirits arose in a similar context. It was at first no more than the attribution of life and assorted powers to the memories of dead relatives, slaughtered prey, and defeated predators, memories with which our ancestors had intense dialogues in their dreams. With the development of civilizations, oneiric activity came to be seen as a magical portal for accessing what today in the African-Brazilian religion Umbanda is called the kingdom of Aruanda, the spiritual dimension in which the ancestors live, the embryo of the world of the gods, immortalized in the memory of multiple generations. In a sample of sixty-eight dreams collected by the anthropologist Franz Boas among the indigenous Kwakiutl people, 25 percent referred to dead relatives or funeral scenes.[54] In the words of an indigenous Pirahã man: "When we dream, we are close to the dead, we are with them."[55]

16

Missing the Dead,
and the Inner World of Culture

The foundational role of the memory of the dead for the development of culture happened somewhat accidentally, since the powerful mechanism for propagating ancestors' habits, ideas, and behavior was affect. The memory of someone who has gone, which is quite visible in chimpanzees who go into mourning when they lose a loved one, became an indelible feature of our species. This did not happen without some contradictions, of course. With the love of the dead also came fear of them. From Egypt to Papua New Guinea, at different moments and places, rituals flourished for neutralizing, pacifying, and satisfying disembodied spirits. In medieval England, the dead were so feared that corpses were mutilated and burned to ensure they would remain in their graves. Among the Yanomami, the burning of belongings is an essential part of funeral rites. The Catholic Church to this day believes the mortal remains of saints to be valuable religious relics.

The propagation of the memes of spiritual beings was therefore driven by positive and negative affects toward the dead. It was the memory of the techniques and knowledge held by deceased grandparents and parents that transformed this process into something adaptive, a really virtuous symbolic circle. It is no exaggeration to say that a fundamental driver of our cultural explosion was how much we missed the dead. The belief in the divine authority for guiding human decisions led to an accelerated accumulation of empirical knowledge about the world, in the form of precepts, myths, dogmas, rituals, and practices. Even if it was supported by all kinds of coincidences and superstitions, this belief was the embryo of our rationality. Causes and effects were being learned through the corroboration or refutation of the effectiveness of religious symbols.

The worship of the dead developed from the Paleolithic, passed through the Neolithic, and culminated in the Bronze Age, with its legacy of majestic tombs and the start of the symbolic written record of all this cultural accumulation. Religions in their countless strands are derived, then, from a technology of psychological and physiological self-regulation that was selected to optimize reproductive capacity and group cohesion,[1] a means of highly adaptive mental functioning whose success is evidenced in the hegemony of theist civilizations right across the planet.[2]

Some 4,500 years ago we find the beginning of the historical record that radically altered the speed of our species' evolution. The birth of literature took place in Afro-Eurasia at the start of the Bronze Age, in the context of the first great civilizational fusion, which involved Indo-European and Semitic peoples. Population growth, migrations, and military conquests started to unite ever-larger groups of people around similar cultural nuclei. As it encouraged the preservation of knowledge from one generation to the next, the coupling of love for knowledge and love between parents and children—memorialized through deification—was transformed into such a powerful force that it literally expelled us out beyond the stratosphere. But just as the main capsule of a rocket keeps going as it ditches all the other disassembled parts, for us to arrive at Apollo 11 we needed to leave behind, a relatively short time ago, a good part of the mental software that we used to begin the hominid consciousness revolution. In order to understand how the gods got us out of our caves, we need to understand how it was that they abandoned us—and we them.

FROM ACHILLES TO ODYSSEUS

The proto-Indo-European languages, which had their origins in Central Asia between nine thousand and six thousand years ago,[3] had by the Axial Age spread across a geographical area reaching from Ireland to India. And in all these places, in all these languages, for all these people, names with similar roots were spoken to connect dreaming and death. If gods are memes of dead ancestors, possessors of all knowledge and masters of all destiny, it is easy to understand why the use of dreams emerged to carry out necromancy and divination. Demonstrably during the Bronze Age, but probably well before this, people began to consult spiritual beings in their dreams. For this reason, the ancients were quite well aware that dreams are not necessarily reliable. While some dreams are well formed, thrilling, and even useful, others are awkward, ill formed, and frustrating.

In *The Iliad,* composed between the eighth and seventh centuries B.C.E., Achilles is visited in a dream by the spirit of his good friend Patroclus, who was brought down in combat by the Trojan prince, Hector. Achilles moves to embrace his comrade, but Patroclus disappears into the ground making strange noises. This dream with such an unsatisfying ending is revealed as merely an unfinished mental construct, nothing but disappointment. In *The Odyssey,* meanwhile, dreams appear as deception as much as they do sources of providential assistance. In Book IV, when Penelope's suitors plan to assassinate her son Telemachus, the goddess Pallas Athena appears in the queen's dream, reassuring her about her son. In Book VI, Athena appears in a dream to the princess Nausicaa, to persuade her to meet Odysseus, who is asleep and who needs her help. In Book XI, when Odysseus enters Hades's world of the dead to hear Tiresias's prophecies, he ends up meeting his deceased mother, who offers him advice. He tries three times to embrace her, but three times he is embracing merely an illusion—what seemed a divinatory dream ends in disappointment. Finally, in Book XIX, when Penelope is besieged by the suitors who have given Odysseus up for dead and he appears dressed as a beggar, she tells him of a dream from the previous night in which an eagle identified with Odysseus kills twenty geese representing the suitors. The fake beggar confirms that Odysseus will return and the next day he fulfills the prophecy, disposing of all his rivals with arrow-fire and spears.

We saw in Chapter 3 how the transition from Achilles's mentality to Odysseus's represents the move toward a consciousness that is similar to what we have today. Achilles has neither nostalgia for the past nor plans for the future. All he wishes for is the glory of the present battle and, in order to attain it, he allows himself to be totally led by Athena's commands. While Achilles is led by the voices of others, Odysseus frequently talks to himself and reverses the causality in his sphere of action. Instead of just reacting to stimuli the way Achilles does, he anticipates situations and makes the future happens as he wishes it.

By understanding how the Trojans feel and think, by understanding their beliefs and their story, Odysseus anticipates the fact that they will interpret the huge wooden horse as an offering made to the gods by the Greeks to allow them to return safely to their homes. Odysseus also predicts that the Trojans will take the horse inside their impregnable city, as an astonishing trophy of war. With this simulated future in mind, Odysseus hides the Greek warriors who are to open the gates of Troy inside the horse.

Even though he does sometimes rely on supernatural help, it is not through divine inspiration that Odysseus wins the war but through his lucid capacity to travel inward to imagine himself in someone else's place. He imagines that the Trojans have a mind not unlike his own and that they ought to react to the offering in a predictable way. Only with theory of mind, by imagining what other people think and feel, can Odysseus lie and deceive, since this requires assuming that others—the Trojans—are psychologically similar to him, even if they do not know what he knows.

The Homeric narrative of the Trojan War, perhaps reflecting a specific siege in the twelfth century B.C.E. in Anatolia, perhaps an amalgam of multiple Mycenaean incursions into Asia Minor, is an important account of the extensive civilizational collapse that marked the end of the Bronze Age and the beginning of the Iron Age. In the space of just three centuries, powerful city-states and whole empires disappeared temporarily or definitively in Afro-Eurasia, including Troy, Knossos, Mycenae, Ugarit, Megiddo, Babylonia, Egypt, and Assyria. The divine plan was shaken up by overpopulation, by more deadly weapons, more frequent wars, naval invasions, land migrations, a fall in literacy levels, deadly plagues, food shortages, famine, and social chaos. The archaic system of belief in the gods, endowed with Paleolithic roots and supported for thousands of years by a superstitious causality, began to crumble.

In this situation of profound social crisis, dream oracles could no longer provide adaptive responses to the countless and ever-more-unpredictable problems of reality, a mixture of those societies' multifaceted, sophisticated problems with the dry old tripartite logic reborn in the new times ahead: kill, survive, and procreate. Kings and generals found themselves deprived of guidance for their actions, as they could no longer hear the wise voices of their deified ancestors. The neural reverberation of divine memes was even further undermined by the dissemination of the written word, capable of traveling in time and space and talking to a reader without that reader even needing to hallucinate supernatural voices. The literature of the end of the Bronze Age provides copious records of people bewailing the gods' silence. These divine voices, which had previously been so ready to command, had now gone quiet, and the members of the human race had found themselves alone in their own minds. It was only after this collapse (c. 1200–800 B.C.E.) that the Axial Age began, and the human consciousness that resembles ours today was awakened. In 326 B.C.E., when Alexander invaded the north of India, Indo-European and Afro-Asiatic languages were already evolving with shared ideas of religion, government, commerce, money, and literature. Since that time, we have gained greater and greater rational control over the world, and dreamers have begun to lose their close connection to oneiric reality. Bit by bit, we have begun to see weirdness and awkwardness where previously there was enchantment and mystery.

THE ORACLE ENDURES

It is not that dreams lost their status quickly and completely. Throughout antiquity, the oracle of night maintained a prominent place in private life and public administration, as Greco-Roman culture amply demonstrates. The exemplary dreams of Julius Caesar and Calpurnia on the night before Caesar's assassination, which had very great significance for the political organization of Rome, were both precognitive in their own terms: his the metaphorical ascent into the heavens to be with Jupiter, an ecstasy and a sublimation of all worldly troubles; hers a terror of concreteness marked by perfect prediction and cruel prophecy, a "theorematic" dream in Artemidorus's rich terminology.

Faith in the divinatory power of dreams is not limited to extinct societies or so-called primitive ones. To this day, in the city as well as in the country, it is hard not to find somebody who will interpret a dream as a warning or premonition worthy of guiding marriages, travels, the buying and selling of buildings, contracts, and monetary bets. In Brazil, the custom of betting on a particular animal in the *jogo do bicho* gambling game, especially popular among blue-collar workers, following an appearance by that animal in a dream is widespread. Searching on

Google for the combination of the terms *sonho*—the Portuguese word for a dream—and *jogo do bicho,* I turned up 350,000 relevant pages. A news story published in one of the main newspapers in the state of Pará, which is widely read in the Amazon region, helps us to understand the excitement:

> Self-employed man Paulo Roberto da Silva, 46, plays the Jogo do Bicho every day . . . In this gambler's imagination, anything can turn into an animal. "I dream about anything, I translate the dream then I come and play. Everything serves as inspiration for my bets, even the clouds, depending on their shape . . . I even won 1,200 Reais once, and I hope to do it again.[4]

This phenomenon is in no way limited to the working classes. In late 1913, when he was traveling by train to visit a relative near Zurich, Jung had a disturbing oneiric vision: he saw all of Europe bathed in a sea of blood, a monstrous flood dotted with corpses. It was only the following year, when the butchery of the First World War broke out, that Jung realized how precognitive that dream had been.[5] Decades later, in an extraordinary déjà vu, Jung would anticipate the rise of Hitler and the disastrous dominance of Nazism through the analysis of the dreams of his German patients.[6] These accounts suggest that cultural history can be predicted, as it develops through the updating of archetypal memes over the centuries. In Jung's words:

> I was sure that something was threatening in Germany, something very big and very catastrophic, and I only knew it through the observation of the unconscious . . . When you observe yourself within, you see moving images, a world of images, generally known as fantasies, yet these fantasies are facts. It is a fact that a man has such and such a fantasy, and it is such a tangible fact, for instance, that when a man has a certain fantasy, another man may lose his life . . . Everything was fantasy to begin with, and fantasy has a proper reality, that is not to be forgotten, fantasy is not nothing, it is of course not a tangible object, but it is a fact nevertheless . . . Psychical events are facts, are realities, and when you observe the stream of images within, you observe an aspect of the world, of the world within.[7]

If the internal world is as real as the external one, then it is necessary to see premonitory dreams as facts of nature—which does not mean that their interpretations are also natural. Denying the effectiveness of the probabilistic oracle is as risky as believing devoutly in its premonitions. The BBC reported that Florin Codreanu, a Romanian living in England, who suffered from recurring nightmares, strangled his own wife upon waking up enraged from a dream in which she was betraying him. Despite citing the dream as the motive for the crime, the man was sentenced to life imprisonment in 2010.[8]

In a piece of research carried out by Carnegie Mellon and Harvard Universities, the majority of those asked said that dreams have a real impact on their daily lives, affecting their decisions and social relations.[9] In 68 percent of cases, this influence was justified by a belief that dreams can predict the future. The interviewees were told to imagine that they had a plane ticket, and asked if they would change their travel plans when faced by four alternative scenarios: a warning of a possible terrorist attack, an intrusive thought while awake about the possibility of a plane crash, a dream on the same subject, or a story in a newspaper along similar lines. It is striking that, compared to all the other scenarios, the largest proportion of those interviewed said there would be the greatest chance of canceling their trip after a dream.

THE INFINITE PUZZLE

Summing up our journey thus far, it is important first to recognize that in the natural world, animals always face the same problems: not dying, killing something to eat, and procreating. In this very tough environment, where each day is a battle and the big problems are always variations on the same themes, dreams evolved as an additional function of sleep, capable of simulating behaviors before testing them out in the real world. For extreme situations, those that are truly dangerous, dreams are creators of the life that escapes death. However, for a human being living in society and furnished with the material conditions of existence, the three big problems are replaced by thousands of small annoyances, limitations, and frustrated desires. In these circumstances, dreaming becomes a much more ambiguous and complex fabric, like several jigsaw puzzles being assembled at the same time, on top of one another, a palimpsest of narratives. This only increases the need to separate and interpret the different narrative threads interwoven in the dream experience.

It is crucial, then, that we recognize the positive potential of the increased awareness that dreaming allows, presenting an unequaled opportunity to explore one's own unconscious. Recounting and deciphering dreams was and continues to be the basis for traditional therapies, whether through individuals who are specialized in this function, such as the Mapuche dream interpreters, or through a widespread socialization of the capacity to experience and explain dreams, as in the case of the Xavante. Dream narratives, plots, and characters are experienced collectively. As in a composite portrait, each story is made up of the recombining of the pieces of the past to try to understand the future. In the analysis of dreams, psychotherapists in general—and psychoanalysts in particular—act as legitimate fellows to the shamans, using more or less the same resources for spinning and unraveling what has happened, albeit with quite different explanations for the phenomena experienced.

If dreams to a psychotherapist are the main internal source of symbols, in many traditional cultures the dream experience does not only refer to another mental reality, but to a reality that is material, concrete, and perceptible. To those people who are embedded in these cultures, the opposition between waking and sleeping does not correspond to any distinction between material and immaterial, or between organic and psychic. The Brazilian anthropologist Antonio Guerreiro, from Campinas State University, explains that among the Kalapalo of the Upper Xingu, the soul that navigates through dreams is equivalent to "each being's potential to be perceived from the viewpoint of other beings (enemies, spirits, etc.), and to relate to them according to their own logics."[10] According to this explanatory key, a dream does not represent a diving into oneself, but the setting off—voluntarily or otherwise—on a journey that is potentially rewarding and threatening. Among the Wayuu people of the Guajira Desert, on the northern border between Colombia and Venezuela, it is common to say before going to sleep: "We will meet again tomorrow if you have good dreams."[11] The words show that dreaming can be dangerous for the Wayuu, as they believe that during their dreams the spirits of the dead roam about the world, foretelling events and causing illness among the unwary.

SURGICAL WORDS

In psychotherapeutic consulting rooms all over the world, the danger that can appear in dreams is revisited in a place that is safe. The Austrian psychiatrist and psychoanalyst Ernest Hartmann was one of the first to defend the idea that dreams in themselves work like psychotherapy, allowing the dreamer to combine thoughts that are normally dissociated in waking life, making connections in a safe place.[12] In countless cultures, however, the moment of dreaming is not safe at all. It is only after sleeping, in morning circles of conversation, stretching over hammocks and yawning widely, that this safety can be attained—in a space where it's possible to talk and listen, to recount over and over until some resignifying has been achieved. Inside the teepee just as on the couch.

After decades of being subjected to relentless criticism, psychoanalysis is beginning to experience the redemption of some of its most important assumptions. Like other methods endorsed by science for lessening traumatic memories, the psychoanalytic method of free association in a safe and relaxed environment encourages a mild recollection of the trauma, something that is of great therapeutic value in reducing the stress and dealing with its consequences. The treatment of post-traumatic stress disorder encompasses various types of psychotherapy, including several relaxation techniques, meditation, habituation to the traumatic narrative, cognitive reinterpretation in an unthreatening context, repetitive sensorial stimulation, and the administering of drugs—all of them aimed at weakening the traumatic memory after it has been voluntarily reactivated.[13]

Even in the case of psychotic patients, combining pharmacological treatment with psychotherapy is more effective than depending exclusively on the drugs.[14] This is because patients acquire some knowledge of their own illness, which makes it possible to develop criticism and suspicion of the hallucinations and delusions that are taking hold of them. With practice, the patient is able to lessen or even block out the impact of the most intrusive disturbances, such as hearing voices. The dialogue between patient and therapist, which in the view of hegemonic medicine is technically no more than a placebo, increases the effectiveness even of treatments that are in principle far away from the symbolic field, such as mild electrical stimulation to treat chronic lumbago.[15] Any patient can tell the difference between a doctor with empathy and a doctor without it. Even if explicitly rejected by or unknown to so many clinicians, the psychoanalytic triad of free association of thoughts, the interpretation of spoken words, and the transference of affects between patient and therapist will interfere in treatments through the human need for understanding and solace.

Jung used the metaphor of surgery to refer to psychotherapy, perhaps because the gained awareness of repressed content, when it is successful, seems to close and cauterize psychic wounds. But just like surgeons with their scalpels and gauze, psychotherapists can do their work well or badly. In the latter case, scars often appear that are caused by the treatment itself. In addition, memories that are highly emotional simply do not get erased. To stay with the analogies between distant branches of medicine, it would be better to say that psychoanalysis is a kind of physiotherapy of the emotions, a massaging of memories, a gaining in awareness of thoughts and of the body, of one's limits and desires, that is capable of rearranging memories and making the mind less inflamed. Or, alternatively, to create an even gentler metaphor: the talking cure is like an untangling of hair to undo the knots.

RECONSOLIDATION AND PSYCHOTHERAPY

The molecular bases for these poetic therapeutic effects, which strictly speaking remain unknown, might have begun to be revealed with the discovery of the reconsolidation of memories, the process by which memories that are already acquired and consolidated can be subsequently altered when they are remembered. The classic experiment in this line of research was carried out by Egyptian-Canadian neurobiologist Karim Nader, then a postdoctoral fellow at the lab of U.S. neurobiologist Joseph LeDoux, at New York University. In the winter of 1999, Nader became aware of some studies from the 1960s that suggested the possibility of modifying memories by reactivating them followed by some kind of manipulation. LeDoux was dismissive: "Don't waste your time, this will never work." Resolving to revisit this heretical idea anyway, Nader trained rats using an auditory signal that was followed by a weak electric shock. This sequence of events made the rats remember that the sound preceded the shock, immobilizing them. Nader then waited twenty-four hours and went back to presenting the auditory signal, only this time there was no shock, but rather the injection to the brain of a substance that inhibits the production of new proteins. The injection took place in the cerebral area of the amygdala that is involved in codifying the fear of specific stimuli. When the rats were tested a day later, or even two months later, they no longer froze upon hearing the sound—they had "forgotten" their relationship to the shock.[16]

In spite of considerable initial resistance from specialists in the area, the phenomenon of memory reconsolidation was replicated in many different ways in several animal models. The research brought Nader some well-deserved renown and the post of professor of psychology at McGill University. Today we know that memories are not consolidated just once, immediately after their acquisition. On the contrary, they become malleable again each time they are recalled, retrieved, reactivated. This renewing of the malleability of memory depends on the same mechanisms of gene regulation and protein production that are activated while awake in a learning setting. Each time a memory is re-remembered, it is partially reconstructed.[17] Even solid, old memories, which have stood the test of time and are considered stable, can undergo alterations to content and to related emotions. Mark Solms noted that revisiting old memories can have positive effects in our lives as they continue to unfold:

> The purpose of learning is not to maintain records but to generate predictions. Successful predictions remain implicit; only prediction errors ("surprises") attract consciousness. This is what Freud had in mind when he declared that "consciousness arises instead of a memory-trace." The aim of reconsolidation, and of psychotherapy, is to improve predictions about how to meet our needs in the world.[18]

To the extent that they revisit and modify existing past experiences, dreams can be seen as particularly potent opportunities to reconsolidate memories.

But this still does not sound like much to account for the amazing impact that a dream can have on the mind of the dreamer. There is still a long way to go before we can understand how the molecular and cellular phenomena triggered by sleep are related to dreams as a psychologically transformative experience of great relevance to the process of the individuation of the self. This journey will take us through the active retrieval of memories from the unconscious, through the increased awareness of our own instincts and impulses (especially those in conflict with social norms), and through a better perception of the chiaroscuro of the mind that we always experience but almost never notice. Dream symbols should not be interpreted as this or that, but rather as this *and* that, given the sheer multiplicity of meanings possible, which are derived not only from the pictorial associations between images but also and perhaps principally from so many semantic, syntactic, and phonetic associations, including intra- and inter-linguistic polysemy. Codified in this broad linguistic space, the shared relationships of ideas and affects allow the construction of an autobiographical experience that combines formal and informal education in the creation of unique individuals. Original perspectives, with real subjects and true points of view. The mental space is not infinite, just very, very vast.

17

Does Dreaming Have a Future?

When the peoples of the Vedic culture believed that Vishnu dreamed the universe into reality, they bequeathed us a powerful metaphor for what we do when we dream, imagine, plan, and realize. Dreaming without intentionality is a human circumstance and necessity, but dreaming with intentionality is a radical life choice. It is a choice that can be experienced in many different ways depending on the dreamer's objective, from the most lofty search for purpose through mystical devotion, scientific research, and immersion in the infinite, to the powerful emotions of an extreme interior sport. Breaking through to the ineffable light in the search for knowledge, whether in intense REM sleep, while spinning nonstop in a Sufi ritual, or under the effects of the psychedelic secretions of the *Bufo alvarius* toad, is a path into the interior of the mind in search of some state that can clarify, inspire, move, transform, or cure.

The methods for obtaining such oneiric trance states can involve fasting, sleep deprivation, sensory deprivation, physical ordeals, or simply falling asleep. This is certainly not the exclusive privilege of Yanomami shamans, Himalayan yogis, or Californian hippies. To go with a familiar example: in the neo-Pentecostal religions that are spread around Brazil and various other countries, mystical trance experiences are highly valued. On FM radio, programs broadcast by the Universal Church of the Kingdom of God advocate vigils, statements of purpose, donations, and atonements in order to reach an ecstatic encounter with the Holy Spirit. It is no wonder these churches are growing so substantially around the globe, since they promise a powerful, pleasant trance to those who are working and suffering in tough, everyday reality.

Advertised by priests all over the world, the promise of an altered state of consciousness appears in many flavors, comprising deities, angels, demons, spirits, rituals, dances, sacrifices, and all the psychedelic plants, animals, and fungi we know: all vibes. Despite their huge variety, all of the mental states reached through these channels are characterized by the same single truth: they evoke what *is not* in order to allow an escape from what *is*. Dreaming occurs through the estrangement of reality, in order to imagine what could be. In a confessional space, in front of a discreet observer, or in the intimacy of a conversation with God, kneeling at the side of the bed, the believer tries to flee from their oppressive reality and yearns for a numinous kind of contact, one that might give life meaning.

When they were on the verge of going hungry in the Canadian Subarctic, the dreamers of the Beaver tribe reported going into an oneiric trance in order to discover where the hunt was to be found. In the Amazonian jungle, Jivaro hunters propitiate the success of their pursuit by drinking ayahuasca. If waking life is the present, then the possibilities of the future and the past belong to the trance, along with everything that was not or that might yet be, the horizon of alternative futures: the world of counterfactuals.

THE DREAM DECIPHERED

Despite important advances in dream science, there is still much to learn about the nature of dreaming and its role in human behavior. Even extremely basic questions about the phenomenon are mysteries that either were only recently solved or that remain unexplained still. Until just a few years ago, some of the most important researchers in the area of sleep and memory believed that the account of a dream does not reflect the actual experience during sleep, but merely a swift elaboration carried out by the already awake brain, immediately after the sleep has ended. This argument was originally formulated in the nineteenth century by the French doctor Louis Alfred Maury in opposition to the ideas supported by his contemporary, the Marquis d'Hervey de Saint-Denys.[1] In 1956, the U.S. philosopher Norman Malcolm returned to the subject, considering it on the basis of the logical incongruence of referring to an unconscious state of consciousness.[2] To Malcolm, a dream was a linguistic deception, a mental phenomenon with no present existence, about which we only know anything because of an account given while awake. Instead of considering the dream account as evidence of the past existence of a dream, it would be more prudent to consider it a phenomenon of waking life itself.

Twenty years later, another U.S. philosopher, Daniel Dennett, currently at Tufts University, revived the argument: if a dream is an event that we can only know of *a posteriori*, how can we rule out the possibility that it does not in fact represent a "subjective experience" during sleep, with some degree of consciousness, but rather the unconscious accumulation of synaptic modifications that only become a subjective experience upon waking?[3]

Dennett believed it was impossible to refute the notion that the formation of a dream occurs exclusively after waking. Even the simultaneity of dreaming and sleep manifested during lucid dreaming was not accepted as evidence, since the objective verification of the existence of the lucid dream also depended on the subjective accounts of already experienced dreams. Echoing the hundred-year-old Freudian idea that a dream could only be known through the verbal account of the dreamer themselves, Dennett became a champion of the most hard-line dream skepticism of all—that which refused to accept that dreams even exist.

However, revolutionary methods for decoding mental images have called this opinion into question. In the last decade, the teams led by U.S. researchers Jack Gallant, of the University of California–Berkeley, and Tom Mitchell, of Carnegie Mellon University, devised algorithms and experimental processes that are able through the imaging of brain activity via functional magnetic resonance to reveal what a person is secretly seeing or thinking.[4] This method, which would have been enough to make the science fiction writer Isaac Asimov's mouth water, is based on extensive collection of data from the individuals when they are repeatedly exposed to various stimuli. Then, machine learning technologies are used to detect relevant information. The vast library of patterns of cerebral activity paired with corresponding stimuli is used in this way to predict new stimuli based on the concomitant neural activity.

This method has resulted in astonishing discoveries, such as the evidence that the semantic representations of different categories of objects presented visually (people, animals, cars, buildings, tools) are mapped out all over the cerebral cortex. This means that the concepts are mapped onto the brain like nationalities in the globalized world:

every country has some people from the main nations. Rather than being overlaid upon one another, the representations of different categories of objects seem to be adjacent to one another. However, when the experimental subject is instructed to look for a specific category in the film that is being used as a visual stimulus, a large proportion of the voxels (three-dimensional pixels that are the spatial unit of measurement in functional magnetic resonance imaging) adjust their response toward the category targeted by their attention. This causes an expansion of the specific category (e.g., man), as well as of those categories that are semantically related (e.g., woman, person, mammal, animal). In contrast, the representation of those categories that are very different from the target category were compressed (e.g., text, drink). Attention paid to a specific category distorts the map of representations as a whole, according to the semantic relations between the objects represented.[5] Intentions, images, and words matter.

Besides revealing quite new aspects of the neural organization of percepts and memories, the results from the new field of brain decoding have profound existential implications, as they breach the inviolability of thought. It is now possible, albeit only in an incipient form, to "read" another person's mind through technology. And this approach goes further: the first application of the decoding methods to dreams was published in 2013 in the journal *Science*. The team led by the Japanese neuroscientist Yukiyasu Kamitani managed to decode categories of mental content during the initial stage of N1 sleep (hypnagogic sleep), which is quite like REM sleep in electrophysiological terms. But N1 episodes are usually much shorter than REM sleep episodes, so N1 dreams are generally short as well, more like isolated scenes than cinema. Using signals from brain regions farther away from the sensory organs, Kamitani and his colleagues managed to decode specific oneiric features (e.g., car and man) 70 percent of the time.[6] While it is still only incipient, the study was enough to be able to test out the hypothesis that dreams are formed immediately upon waking. The results showed that the greatest correlation between neural signal and mental content happened around ten seconds *before* waking, dropping subsequently (fig. 15). In other words, dreams are not formed after sleep but during it.

More recently, Giulio Tononi and his team obtained similar results when investigating electrical brain waves. They decoded dreams successfully, separating them according to the activation of the cerebral

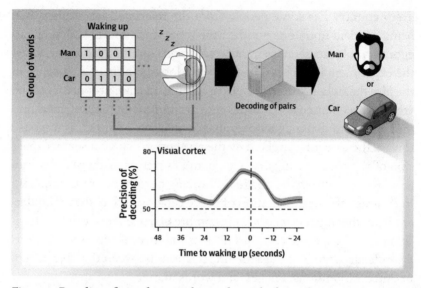

Figure 15. Decoding of visual images during sleep. The figure shows a peak of decoding in the visual cortex in the period of ten seconds before waking. Note that the greatest accuracy of the decoding occurs before awakening.

regions involved in the representation of specific mental categories, such as faces, places, movements, and speech.[7] The advent of neural decoding finally made it possible to discover general aspects of other people's dreams, leading to an expectation of doing away entirely with secondary elaboration—the account given of the dream, instead of the dream itself—to access the oneiric content "directly." In theory this would make it possible to access the raw material of dreams, totally free of repression, censorship, or ancillary associations.

The decoding of the specific sequence of images experienced during a dream seems to reveal a scientific object that really is new to science, perhaps comparable to the advances that came from the first chemists to isolate pure substances, or from the invention of methods for polishing lenses for telescopes and microscopes. The advances in neuroscience sound like the beginning of the end for the supposed unreliability that has always been a feature of dream accounts, a source of justifiable skepticism regarding the self-serving dreams of Julius Caesar, Constantine, Frederick, Kekulé, and so many others. In the future, it will be harder and harder to justify coups d'état, religious conversions, political ambiguities, and dubious originality by simply making reference to a convenient dream. An age of dream transparency might perhaps be approaching.

But we must be patient, since these advances are all very recent. There is still discussion over whether the pioneering decodings of dreams do indeed open the way for us to understand the phenomenon objectively. It is worth remembering that the method itself requires the dreamer to recount their dream, creating a secondary elaboration that will serve as a sample for the future evaluating of the results of the brain imaging. In addition, the decoding requires the construction of a vast bank of visual images and the respective cerebral responses, with a verbal tagging of each stimulus carried out by researchers and not by machines. A large number of these image-response pairs need to be presented to a computer network to make it capable of recognizing and classifying patterns that are specifically related to concepts. If the whole experiment seems somewhat circular, that's because in fact it is—which should give the philosophers something to ponder for several decades yet.

How Freud and Jung would have smiled if they'd lived to see these discoveries and new ideas. And what expressions of astonishment we would see on the faces of an Akkadian priestess or a Siberian shaman from the Bronze Age if they were only able to witness, with their own eyes, a dream revealed by functional magnetic resonance imaging. Their eyes would surely shine, and then perhaps their eyelids would close for them to travel into a really crazy dream of their own.

A NEW PSYCHIATRY IS BORN

The new science of the mind lays the foundations for a new psychiatry, turned toward the future but connected to the past, pharmacologically better equipped and much more alert to tradition and the preparation of the therapeutic setting. It is increasingly obvious that the professional care of mental illness needs to incorporate and honor the knowledge from traditional shamanic practices. Sleeping and dreaming improve the body's health and increase neural plasticity. This observation converges with the recent evidence that classic serotonergic psychedelics are among the substances that best emulate the dream state,[8] strengthening processes of primary consciousness.[9]

We need to recognize the crisis of the over-medicalization of psychiatry. Despite being commonly recommended for daily treatment

over months, years, and decades, the antidepressants found at phar-
macies show only very modest positive effects, little better than those
obtained with placebos and with an effectiveness that is only confirmed
for the first two months of the treatment.[10] Faced with the worrying
side effects of these drugs, which include the risk of chronic depression that is impervious to treatment,[11] mainstream psychiatry has allied
itself with the interests of the pharmaceutical industry and washed its
hands.[12]

This disappointing picture is in contrast, for example, with the effects of psilocybin, the main psychoactive compound from the *Psilocybe
cubensis* mushroom, which reduces depression and anxiety for several
months when administered in two doses during sessions of psychotherapy.[13] Depressed patients impervious to other treatments show an
increase in feelings of "nature relatedness" and a decrease in the expression of "authoritarian political views,"[14] with a significant improvement in the recognition of other people's emotions.[15] It is important to
note that the quality of the psychedelic experience (ranging from delightful to frightening) and also its intensity (subtle to overwhelming)
determine the results in the long term.[16] The course of the journey
determines the port of arrival.

When the psychic suffering is related to past traumas, producing
post-traumatic stress disorder, for example, the best clinical solution
would seem to be psychotherapy assisted by MDMA.[17] This is the active
ingredient in ecstasy, which provokes an intense release of serotonin,
noradrenaline, and dopamine produced by the brain itself when it is
not contaminated by other substances. The psychological effect of an
appropriate dose of MDMA, with the right people, in the right place,
with good, warm lighting, and music likewise, is extremely pleasant.
The effect corresponds to a feeling of anxiety disappearing and an
intense love for people, an immense happiness at being human that
is revealed principally through touch. The effect can last several hours
and can persist in a subtle way for several days after the substance has
been ingested.

MDMA was used for couples therapy in the 1970s, but it was banned
in 1985 by U.S. president Ronald Reagan despite being safer than almost
all the main psychoactive drugs.[18] Unlike other psychedelics, MDMA
does not provoke great changes in perception or hallucinations in most
people. When it is administered to traumatized patients, as in the case

of the thousands of U.S. veterans of the wars in Afghanistan, Iraq, and Vietnam, MDMA showed impressive positive results. In May 2018, the prestigious journal *The Lancet Psychiatry* published the results of a rigorous clinical trial of the effects of MDMA on twenty-six patients suffering from post-traumatic stress for at least six months, including war veterans and first responders. The randomized study, double-blind and with dose-response assessment, was carried out by a team led by U.S. psychiatrists Michael and Ann Mithoefer and by the doctor in public policy Rick Doblin, founder and executive director of the Multidisciplinary Association for Psychedelic Studies (MAPS), one of the main organizations promoting the legalization and regulation of the medicinal use of psychedelics. The results showed that two sessions of treatment with MDMA and psychotherapy were enough to significantly reduce the symptoms of post-traumatic stress, even when measured a year after the treatment.[19] Unquestionably, help is on the way.

Although psilocybin, MDMA, and other pure molecules are very close to being accepted by traditional psychiatry, perhaps the most effective of the psychedelic antidepressants will, one day, be the complex mixture of molecules present in the Amazonian beverage ayahuasca. Preliminary studies have shown that this tea of traditional use reduces depression quickly and lastingly, starting forty minutes after its ingestion and capable of being sustained for two weeks after a single dose.[20]

A rigorous experiment, randomized and placebo-controlled, on thirty-five patients admitted and monitored for several days by EEG, functional magnetic resonance imaging, and a range of psychological tests, recently confirmed the significant antidepressive effects of ayahuasca. It is no easy matter to conduct a clinical test using psychedelics in a Brazilian public hospital on low-income patients. On the one hand, it requires dealing with negative symbolic impacts of the proximity to human suffering, which can lead the quality of the psychedelic experience toward negative emotions. On the other, the experimental effects can be masked by a surprising placebo effect, with a significant improvement in the depressive symptoms from the simple fact of the patient feeling well cared for, in an environment that is more wholesome than at home.

To coordinate this research, the neuroscientist Dráulio de Araújo, from the Brain Institute of the Federal University of Rio Grande do Norte, needed to set up a real task-force of specific talents and skills.

The multidisciplinary team that carried out the study ranged from Fernanda Palhano, a neuroscientist and the lead author of the work, to doctors Sérgio Mota Rolim and João Paulo Maia, who were in charge of the polysomnography and the psychiatric screening. The results justified all the effort, as the antidepressive effects of a single dose of ayahuasca, which could be detected just minutes after ingestion, were sustained for at least seven days, which did not occur with the placebo.[21] Confirming the importance of the context of use and of the psychotherapeutic handling of the experience, the researchers ascertained that the more intense the psychedelic experience, the stronger the antidepressive effect.[22]

Such quick and long-lasting effects could not exist without the mobilizing of molecular and cellular mechanisms capable of transforming short-term synaptic changes into long-term morphological changes. The first demonstration of the huge plastic potential of psychedelics dates from 2016. Driven by the Serbian biologist Vanja Dakic and supervised by the Brazilian neurobiologist Stevens Rehen from the Federal University of Rio de Janeiro and the D'Or Institute, a team of which Araújo and I were a part showed that substances contained in ayahuasca increase the levels of proteins related to synaptogenesis and neurogenesis in cultures of human neurons.[23] Another study led by the Brazilian neurobiologist Richardson Leão and the doctoral student Rafael Lima showed that a single dose of 5-MeO-DMT stimulates cell proliferation and neuronal survivability in the hippocampus of mice.[24] A third independent research group, coordinated by the U.S. chemist David Olson, from the University of California–Davis, demonstrated similar phenomena *in vitro* as well as *in vivo*, after treatment with LSD or N,N-DMT.[25] This means that the ingestion of psychedelics flings open the doors of neural plasticity, allowing the transformation of a subjective experience lasting a few hours into a psychic cure that can last months and years. These data become even more interesting if we think about the use of classic psychedelics and cannabis for treating the abuse of more dangerous substances, such as alcohol, tobacco, crack, and cocaine.[26]

In considering the new scientific discoveries that support the benign use of psychedelics, we must keep in mind that their use is only safe when certain important precautions are taken. Just like extreme sports, psychonautics—traveling one's own mind using psychedelics—offer profoundly transforming and thrilling experiences, which can give new meaning to life if practiced appropriately by people who are well prepared. Psychedelic trips require as much technique, know-how, and caution as paragliding among the clouds or diving to the depths of the sea. Just as with extreme sports, the initiation of novices without the supervision of qualified guides should be strictly avoided. Just as with extreme sports, people who are in specific high-risk groups should abstain from psychedelia. And just as in those sports, the context of the use substantially determines the course of the trip.

Whether naturally or pharmacologically induced, dreaming increasingly appears to act like therapy for brains that are ailing from an excess of stress or the abuse of any kinds of substances. It is also a huge learning opportunity for the sick and the healthy alike. In April 2017, three thousand people from different countries gathered in Oakland, California, to take part in the Psychedelic Science congress, promoted by MAPS and the Beckley Foundation. Unlike in previous iterations of similar events, in which hippies dressed as J.R.R. Tolkien characters jostled for space with a few researchers and their students, what was striking was the large contingent of scientists present, some of them highly renowned, in addition to journalists, documentary makers, and foundations and firms interested in funding research on the medicinal use of psychedelics.

Most important of all, the gathering was an opportunity to reflect on the terrible schisms of the twentieth century, which criminalized sacred fungi and plants, stigmatized colleagues, and finally did damage to therapeutic effectiveness and the ideological tolerance between the different branches of psychology and psychiatry. In a session on the future of psychedelic psychiatry, the U.S. psychiatrist Thomas Insel, who spent fourteen years as the director of the National Institute of Mental Health (NIMH) in the United States, publicly acknowledged mainstream psychiatry's failure to find a solution to psychic suffering, and recognized the huge potential of psychedelics when used in the appropriate setting.

His words, which were very revealing about the pharmaceutical hypocrisy that specialized in selling illusions, still echo in my mind:

Many of us [psychiatrists] have the experience that we're much more helpful today than we were twenty or thirty years ago to patients with similar kinds of problems. And yet, the data don't support that. The suicide rate is higher now, trending higher than it was ten, twenty, thirty, forty years ago . . . Measures of morbidity, as opposed to mortality, are actually higher, not lower. So by public health measures, we're not doing very well . . .

We need to wake up to the idea . . . that the complexity here is going to demand a networked approach, a comprehensive approach . . . I'm really impressed actually by the approach here. People don't say simply "we're going to give psychedelics," they talk about "psychedelic-assisted psychotherapy." You know I've never heard anybody talk about antidepressant-assisted psychotherapy . . . I think it's a really novel approach . . . to really make a change in somebody's life . . . [but] how do you get that through a regulatory pipeline, what would the FDA say about this, or EMA?* I mean, they won't even touch psychotherapy alone.[27]

Would this really be a different approach, something really new? It is more like a millennially ancient innovation, a retrieving and above all a reinventing of ancestral knowledge about the enormous importance of the context in which psychedelics are used for determining their effects. Could the international health system's lack of curiosity in doing in-depth investigation into the benign properties of psychedelics have something to do with the low profitability of treatments that are based on a lot of human contact and little medicine, compared to the much more lucrative model of little human contact and a great deal of medicine?

THE REBIRTH OF DEPTH PSYCHOLOGY

The birth of the new psychology also coincides with the restoring of its historical origins. With the perspective of time, Freud and Jung become more and more secure in their roles as real pioneers in human ethology, honorable fulfillers of the prophecy made by Konrad Lorenz right after the end of the Second World War: hard as it was to accept their theories, it would be impossible to ignore their discoveries.

---

* The Food and Drug Administration, and the European Medicines Agency.

It is instructive to list those that have so far had scientific confirmation. Not only do the id, ego, and superego correspond to distinct cerebral processes,[28] but this tripartite conception of the mind also inspired the original conception of artificial intelligence.[29] Psychotherapy through words, a process that is introspective and reflective, is clinically effective and indispensable in most cases.[30] Dreams cannot be explained away by limiting them to REM sleep alone, and they reflect traces of memories acquired while awake.[31] Memories can be suppressed.[32] Without the activity of the dopaminergic circuits involved in desire, there can be no dreams.[33] Dream accounts are particularly informative of the patient's psychiatric state.[34] Sexuality begins in childhood and can persist throughout life.[35] Traumas can leave an indelible mark on future behavior, including on that of offspring.[36]

We are a mixture of life and death instincts, filled with contradictory emotions and desires. For this reason, what matters most is not what we think but what we do. Dreams emerge from the unconscious in such a way as to describe present situations and possible alternative futures, and are able also to express collective patterns of thought. Over the course of our lives, we go through well-characterized phases in our relationships with our own bodies, with those people closest to us, and with the objects in the world, marked by successive stages of discovery, development, maturity, and senescence of mental representations. Recounting without inhibitions helps us to navigate the course of frustrations and mitigate our pain—not only through the countless psychotherapies of self-knowledge, but increasingly, explosively, by the feverish use of Facebook, Instagram, Twitter, blogs, vlogs, and a whole new universe of ways of telling stories. The cultural ratchet that we began while sitting around the fire, our unruly production of narratives, is flaring up and spinning ever faster.

Freud and Jung created the bases for all this understanding. In their surprisingly coherent works, they threw light on the recesses of our own behavior through the masterly exercise of induction, deduction, and abduction. Their inclusion in the pantheon of the great scientists of humanity does not only require understanding and valuing their legacy, but also defending them against the various accusations made against them with more or less justification, many of them of a moral nature. If Freud and Jung were measured with the same ruler that is applied to other geniuses of humanity, defending them would be easy, since excluding all the others would be simply disastrous. For lust and obscene behavior, we would lose Mozart and Caravaggio. For causing harm to his enemies and liking money, goodbye, Isaac Newton. For mysticism, we would bid farewell to Johannes Kepler and Hans Berger. For changing their minds and making theoretical adjustments, no more Albert Einstein or Stephen Hawking. For being an apologist for the use of drugs, we would have to do without Aldous Huxley and Carl Sagan. We must separate scientific discoveries from the imperfect behavior of their discoverers, which are inherent in the human condition. If Freud and Jung were not titans of the science of the mind, who was?

THE SOCIETY OF THE MIND

Looking further ahead, the development of the method of decoding dreams should shortly make it possible to test the hypothesis that dreams extrapolate the dreamer's specific point of view. It is quite possible that we do not have just one dream at a time, but rather a whole host of parallel dreams at any one moment, inhabited by different autonomous representations that we carry within us, the "creatures of the mind" that seem to take on their own lives when we dream. Marvin Minsky proposed the idea that a human personality is not unitary, but comprises a society of memes that inhabit a virtual space created by the brain. The British novelist and philosopher Aldous Huxley agreed:

> Like the giraffe and the duck-billed platypus, the creatures inhabiting these remoter regions of the mind are exceedingly improbable. Nevertheless they exist, they are facts of observation; and as such, they cannot be ignored by anyone who is honestly trying to understand the world in which he lives.[37]

The creatures of the mind that sometimes strike us powerfully as divinities and at other times disappoint us as incorrect composite sketches are what Jung called imagoes: mental images of varying degrees of complexity, individual representations with different levels of verisimilitude and independence. Jung left a detailed record of his relationship with a dream figure called Philemon, an Egypto-Hellenic pagan who first appeared to him in a dream in 1913, and who came to be a gnostic guru to the young psychiatrist. In his words,

Philemon and other figures of my fantasies brought home to me the crucial insight that there are things in the psyche which I do not produce, but which produce themselves and have their own life. Philemon represented a force which was not myself. In my fantasies I held conversations with him, and he said things which I had not consciously thought. For I observed clearly that it was he who spoke, not I. He said I treated thoughts as if I generated them myself, but in his view thoughts were like animals in the forest, or people in a room, or birds in the air, and added, "If you should see people in a room, you would not think that you had made those people, or that you were responsible for them." It was he who taught me psychic objectivity, the reality of the psyche.[38]

Mental fauna is quite an apt description of the myriad objects and social relations that are mapped out in our minds, including the simulating of other people's behaviors and a surprising autonomy of the characters, who are echoes of the much more hierarchical fauna of the recent past, when the ancestors' word was law and the patriarchy would not brook any objections. Between people, beings, and divinities, alive or dead, we carry in our heads, in an explosive outburst of images, emotions, and associations, the whole heraldic legion of our past, from the Upper Paleolithic's mysterious Lord of Beasts to Godzilla, from Achilles to Muhammad Ali, from Enheduanna to Barbara McClintock, from Innana to Amy Winehouse, from our grandparents to our children.

It is with imagoes, with all of them and only with them, that we appear in dreams, though each one is merely the filtered and edited fraction of the totality of that externally existing person or character. It is not only the ego that inhabits the temporoparietal cortex, it is our internal fauna. During our waking lives, the circuits of the prefrontal cortex carry out the inhibitory control that filters out all the discordant voices of our mental democracy to generate a single action. But during sleep, the brakes go, the cages are opened, and all the wild animals come out for a stroll.

Following this theory, the feeling that we have a single dream arises from the presence of the dreamer's self-representation in just one dream at a time, just as the presence of an actor on a specific set doesn't prevent the studio from shooting several movies simultaneously. To paraphrase a Lakota saying about memory, a dream is like walking a trail at night with a lit torch. The torch only throws its light up to a certain distance . . . and beyond that is darkness.[39]

I remember dreams in which the characters did not simply enter or exit the scene, but where entire casts and settings were changed abruptly with great commotion, as if my dreaming self had left its own dream and entered another, a short circuit of narratives that are made of the same material, the pure electrical reverberation of memories, but with the particular distinction that the second dream seemed to have been started and developed in the dreaming self's absence, as if it already existed before the moment that the self invaded the next-door dream.

For good or ill, there have never been so many opportunities to replicate memes, which can reach millions of people all over the planet in a matter of seconds. Today, when a person dies, it is normal that countless impressions of that person survive in photos, texts, sounds, words, and narratives, partial representations that could perhaps persist over time in the vast collective unconscious of the digital cloud and its users. We are inventing eternal life, not only of people of flesh and bone but also of characters. In the bazaar of digital and cerebral representations, the ancient Sumerian goddess Inanna has almost completely disappeared. In some minds she still sells incantations and begs for attention at the doors of the temple of Babylon. Perhaps, among a handful of the more learned scholars, the goddess does even still blaze bright wherever she goes, calling up her celebrated manifestations like Ishtar, Aphrodite, and Venus. But in the majority of minds she no longer even exists, and it is her inheritors who are flourishing: Marilyn Monroe, Madonna, Beyoncé . . . And these representations clearly interact and compete with all the other memes, from Mickey Mouse to Pelé, from John Lennon to the Dalai Lama, in a transcultural accumulation of crossed references so unparalleled as to defy understanding.

We are still faced with the challenge of constructing dreams in robots. We already know how to simulate in computers some of the mechanisms unleashed in sleep, but we remain a long way from building androids capable of dreaming of electric sheep—as in the title of the book that inspired the movie *Blade Runner,* a dystopian narrative in which the distinction between man and machine is no longer clear.

There is extraordinary promise in the combination and maximization *in silico* of induction (vast databases), deduction (incredibly fast calculations), and abduction (probabilistic simulations). In the race to develop artificial intelligence with corporate personality to govern the planet in the first century of the new millennium, it is likely that we have already synthesized our new gods—even if we haven't noticed it yet. As in Mbundu beliefs, *souls will live in things*.

18

Dreaming and Destiny

When the clocks of the midnight hours are squandering
an abundance of time,
I shall go, farther than the shipmates of Ulysses,
to the territory of dream, beyond the reach
of human memory.
From that underwater world I save some fragments,
inexhaustible to my understanding:
grasses from some primitive botany,
animals of all kinds,
conversations with the dead,
faces which all the time are masks,
words out of very ancient languages,
and at times, horror, unlike anything
the day can offer us.
I shall be all or no one. I shall be the other
I am without knowing it, he who has looked on
that other dream, my waking state. He weighs it up,
resigned and smiling.

—JORGE LUIS BORGES[1]

With regards to prognosis, . . . dreams are often in a much more
favorable position than consciousness.

—CARL JUNG[2]

Where is God, even if he doesn't exist?

—FERNANDO PESSOA, *BOOK OF DISQUIET*[3]

The coming decades will bring an integrated understanding of what dreaming can revert to being, or can become: a sophisticated psychobiological gearbox capable of promoting continuous behavioral adaptation, deployed as needed; when it is well calibrated, a powerful computer of possibilities, an oracle that sums up the vast landscape of accidents or near accidents so as to assess which way the compass of destiny is pointing—not destiny as an inescapable, predetermined future, but the place or state to which everything converges. Destiny is the place toward which the wind is blowing, the river is flowing, where desires and circumstances lead. Our marvelous and multifaceted brain machine for the extracting of probabilities, built by genes and memes over the course of the evolution of the species, feeds on conscious emotions and preoccupations, but also on our almost insatiable capacity to be interested in the world.

A dream expresses the destination, but it does not guarantee the arrival, like somebody who might be traveling on the correct course but might yet stop earlier, or speed up, or choose to follow alternative routes. Our destination-destiny is where we are headed, but it is not necessarily where we go. Well-dreamed dreams glimpse our destinies through simulations of the possible journeys and outcomes. Dreaming is like feeling one's way around the dark room with a sliver of light, where the walls are the future itself.

We have descended from dreaming peoples since the dawn of time. If in urban civilizations dreams have stopped being essential to the functioning of society, in many indigenous cultures, this transition never occurred. To this day, dreams live in and illuminate the minds of hunter-gatherers, contemporary representatives of the way of life adopted by almost all our ancestors. Understanding the dream perspectives of these peoples is very important if we are to explain the path that has brought us here, and the challenges that face us.

AMERINDIAN DREAMS

Generally speaking, across time and space, Amerindian peoples have traditionally acknowledged the capacity of dreams to predict the future, whether in a form that is regularly accessible to the common dreamer or by means of revelatory dreams at especially significant moments in a life, or even through shamanic dreams stimulated by rites of passage, cure, or spiritual guidance. They cultivate dreams of great potency and lasting significance, capable of inspiring, initiating, advising, teaching, and maturing people. It is through these formative dreams that the young decide to follow the different paths that adult life offers them, such as that of shaman or hunter-warrior.

The first European accounts of the people of the New World testify to the social significance of the Amerindian dream event. Hans Staden, the sixteenth-century German soldier who was shipwrecked on the Brazilian coast and kept captive by the Tupinambá, described how before going off to war, the anthropophagous Native American people were urged by the medicine men to examine their dreams with great care. If they had visions of their own flesh being roasted, they would give up on fighting and remain in the village. But if they had visions of the enemy being roasted, they would take up arms, celebrate, and wage war.[4] Seventeenth- and eighteenth-century Jesuit missionaries reported that the Iroquois of the northeastern United States and southeastern Canada understood dreams as enigmatic voyages to satisfy the desires of the soul.[5] In order to comply properly with the oneiric revelations, the Iroquois dreamer would recount them publicly until there was a metaphorical interpretation capable of guiding the best action to be taken. Alongside psychoanalytic theory, these beliefs would influence Jung's notion of the big dream.

Almost three hundred years later, among Ecuador's Achuar Jivaro people, there are records of a belief in dreams as metaphors concerning the future, which are substantially determined by relationships of predation. Dreams are classified by the Achuar into three basic types. Dreams of good omen for the hunt are imagetic and silent, and should be interpreted in secret so as not to scare away the prey. This interpretation ought to acknowledge equivalents and inversions, so that a dream of fishing, for example, might be interpreted as indicating a good opportunity to hunt birds. A good-omened dream is a necessary but insufficient condition for the hunt: it does not guarantee success, but it suggests to the dreamer how they should act in order to attain it. The second type of dream consists of ill omens for the dreamer or their relatives. These, too, are imagetic and silent, but very frightening, in that they present enemy people in the form of animals. The third type is the "truthful dream" of ancestors and spirits, characterized by the occurrence of verbal messages. In these dreams, it is possible to invoke specific spirits to carry out tasks that are suited to their characteristics. In order to induce this demiurgy of creatures of the mind, the dreams are prompted by a range of abstinences and by the ingestion of tobacco and psychedelic plants.[6]

The Aguaruna Jivaro of Peru use the same word to refer to dreams as to the trance induced by ayahuasca. They believe that under its influence it is possible to contemplate events that are under way, which have not yet been consummated but with different probabilities of happening. The dream, then, is not a premonition of an inexorable future for the Aguaruna, but an opportunity magically to shape the future, through intention and most of all through dream action.[7]

Among the Pirahã, who are native to the Amazon region, people dream in order to collect songs, to wage wars, or to form alliances with the spirits.[8] In the words of Brazilian anthropologist Marco Antonio Gonçalves, from the National Museum at the Federal University of Rio de Janeiro, "If a dream can produce an occurrence, an occurrence can produce a dream. Or rather, what happens in the dream will happen in the world as a repetition, and what happened in the state of wakefulness will happen in the dream as a representation."[9]

Among the Waura of the Upper Xingu, a dream is considered a phenomenon similar to a trance, to sickness, to rituals, and to myths. In these states, the soul goes on a journey and manages to get in contact with beings that are extra-human, arcane, and monstrous, very close to animals. Out of the difficult negotiation with these beings can come useful knowledge, such as the magnificent Waura repertoire of geometric drawings received in dreams.[10]

Among the Parintintín of the Madeira River basin, in the southeast of the Amazon region, dream narratives are recounted in the morning, with a view to predicting the future. Just as in the case of myths, specific grammatical forms are used for the relating of oneiric experiences.[11] Among the Kalapalo, meanwhile, no specific word for dream would seem to exist, and dreams are interpreted as expressions of the desires of the person dreaming, of their goals and future possibilities. The Kalapalo distrust the verbal elaboration of dreams, but believe in the truth of the oneiric picture. This is why a great deal of effort is expended on the search for the best words for recounting dream actions.[12]

Among the Mehinaku of the Upper Xingu, dreams are also an object of narrative and everyday interpretation immediately upon waking, while the dreamers are still in their hammocks and can tell their nearest neighbors about the journeys taken that night by the eye soul. Dreams can have direct relevance to the future but they cannot determine it, though they do supply clues as to how best to proceed in order to obtain desired effects.[13] The Mehinaku also value the metaphorical interpretations of dreams. Dreaming about flying ants, for example, can be interpreted as referencing the death of relatives, since these ants have short life-spans.

THE XAVANTE DREAM THEIR FUTURE

A couple of hundred miles southeast of Xingu, among the Xavante—who in spite of the small geographical separation are linguistically very far removed from the Mehinaku—dreams play an even more central role in the social life of the group. The history of how the Xavante used their dreams to survive the conflicts with the whites, coming to be one of the most numerous South American indigenous groups today, with more than eighteen thousand people,[14] justifies a detailed presentation of their case.

In Xavante culture, dreams are not the privilege of medicine men and shamans, as everybody is able to have divinatory dream visions that serve three main functions. The first is related to hunting, war, and sickness; the second is to do with dreams as a voyage to explore the habits of other peoples; the third consists of the revelation of songs, laments, dances, and rituals that are destined to be cultivated by the whole community.[15]

Dream revelations are not passive events to the Xavante. On the contrary, a lot of concentration is needed to bring them into waking life. Magical dreams are awaited and ritually propitiated with great excitement:

> You must always focus your attention on the things you want to dream, you must concentrate, in music or some festivity. You must not go to sleep unprepared . . . You cannot just wait, you have to have a hope, from trying so hard . . . The spirits and some people who lived together in the old days, before the contact, the village—will see that you are committed, and later you

will dream of some beautiful music or you will receive some music for a party . . . [16]

Dream practice is necessary for the good functioning of Xavante society. "In sleep I dream, I sleep and I dream. The others sing. I dream in order to make those who sing my dream happy."[17]

According to the Kayapo writer Kaká Werá Jecupé,

> among the old Tapuia peoples, the ones who most preserve the Dreaming tradition are the Xavante. The dream is the sacred moment when the spirit is free and it can carry out a number of tasks: purifying the physical body, and its home; traveling to the home of the ancestors; often flying above the village; and sometimes, through [the] Spirit of Time, it goes to the edges of the future . . . A Xavante village is semicircular . . . At the centre is the clearing for activities: ceremonies, parties, advice circles and dream circles. It was in this clearing that, from a dream, the beginning of the story of the taming of the whites was told.[18]

The Xavante are among the oldest peoples to inhabit South America. The importance ascribed to dreams among the Xavante and other peoples who came to occupy the heart of the subcontinent affiliates them with the general denomination "Dream Tradition," in contrast with the users of psychedelics in the Amazon basin ("Moon Tradition"), or the indigenous populations migrating along the Atlantic seacoast ("Sun Tradition"). Since time immemorial, the Xavante have inhabited Brazil's Central Plateau, where the state of Goiás is located today, the proud masters of a land populated by jaguars, armadillos, tapirs, anteaters, toucans, parrots, and macaws. However, from the middle of the seventeenth century, *bandeirante* ("flag-holder") pioneers in search of slaves, gold, and emeralds started to invade their territory on the right bank of the watercourse whose name sums up the nature of that contact: the River of the Dead. There were a hundred years of bloody conflicts with prospectors and military troops, who sought to "tame" the Xavante by forcing them into submission.

And then something surprising happened: the indomitable Xavante disappeared. Might they have discussed this radical change of strat-

egy in a dream circle? No historical record of their decision survives.
But the fact is that between 1844 and 1862 the group set off westward,
crossing over to the west bank of the Araguaia River and migrating
toward the Roncador Mountains in what is today the state of Mato
Grosso.[19] A search party ended up being sent to locate them, but all in
vain. They disappeared into the vast scrubland of the Central Plateau,
becoming invisible among its trails and tablelands. Over their long diaspora of cultural attrition and exodus to ever more remote regions, the
Xavante had become experts at isolating themselves. Whether through
distance, through their aggressiveness, or through the magic obtained
in dreams to which they attributed their invisibility, the Xavante managed to spend the century undisturbed by the whites.

However, as time passed, the border between the two worlds shifted
once again. In the 1930s, violent conflicts resumed, but now there
was much less space to escape to. In 1938, the dictator Getúlio Vargas
launched the "March to the West," an official government campaign
to occupy central Brazil. In search of patriotic representations of social
purity, Vargas selected certain indigenous people as symbols of the
national soul. The Indian Protection Service (SPI), which had so often
colluded in the invasion of indigenous lands and in the genocide of
their people, in the latter years of the Vargas dictatorship experienced
a temporary return to the romantic days of its founder, Marshal Cân-
dido Rondon, who from the late nineteenth century had spread telegraph lines across the Brazilian backlands without resorting to violence
against the natives. Vargas paid a visit to the Karajá on Bananal Island to
record propaganda footage; he flew over the territory of the Xavante,
and ordered expeditions to be sent out to make contact.

At first, the encounter was not easy. At the end of 1941, the engineer
Genésio Pimentel Barbosa headed a team made up of employees of the
SPI and Xerente interpreters to create an attraction post close to the
Xavante, on the right bank of the River of the Dead.[20] While the ini-
tial offers of gifts were accepted, on November 6 the Xavante clubbed
Pimentel Barbosa and several members of his team to death.

Fortunately, at that moment the SPI decided not to use violence.
Engraved at the entrance of the cemetery that exists today on the site
of the attack, a slogan expressed a different attitude: "Die if necessary,
never kill." In 1943, the Brazilian government set up an official mission to map out the areas occupied by the Xavante and other indig-

enous groups, the famous Roncador-Xingu expedition. In 1946, the *sertanista*—an expert in the Brazilian interior—Francisco Meireles led a horseback expedition across fields and swamps, crossing tracts of scrubland filled with buriti palms until they came very close to the imposing Roncador Mountains. The many sounds of the deep forest did not include human voices . . . They waited a day and a night, they made signals with fire and flares, waited some more, but no one showed up. They left gifts and returned to the River of the Dead.

And then, a few days later, the objects that had been offered as gifts were accepted. Once again, the Xavante's strategy had changed: they were now seeking peaceful exchange with the whites. After tensions followed by arrow-fire and furtive contact on the banks of the River of the Dead, the Xavante lowered their clubs and in exchange received machetes, axes, fishhooks, domestic utensils made of steel, firearms, munitions, clothing, mirrors, and medicines. The indigenous figure who was essential for legitimizing this project of mutual taming was the chief Apoena ("he who sees far"). According to Xavante tradition, Apoena was to execute a strategy foreseen in dreams by his grandfather, relating to the start of a new cycle in the spiritual world. Fight and flight were no longer viable solutions, something new had to be done. In 1949, Meireles was finally received by Apoena in his village. The exchange strengthened the indigenous chief in the Xavante's internal struggles and defined a new model of contact, focused on a controlled integration with the economy of the whites, which made it possible to mix the consumption of supplies provided by the government with maintaining their own semi-nomadic lifestyle.

However, despite Apoena's diplomatic efforts, it took a long time for the Xavante lands to be delineated. The creation of the Xingu Indigenous Park to the north of the Xavante lands, a product of the efforts of *sertanistas* Orlando and Cláudio Villas-Bôas, seemed to the soldiers and businessmen of the Brazilian metropoles a more than ample concession. Squeezed by the growing land-grabbing pressure from colonists and politicians, the Xavante only started to see their lands made official—and even then reduced in scale—at the end of the 1960s. The increased contact with the whites who were in the middle of a population explosion (due to white immigration from southern Brazil) lured Xavante out of their villages to the cities or religious missions. Incursions and low flyovers by airplanes became common, illness and hunger spread, and the Xavante population began to decrease. There was a real risk of disintegration of the group.

It was then that Apoena showed his long-term vision once again. As an appeasement strategy inspired by a dream and agreed to by his community, Apoena sent eight of his grandchildren to live in the booming agribusiness city of Ribeirão Preto with white families connected to one of the rare friendly farm owners in the area. The aim was to incorporate the practices of white culture, but to infuse white culture with Xavante culture, too. As a complement to this strategy, at the same time as they were sending apprentice-ambassadors to the mysterious outside world, the Xavante closed the borders of their territories in order to stall the process of acculturation. This was 1973, the brutal height of the military dictatorship, and yet it worked. The closing of the borders allowed Apoena's grandchildren time to grow up as real and valued members of the families that had adopted them, creating bonds of appreciation and solidarity that have protected the Xavante ever since. Apoena himself died in 1978, proudly maintaining his Uto-pian peacemaking dream.

Did Apoena really dream all this? There is no anthropological docu-mentation that can answer that question, and maybe it doesn't even matter. More than knowing whether or not it was in a sleeping dream that Apoena saw his political actions, it is essential to understand that the story that supported them was told and retold as a dream until it had spread right through the Xavante community and far beyond—just as it's doing now, at this moment, as you are reading. Through the reiterating of the narrative, the individual desire was transformed into a collective desire for cultural survival.

Today, Apoena's grandchildren play a crucial role in the relations be-tween the Xavante and the outside world, filtering information, defend-ing their rights, and preserving their cultural identity. The new leaders are university-educated and they record their ancestors' traditions on video and audio, creating documents that are watched by mesmerized non-Xavante people young and old, replicating their culture through the technology of the whites. Their villages, which are filmed on por-table digital cameras, are now centers for spreading Xavante culture right across the planet.

The important Wai'á rini ritual, performed every fifteen years and until recently kept secret, was recorded in a detailed film by the Xavante filmmaker Divino Tserewahú. In this ritual, boys try to make themselves pass out through dances, rituals, staged conflicts, and exhausting physical trials that include races, water deprivation, and staring into the sun. When they do finally pass out and go into a trance, they receive visions and are initiated into adult life under the guidance of their ancestors. The film describes how they acquire powers to cure, to sing, and to interpret dreams:

> Dreaming is very important in the life of Xavante men. Through suffering and fainting during the ritual he is able to see what will happen in the future. When he tells what he has dreamed, it really happens. He can also encounter the dead through his dreams. This is why it is important to suffer and faint a lot during the celebration of the Wai'á rini. He who suffers the most, dreams the most, and has the most power.[21]

Dreams continue to be vital to the Xavante's policies of external relations. When they are called to Brasília to discuss questions of territorial boundaries, the elders gather to discuss the subject and then they try to dream of their ancestors and creator gods to ask for counsel. Sometimes they are transported to Brazil's capital to experience the proposed meetings in dreams. When the result is not good or when they do not have confidence in the white negotiators, they don't even bother to make the journey in the real world.

MORE REAL THAN REALITY

The use of dreams for political guidance when faced with powerful enemies has also marked the history of the Mapuche people of Chilean and Argentinean Patagonia. Ever since the Spanish invasion in the sixteenth century, through the national independences and up to the present, the Mapuche have resisted the expropriation of their territories by means of wars, rebellions, and messianic struggles.[22] Throughout this process, there has been a forceful conflict between the individualism of the European colonists, capable of a ruthless pragmatism, and the Mapuche collectivism based on egalitarianism, community socialization, reciprocity, and civility. The seizing of indigenous land was and continues to be the keynote of the conflict. Up until the start of the twentieth century, on both sides of the colossal southern mountain range, good money was still being paid for the head of an Indian.

Faced with a nightmare on this scale, it is not surprising that the Mapuche resistance has come directly through dream work. To these people, a dream (*pewma*) is a journey that the soul undertakes while asleep, a pan-Amerindian concept found in most ethnographic studies of the continent. Traditionally the Mapuche distinguished between dreamers, who were responsible for receiving messages from the spirits, and dream interpreters (*pewmafes*), typically women, extensively prepared to decipher oneiric accounts.[23] Between 1910 and 1930, the Mapuche leader Manuel Aburto Panguilef, one of his people's great prophetic dreamers, led an independence movement with significant dream guidance. Panguilef—which means "swift puma"—organized several conferences in which the Mapuche were able to sing, dance, pray, recount dreams, and discuss politics in their own tongue.[24] In 1921 the Federación Araucana was created, which Panguilef chaired until it was dissolved in the 1940s. Despite the indigenous beliefs' syncretism with Christianity, the Federation advocated keeping a distance from the customs of the whites and an adherence to Mapuche traditions, as well as the use of their own tongue, Mapudungun, instead of Spanish. In 1931, Panguilef proposed the creation of an autonomous Mapuche republic, but the power of his messianic leadership subsequently waned, giving way to a number of other resistance movements, almost all of which were violently suppressed.

On the eve of the atrocious military coup that toppled the Chilean president Salvador Allende, on December 11, 1973, the Mapuche leader Martín Painemal had a premonitory dream:

> I dreamed, at that moment I saw millions of birds that were at war. The birds were tearing one another apart. It was out of control, thousands and thousands of birds destroying one another like in a war. The birds were broken apart to bring down Allende. I dreamed before it happened, I kept thinking about it and understood that's what it was, it was a warning.[25]

Alerted to imminent disaster, Painemal took a number of precautions to escape persecution by the pro-coup forces. He went into hiding and survived.

JOURNEYS AND MAPS

With many variations on the same theme, Amerindian cultures typically assign dreams a crucial place in which time is condensed, with past, present, and future all together in a vast, ongoing continuum. In the soul's wanderings through the horizon of possible futures, the dreamer and especially the shaman try to diagnose the situation that is underway and take control of the dream so as to reverse the causality of events. Instead of simply seeing what has happened or what will happen, they try to create a new reality through their own actions.[26] One exemplary narrative tells of a shaman who goes off in search of a cure or a solution through dreams.

A Juruna myth describes how a young man called Uaiçá went off to hunt and found a tree with a number of dead animals surrounding it. He fell asleep and dreamed about animals from the forest, about people singing, and about Sinaá, the Jurunas' jaguar-god ancestor, with whom he spoke at length. Uaiçá woke up as the sun was setting and returned home. The following day he decided to fast. For some time he would return daily to the tree, and the dreams were always repeated, until Sinaá himself ordered the visits to stop. When Uaiçá awoke, he made a tea out of the bark of the tree, drank it, got intoxicated, and thus began a shamanic process of acquiring powers: he caught fish with his hands, he drew illnesses out from people, and he had eyes in the back of his head. When he slept, he traveled to Sinaá and returned from the world of dreaming bringing everything the Juruna wanted. He became an important medicine man.[27]

Kaká Werá Jecupé provides valuable testimony about the social importance of Amerindian dreams:

The dream is the moment when we are stripped of the rational structure of thinking. We are in a pure state of spirit, in *awá*, complete being. It is a moment in which we become connected to our deepest reality. In a dream your spirit literally travels and can be directed to any place or time you choose. This requires training, like learning to speak . . . There exists among some peoples a thing that is done in the mornings called dream circles. They gather fifty people into a circle and they start recounting their dreams. And those dreams start to give a steer for the village's daily life . . . The peoples consider dreams a moment of liberation for the spirit, when the spirit sees everything from all angles.[28]

Big dreams are yearned for, obtained, and revered by all of indigenous America, from the farthest south to the farthest north. In 1981 the British anthropologist Hugh Brody described impressive dream hunts among the indigenous Dane-zaa, the Beaver people of the Canadian Subarctic. This is an ancient tradition that was already being lost at that time, known only to the elders. In these special dreams the hunter would undertake a journey to find out where the animals were that he was looking for—and he had an opportunity to choose which exact animal should be sacrificed. On waking, he would produce a map that showed its location. The dream map described by Brody had been kept folded up for many years. It was as large as their tabletop and it was covered in thousands of tiny marks, each of them firm and colored. The white visitors were invited to gather around the table to examine the map.

[The Aborigines] Abe Fellow and Aggan Wolf explained. Up here is heaven; this is the trail that must be followed; here is a wrong direction; this is where it would be worst of all to go; and over there are the animals. They explained that all of this had been discovered in dreams.

Aggan also said that it was wrong to unpack a dream map except for very special reasons.[29]

THE TIME WITHIN TIME OF THE ARRERNTE

The idea of dreams as portals for communicating with another world might have reached its apex among the Arrernte Aboriginal people of central Australia, who descend from the first human immigrants to explore that arid territory at least 65,000 years ago. They believe in the Alcheringa, a primordial spiritual plane that existed before the dreamer was born and will continue to exist after their death, inhabited since the beginning by all ancestry in an overlapping of past, present, and future that Jung would not have hesitated to call a collective unconscious: the collection of memes of an entire culture. The experience of the Alcheringa seems to reach such high levels of vividness that the

Arrernte believe that they actually live in that world. For them, dream life is more real than waking life. Among different subgroups, which are culturally diverse but genetically similar, the different words used to mean "dream"—*alchera, bugari, djagur, meri, lalun, ungud*—are synonyms of the primordial time of the creation of the world, which was popularized in the West under the English name *Dreamtime*.[30]

In this fundamental dimension of existence, time is not experienced as "one-thing-after-another," but as "all-at-the-same-time-now," or "time-outside-of-time." In some groups, this is a time before the present, in others a time within this one, in others a time parallel to this one.[31] In the Alcheringa are preserved all the initiation secrets, the cosmogony and ontology of the Aboriginal peoples, as well as the vast repertoire of practical knowledge that enables them to inhabit such an inhospitable environment, with such scarce resources and such deadly predators. The Alcheringa is the source of hunting, cooking, and painting techniques, but also of the maps of sacred trails that make it possible to move safely around the largest island in the world, using specific geographical features to mark out excursions and routes. Also from the Alcheringa come the teachings of the elders to the younger people in the form of songs, dances, and stories that explain where to find water, hunting grounds, shelter, and materials for constructing wooden and stone tools. Totemic secrets are revealed in the Alcheringa, which is why it is said that a particular person specifically has a kangaroo dream, a honey ant dream, a shark dream, or a badger dream. Some mysteries are only transmitted in old age, when the dreamers have become sufficiently mature.

The Alcheringa is where the mythical identification with the past of the ancestors occurs, a renewed source of the present and canonical reference-point for situations and attitudes that are never exactly new, since they are repetitions of arcane patterns. In this way, dreams make it possible to speak with the ancestors and other spirits in search of knowledge and guidance, in an encounter that is likely as thrilling and inspiring as it must have been for Axial Age Greeks dreaming about the Homeric heroes of the Bronze Age.

In Arrernte culture, nature is an incredibly vast temple and life is a continuously numinous experience, since spirits endowed with intentionality live inside the flora, the fauna, and the mineral world. Their animism is intense and archaic, possibly the oldest uninterruptedly practiced religion on the planet. Being immersed in it, the Arrernte identify freely with any natural object, both in dreams and in their waking lives. The Alcheringa allows them to live a totally different life in their sleep with spirits of all kinds, including animals and plants and multiple generations of forebears, in an experience that is so full and heightened that returning to waking life is like returning to a dream, and falling asleep is like waking.

LEAVING THE BODY

Tibetan monks, meanwhile, understand dreams as mere constructs, illusions subject to the manipulations of the dreamer's will, the limits of their technique, and their intentions. They consider every falling asleep to be a preparation for death, and they practice *milam*, or dream yoga. This discipline allows them to attain states of heightened lucidity in which they learn to control their dreams without any difficulty or fear, knowing that what they are is a strictly interior reality.

Milam is learned in successive stages, with small variations depending on the different traditional lineages. To start with, the dreamer needs to learn to recognize that they are dreaming, that is, they need to learn how to become lucid within their dream. At first, it is very hard to establish the consciousness that one is dreaming, and it is common to attain oneiric lucidity but then immediately lose it again, simply forgetting the suspicion that none of that is real. The capacity to sustain this doubt about reality is essential for reversing the causality of dream accounts, such that the oneiric events stop simply happening to the dreamer and start to be prompted by their will.

In the second phase, the dreamer must free themselves from any fear caused by the content of the dream, becoming aware that no event there—terrible as it might seem—can cause any real harm. This learning is necessary because the threshold of dream lucidity harbors a whole host of astonishing things that can present themselves to the unsuspecting dreamer, an army of scares that have been incubated, nourished, elaborated, and treasured up over a lifetime. Typical exercises during this stage include setting fire to one's own dream body to confirm that it does not cause any pain or leave any marks.

In the third stage, the dreamer must consider the fact that all things, whether in dreams or waking, are in a permanent state of mutation and are mere illusions, impressions that are fleeting and void of substance.

When a beloved being appears or disappears in a dream, it is essential to know that person is merely a shell, an incomplete apparition, an imperfect collection of representations. Whether blissful or repellent, dream images are mere chimeras.

The deep acceptance of this fact allows the practitioner of *milam* to begin the next stage, in which the dreamer learns to transform dream objects at will, altering their size, weight, or shape in a controlled fashion. In the mental space of the dream, natural laws are no more than conventions, acquired from waking experience and entirely violable by the active imagination. Even if in dreams most oneiric objects and people fall to the floor when they are knocked over, they are not in fact subject to the law of gravity, and can float if the dreamer so wishes it—or rather, if they know how to wish it. The expanding of this remarkable skill also allows the dreamer to determine the setting for the dream, as well as its characters. Progress in this stage of *milam* requires an increase of the dreamer's force of will, as it is through robust intentional desire, through the force of voluntary will, that the dreamer is freed from the role of the dream's character to become its creator.

Once they have mastered the art of shaping oneiric objects and scenes, *milam* practitioners are trained in the art of transforming their own bodies, increasing or decreasing their size, altering their shape, or even extracting themselves completely from a scene without causing the dream to end. In this stage, the difference between the dreamed body—a mere representation of the self with a particular point of view within the dream activity—and the dream as a whole, the construction of a mind that houses the self but is much more than it, becomes explicit.

Finally, at the most advanced stage, the dreamer must learn to unite their dream with the "clear light of the void," visualizing a Buddha or some other divinity in a state of oneiric lucidity. Of course the transcendental significance of this stage is beyond the comprehension of anybody who has not reached it. But it is not necessary to know this meaning to understand that *milam* is a path of self-awareness that expands the mental capacities of the dream.[32] With a few important differences, the Hindu *yoga nidra* offers a similar journey of self-discovery through the exercise of unfoldings of the body in the transition between sleep and waking.

INWARD AWAKENING

While traditions around the Himalayas or in South America use meditation to attain visions, many other cultures around the world advocate
suffering, fasting, and penance to similar ends. From the Sun Dance
on the United States prairies to the Xavante faintings, from medieval
Catholic ordeals to the walking over hot coals of a Hindu sadhu, leaving
the body and seeking out visions is also achieved through the acceptance and overcoming of pain. What revelations might have taken hold
of Giordano Bruno when he silently faced the flames of the Inquisition? Would he have been feeling the same pain that other people feel?
Or would he have been in a lucid, mystical trance, his internal reality
completely dissociated from the external world, his hallowed mind
someplace far beyond its own immolated body?

Just as pain is a route to a trance, dream states can also be achieved
through pleasure. Imagine the delicious altered state of consciousness
reached by some Islamic Sufis with their repetitive whirling and hyp-
notic music. The attitude of turning oneself toward the inside of the
mind is at the heart of many techniques that involve meditation, visualization, posture, mantras, recitations, and chants. Today there are
already programs of sound stimulation aimed at producing alterations
to the visual experience. It is also possible to achieve visions through
massage and tantric sex. And there are even more incredible, enchant-
ing visions that arise from breath control, whether through the practice of Prānāyāma or other traditional Eastern methods, or through
Western techniques developed in recent decades, such as the holotropic
breathwork of Czech psychiatrist Stanislav Grof.

Bruce Lee, the great master of kung fu, emphasized the importance
of a state of enlightenment: "*Satori*—in the awakening from a dream.
Awakening and self-realization and seeing into one's being—these are
synonymous."[33] While it is so deeply rooted in the East through Buddhism, Zen, Taoism, and Tantra, introspection in the West still meets
with resistance and skepticism. We are almost completely blind when
it comes to our internal organs and processes. Try moving your left
thumb. Easy, isn't it? Now try activating your right hippocampus . . .
The insensitivity to almost everything that happens normally within
our bodies might be the human being's default psychological state, but
techniques such as Chinese Qigong or Hindu asanas alter this experience enormously. Their practice makes it possible to hear one's own

heart beating, control one's body temperature, and feel one's guts. These are matters on the threshold between science and metaphysics, places where scientists have not yet spent sufficient time to map without prejudice these phenomena that they do not yet understand. The few scientific studies of the subject suggest that these skills are real.[34]

If a process of inwardly waking up can be viscerally physiological, there is nothing to stop it from being deeply symbolic, too. Both in *milam* and *yoga nidra*, all of the actions and non-actions of the practitioner occur in a mental state of internal freedom, what is known to science as lucid dreaming. It is a state normally associated with the later stages of REM sleep, moving toward the morning, when the body has already slept enough and so goes into a very special state in which there is little pressure to sleep, along with elevated stocks of releasable neurotransmitters, and abundant REM. It is at this moment, when the brain is dreaming vigorously but is already ready to wake up, that sometimes, almost miraculously, it wakes up into itself.

LUCID DREAMING

Amerindian peoples, Australian Aboriginal peoples, Tibetan yogis, and Christian monks are all masters of dream navigation. Being conscious that one is dreaming is a necessary condition for beginning transformational journeys. Among the Mapuche, dreaming that you are dreaming suggests a great vitality of the soul. These are dreams of profound emotional impact, that strengthen the sense of autonomy and greatly empower the dreamer.

The normal course of REM sleep typically leads to two opposing situations: waking up quickly and then right away returning to the dream state, or waking up and sustaining the vigil. However, persistent practice allows the dreamer to balance on the subtle threshold between REM sleep and waking, expanding their consciousness in such a way as to master the process of mental simulation that characterizes dreaming. This kind of dreaming is very impactful and adds a fresh dimension to mental life—not more or less than what there was before, but on a new and totally different axis. This is the dream of heightened lucidity, in which the dreamer knows they are dreaming and can exercise total or partial control over everything that goes into the dream narrative.

When a dream is transformed into a portal to voluntary oneiric action, it becomes a privileged space for learning, practicing, loving, traveling, and reflecting. It also becomes a space that is conducive to finding and interacting with the creatures of the mind: relatives, friends, ancestors, spiritual beings, gods, and even God Himself. In some strands of New Age Christianity, it is believed that the main phenomena of Christian gnosis can be realized in the space of the lucid dream, giving practitioners a concrete path toward reaching elevated mystical states, described as "seeing the light."[35] While a lucid dream most usually develops from a non-lucid dream state, traditional and contemporary accounts agree that it is also possible to reach oneiric lucidity from waking.

Recognized by Aristotle, Galen, and St. Augustine, the self-aware dream was the subject of a bulky philosophical treatise by the Marquis Léon d'Hervey de Saint-Denys, entitled *Dreams and the Ways to Direct Them.*[36] Prompted by the marquis's ideas, by the term *rêve lucide* ("lucid dream" in French), and by his own experiences, the Dutch psychiatrist Frederik van Eeden provided a scientific report of the phenomenon in 1913:

> I can only say that I made my observations during normal deep and healthy sleep, and that in 352 cases I had a full recollection of my day-life, and could act voluntarily, though I was so fast asleep that no bodily sensations penetrated into my perception. If anybody refuses to call that state of mind a dream, he may suggest some other name. For my part, it was just this form of dream, which I call "lucid dreams."[37]

While van Eeden's report is powerful, not many people were convinced by it. If the study of common dreams is already fragile because of its dependence on the reporting by third parties, how are we to take seriously a superdream in which the dreamer claims to have complete self-awareness? For decades, the skeptics of lucid dreaming disseminated their interpretation that it was really a state of resting wakefulness, in which the body is immobile but awake. It was not until the 1970s that an empirical response to this objection appeared, which established convincing physiological bases for the scientific study of the internal space provided by lucid dreaming. In 1978, the British psychologist Keith Hearne demonstrated in his doctoral thesis that it is possible to signal the entry into lucidity through the eyes, which are normally active during REM sleep and are therefore a "window to the (dreaming) soul."[38] The same was shown during the doctoral work of U.S. neuroscientist Stephen LaBerge, completed in 1980 at Stanford University under the guidance of William Dement.[39] In both cases, the researchers made use of the eye movements that occur during REM sleep, despite the total muscular atonia of the rest of the body, to get around the dogmatic assumption that it would be impossible to signal the occurrence of a lucid dream without immediately waking up. Working with research volunteers who had been trained to enter dream lucidity easily, the researchers asked them to perform pre-agreed eye movements to announce the start and end of each episode of lucid dreaming. The accompanying lack of body muscle tone and of high-frequency brain waves confirmed that this was REM sleep and not waking. In other words, the volunteers were managing to move their eyes voluntarily even while they were dreaming. One point to the yogis.

Over the course of the 1980s, LaBerge carried out a number of different studies that were fundamental to the understanding of lucid dreams. He showed that lucid dreaming is a voluntary skill that can be verbally suggested, practiced, and stimulated by sensory signals. He also showed that in this state it is possible voluntarily to control one's breathing and that lucid dreams typically occur during periods of REM sleep governed by the sympathetic nervous system, creating a "super REM sleep" with high cardio-respiratory rates and abundant eye movement.[40]

LaBerge and Hearne's discoveries, though neglected at first, have in the last twenty years been confirmed and greatly built upon. We know today that a lucid dream is an intermediate state between waking and REM sleep, a hybrid state in which attention is turned "inward" as in sleep, but with the intentional consciousness that characterizes being awake. Though it is rare, lucid dreaming occurs spontaneously at least once in most people's lives, especially in women, with a drop in frequency after adolescence.[41] Most people would like to have another dream like it, but almost nobody knows how to repeat the experience. The U.S. psychologist Benjamin Baird recently got together with Stephen LaBerge to show that a lucid dream can be induced by galantamine. This substance increases the neuronal responses to acetylcholine, the increased release of which occurs during REM sleep. What results is an extremely vivid dream, marked by concentration, focus, and the voluntary making of decisions.[42]

THE NEURAL CORRELATES OF LUCID DREAMING

The thing that allows the well-trained lucid dreamer to attain control over the oneiric narrative is the volitional governing of the imagination, the directed wish that rules over the dream's actions and scenes. Without any fear, and without letting themselves get carried away by their own ability to create the dream narrative, the practitioner of this controlled madness is able to master the vertigo of accessing their unconscious and navigating it at will. The use of executive function during a lucid dream suggests that the prefrontal cortex, which is generally inactive in REM sleep, must be activated during a lucid dream.

In accordance with this hypothesis, J. Allan Hobson and the German neuroscientist Ursula Voss showed in 2009 that lucid dreams are accompanied by an intensifying of high-frequency brain waves in the prefrontal cortex, in comparison to non-lucid REM sleep.[43] Combining the EEG records with functional magnetic resonance measurements, the German neuroscientist Martin Dresler from the Max Planck Institute of Psychiatry in Munich showed that lucid REM sleep, unlike non-lucid REM sleep, shows more activation in the brain regions related to decision-making and intentionality (prefrontal cortex), vision (occipital cortex and cuneus), reflexive consciousness (precuneus), memory (temporal cortex), and space (parietal cortex).[44] The same researchers showed that a motor task carried out during lucid dreaming—opening and closing the fists—triggers activity in the sensorimotor cortex that is normally activated by the same action when performed awake.[45] This was the first visualization of the neural representation of a piece of dream content.

In order to test out the hypothesis that an artificial increase in prefrontal activity during REM sleep can prompt the transition to a lucid dream, two different teams of scientists carried out experiments stimulating the prefrontal cortex during REM sleep. In 2013, the Lithuanian Tadas Stumbrys and the Germans Michael Schredl and Daniel Erlacher showed that reports of lucidity increased after the stimulation, but only among experienced lucid dreamers.[46] In 2014, Voss and her team showed an increase in lucidity during transcranial stimulation at high frequencies, even in people without previous experience of this sort of dreaming.[47] While the question of how much oneiric lucidity depends on practice or innate propensity continues to be debated, there are no longer any scientists who can bring themselves to deny its existence.

MONKS AND NEURO-JEDI

If lucid dreaming is now a well-established fact, what can we say about its supposed usefulness for practicing skills that are relevant in real life? Might it be possible to re-create in a laboratory the essence of those epiphanic dreams of Crazy Horse, in which he "dreamed himself into the real world, and he showed the Sioux how to do many things they had never done before"?[48] Might it be possible to use a lucid dream as a virtual space for constructing special skills, like the character Neo in the movie *The Matrix*, who learns kung fu without any fear of getting hurt? Could computer programmers write code in their dreams? Might it be possible to practice in your sleep?

These questions are far from being answered, but scientists are making progress in that direction. A study of 840 German athletes and their relationship to oneiric lucidity showed that 57 percent experienced at least one lucid dream in their lives, and that 24 percent were frequent lucid dreamers, with at least one episode per month. The most interesting piece of data is that 9 percent of the athletes able to have lucid dreams claimed to use this state to practice their sporting skills, as they believed it would help them to improve their performance in real life. The researchers decided to follow the trail supplied by the athletes in a more prosaic but no less interesting way, looking at the dream practicing of motor skills as simple as throwing coins into cups that are farther and farther away, or hitting targets with darts. The results showed that practice in sleep leads to a significant increase in accuracy in real life.[49]

There has also been research comparing the perception of time in lucid dreams and the perception of time while awake. For mental tasks that do not involve movement or physical effort, there is equivalence between dreamed time and real time, but in the case of motor tasks such as walking or doing exercises, the time taken to carry out the dreamed task can be up to 40 percent more than the time it takes while awake. We still do not know whether this increase in the duration of motor tasks when they are carried out during a lucid dream reflects a possible slowing down of motor processing during REM sleep, or the lack of muscle signals that can feed the movement being dreamed back to the brain. However modest the tasks that have been investigated hitherto, we still retain the promise of an unlimited arena for mental training. A recent study carried out by LaBerge, Baird, and the neuroscientist Philip Zimbardo, of Stanford University, showed that the eye movements produced during lucid dreams are more like those of open-eyed perception than those of closed-eyed visual imagination. The scientific evidence that lucid dreaming is, in fact, a state of being inwardly awake is growing.[50]

A SEDUCTIVE INVITATION

The most significant experiences in life can only be appraised by the person who lived through them directly. Explaining the experience of having children to somebody who has never had them is an existential impossibility. In the same way, it is almost impossible to convey the thrill and adventure of oneiric lucidity. Lucid dreams are typically very pleasurable in their revealing of the absolutely vast internal space of mental representations, a conscious expression of the huge trove of memories from the whole mind, a place where it is possible to satisfy almost any desire. If you have never had a dream of this kind, now is a perfect moment to learn to have them.

There is no one single methodology for practicing lucid dreaming, but some protocols can help. The first step is to resume the dream diary recommended in Chapter 1, which is very good practice for recalling and relating one's own dreams. In addition, it is also important to apply techniques that can increase perception of the dream state, such as the habit of asking oneself frequently, during the course of the day: "I wonder if I might be dreaming?" This question could accompany the sight of a specific object, such as your own hand. A brief period of auto-suggestion before going to sleep also helps to encourage lucid dreaming, through the mental conception of the experience you want to incubate. Even more useful is waking up in the small hours of the morning to carry out that auto-suggestion on the threshold of the final episodes of REM sleep of the night. With an increased understanding of what happens in the brain during lucid dreaming, it is likely that more and more people in the world will have access to this powerful form of self-control. One further thing that might be useful for attaining lucidity is the use of those electronic masks sold on the Internet that can signal the start of REM sleep with lights and sounds, thereby creating a low-intensity external stimulation that could facilitate the transition to oneiric lucidity. It is quite possible that transcranial stimulation devices will be launched onto the market to help to induce lucid dreaming in even the most obstinate skeptics.

Once the process of entering a lucid dream has been mastered, what can you do then? Almost anything. Be reunited with loved ones, live out a great romance, go on dangerous adventures, travel to the edges of the imagined universe, practice maneuvers that are risky in real life, and freely fulfill your desires with no guilt and no obstacles. How one interprets the oneiric lucidity phenomenon depends on one's point of view. To mystics, this sort of dream is a doorway to explore the world of the spirits, a state that makes it possible to unfold and project the astral body in order to undertake journeys in other planets and dimensions. To materialists, the dream is the key to navigating the broad ocean of the unconscious, the totally individual collection of a lifetime's memories and their combinations.

In addition to the fascinating experiments centered on the lucid dreamer's capacities and limitations, there are pioneering advances being made in experimentation on the cognition of those characters who appear during lucid dreams, those magical beings revered by Tibetan monks, Jungian psychoanalysts, and Xavante medicine men. Field studies and lab studies have shown that these characters are able to write, draw, rhyme, supply words that are not known to the dreamer, and even propose creative solutions to metaphorical puzzles, but they possess one curious weakness: they have great difficulty with problems of logic and arithmetic.[51] It is as if the creatures evoked during a lucid dream were mentally limited by the human difficulty in dreaming about letters and digits.

IN WHICH DIRECTION IS DREAMING EVOLVING?

The materialist perspective on lucid dreaming is an updated version of a moral dilemma from sixteen centuries ago. St. Augustine exempted people from responsibility for the sins committed in their dreams, because he considered that these were not under the dreamers' control but rather something that happens to them. The possibility of acting with intentionality to affect the course of the oneiric narrative calls this argument into question. A lucid dream allows us to kill other characters and commit any kind of repulsive act. What would be a heresy to countless ancestral traditions is transformed, in the minds of young hedonists with no guilt and no responsibility, into an amusement park as amoral as any session on a video game.

There are many ways of trivializing lucid dreaming. The emulating of specific characters merely in order to have abundant sex or, shocking as this is, to inflict torture or commit murder, illustrates the extremes of this privately cultivated self-stimulatory commotion. Navigating one's own unconscious as though it were a computer simulation would not be approved of by the Aboriginal people of Australia, nor by psychoanalysts, who believe in the importance of maintaining the integrity of the spirits or of their mental representations, respectively.

Psychoanalysts tend to see in the hedonistic use of lucid dreaming a reckless misappropriation of the dream function, since the gratification of the fantasy of internal control at the expense of action in the world can stimulate harmful features of the personality. According to this view, lucid dreaming would act as a narcotic, rewarding the subject not for attainments in the real world, but for purely imaginary successes. The fulfilling of desires without consequences in reality would be particularly perverse, as it would disconnect desire from responsibility, occluding the escape valve for normal tensions offered by the unconscious.

Yogis and neuroscientists tend to be more optimistic about the benign potential of dream control, but this potential depends on the choices made by the dreamer in the act of dreaming.[52] If the lucid dream is a sophisticated way of deliberately reprogramming the brain, its effects should depend on the images and actions chosen to make up the experience. In other words, if a lucid dream reverberates memories and regulates gene expression—as happens in non-lucid REM sleep— then its effects should be similar to the effects of performing those actions in the real world.

This consideration marks out an ethics—or should we say a mental hygiene?—of lucid navigation. It also celebrates and honors those dreams of diagnosis and cure, from the cult of Asclepius to the *pewmafes*, the Mapuche dream interpreters. There is preliminary evidence suggesting that lucid dreams can suppress recurring nightmares and chronic pain. On the other hand, the suggestion made some years ago that lucid dreams can treat psychoses is not sustainable. Oneiric lucidity is psychologically safe for people without psychotic symptoms, but in psychotic patients this practice can reinforce delusions and hallucinations, making internal reality resemble external reality even more.[53] In the words of Mariano Sigman,

> Lucid dreaming is a fascinating mental state because it combines the best of both worlds, the visual and creative intensity of dreams with the control of wakefulness. It is also a goldmine for science . . . Perhaps lucid dreaming is an ideal model for studying the transition between these primary and secondary states of consciousness. We are now in the first stages of sketching out this fascinating world that has only recently appeared in the history of science.[54]

A DOOR TO THE FUTURE OF CONSCIOUSNESS

In which direction are we evolving? Where will our consciousness end up? Could the lucid dream be the embryo of a new human mind? Meeting our ancestors, which is of the greatest importance to so many civilizations, is only one of the many impossible experiences made possible in this state. The movement inward creates a space for scientific discoveries, through the perceptual intuiting of things that today can be expressed only mathematically, such as the possibility of many dimensions of reality existing beyond the four we know about. Could Giordano Bruno's dream flight outside the solar system and into the cosmos have been lucid?

If the invasion of waking life by dreams was crucial for the evolution of our way of thinking, the evolution of our minds from here on might be linked to the ability to wake up within a dream—and thereby to expand the known states of consciousness. Beyond the neural mechanisms of this waking, which we have only recently begun to unlock, it is impossible to deny that this state represents a frontier of great wonder in our dizzying cognitive evolution. At a time when people are depending more and more on digital virtuality to store up memories and simulate ideas, the virtual navigation through the world of symbols, at the conscious disposal of the dreamer in its pseudo-infinity of recombined representations, signals a renewal of our future in flesh and bone. The exploration of dream lucidity will open up new pathways for human creativity, invention, and discovery, with extremely rich possibilities yet to be explored.

We can no longer say that we are as we were. Those people who were born, as I was, before the advent of the Internet can and should think of themselves as cyborgs 1.0, humans who learned in the past decades, with varying degrees of effort, to delegate almost all their memories and basic everyday activities to machines. Our children and grandchildren, the cyborgs 2.0, were already born in the brave new world in which computers and the Internet are as normal as an apple tree. (Actually, to most children around the world, very much more normal.) This generation and the next will have no trouble accepting the innovations that will embed the technology of virtual communication in their own bodies, in electronic contact lenses or nano-implants of various kinds that will allow them to access files and navigate the net as though by telepathy. But the question has to be asked: equipped with such technology, will we know how to survive ourselves?

Among Native peoples in North America, the principle of the "seventh generation" posits the idea that every individual or collective decision should be considered not only for its effects on the present moment, but also and principally on the future, symbolized by seven consecutive generations. Described in this way, it's shocking that this is not a universal principle. Thinking about the long-term consequences of our actions as far as our imagination can go, we can simulate chains of action and reaction that often reverse the intentions of the original act. This reasoning is at the basis of the "Great Law of Peace" of the Iroquois Confederacy (the oral constitution of its six nations), and is currently used to guide the pan-indigenous struggles in the United States, Canada, and Mexico. If we do not imagine the future, we run the risk of compromising it irremediably.

The United Nations' Intergovernmental Panel on Climate Change (IPCC) reported in October 2018 that the planet is headed for an increase in surface temperature of 3–4°C (approximately 5–7°F) by the end of the twenty-first century. We can expect major climate fluctuations, extremes of cold and heat, huge storms, droughts, and floods. The acceleration of the rise in sea levels echoes the floods of the Sumerian Ziusudra and the Hebrew Noah. According to the panel, a radical change to our economics, a change "unprecedented in scale," is needed to avoid global climate chaos. Drastic geopolitical changes await us, as the Arctic tundra will become fertile, while the continental masses of the Southern Hemisphere, which are concentrated close to the equator, are likely to experience inexorable desertification.[55]

The United Nations' warning that the time to avoid disaster is running out reverberates profoundly among the hunter-gatherers of the Amazon, who with their tenacious diversity represent one of our species' most successful ways of living. A lot of consciousness-expanding will be necessary for us to escape the trap of symbols that we have invented, this dangerous mixture of high technology and low instincts. Under the military-ridden government of president Jair Bolsonaro, Brazil has seen a spike in deforestation and in the killing of indigenous leaders. In his seminal book *The Falling Sky*, the shaman Davi Kopenawa warns us:

> Our spirits are already talking about this, even if the white people are convinced all the words are lies. The *xapiri* and *Omama*'s image try to warn them: "If you destroy the forest, the sky will break and it will fall on the earth again!" They do not pay any attention to them, because they do not drink the *yãkoana*.* Yet their skill with machines will not allow them to hold up the falling sky and repair the spoiled forest. They do not seem to worry about disappearing either, probably because they are so very many. But if we peoples of the forest are no longer, the white people will never be able to replace us there, living on the old traces of our houses and abandoned gardens. They will perish in their turn, crushed by the falling sky. Nothing will remain. It is so.[56]

The grim warning of the imminent falling of the sky on our heads echoes the atavistic fear of collective annihilation. Davi Kopenawa's words echo the Sumerian nightmare of Dumuzid, similar in turn to what must have been experienced by the Xavante fleeing into the Bra-

---

* A psychedelic snuff made from *Virola theiodora*, a plant rich in 5-MeO-DMT. Kopenawa uses the verb "to drink" *(koai)*, but the *yãkoana* is in fact a powder that must be inhaled.

zilian backlands, by the Lakota wandering the icy prairies waiting for the murderous bluecoat mob, by the Mapuche with a price on their heads wandering through the cold winds of Patagonia:

> I constantly saw *garimpeiros* (gold prospectors) attacking me when I dreamed. . . . They told them: "We have to get rid of this Davi who claims to prevent us from working in the forest! He knows our language and he is our enemy. We have had enough of him, he bothers us! These Yanomami are dirty and lazy. They have to disappear so we can search for gold in peace. We have to smoke them out with epidemic fumes!" . . . The army was also hostile at the time. They wanted to carve our land into pieces to let the *garimpeiros* in. Then I would see the images of soldier spirits, with their steel helmets and their war planes trying to take hold of me to lock me up and mistreat me. Yet my *purusianari* spirits . . . came down into my dream to battle the white soldier spirits. They tore out their paths and carried them off into the sky's chest. Then they abruptly cut them, and all of them were hurled into the void.[57]

There is no guarantee that the future of the human dream will not be—as it probably was at its beginning—a grim nightmare. In the chaotic Brexit of 2020, the line from the punk band the Sex Pistols echoes presciently: "There's no future / In England's dreaming." In these times of viral, economic, and environmental dystopia, with the jolt that the Trump presidency brought to an already quite poisoned and warring planet, and now that we have been hit hard by the COVID-19 pandemic and by the crude surfacing of our most neurotic and perverse shortcomings, every day that dawns is a sigh of relief, a renewal of hope like that experienced daily by the Mayan people in the last millennium, fearful with each sunset that the sun would not interrupt the night to rise again the following day. If we survive the ongoing holocaust, it is likely that we will still lose sleep, as global warming is likely to reduce the hours of nighttime sleep substantially.[58] Lest we manage to dream our way out, the future promises insomnia.

The vertiginous growth of our capacity for virtual communication is also a cause for major concern, as it drains away the time for real interactions and entangles us in the absolute relativism of opinions.

Measurement of the impact of rumors on Twitter between 2006 and 2017 show that the most widely shared posts are precisely the most fictional ones. Algorithms, software robots, "souls without bodies," fully up and running, have already won elections on extremist platforms in the United States, in the United Kingdom, and in Brazil, through a massive and automatic propagation of fake memes that infect people until they believe that it was they themselves who wove together these mendacious narratives. The excess of information and lack of judgment means that we run the risk of losing trust in accumulated knowledge and experiencing a new Tower of Babel, a cackling of dissonant voices with no harmonization possible. It is natural for a monkey to hurt itself when it starts playing with the new toys it has invented. Adolescent monkeys give out false alarms all the time, and they get ignored. Talking to thousands of people at the same time is an incalculable power that we have not yet learned to use properly. The disastrous mismanagement of the COVID-19 pandemic by presidents Trump and Bolsonaro and prime minister Boris Johnson would not have happened without the capacity to instantaneously spread lies worldwide. For the pandemic of fake news to be quelled, the mean lying monkeys must no longer be heard.

In order to avoid our cultural ratchet proceeding uncontrollably toward global collapse, we need to broaden our perspective. We must urgently recover the capacity to imagine the worst consequences of our most ingrained habits. The science of biologists, chemists, and physicists needs to walk arm in arm with the wisdom of shamans and yogis, not to be massed against it. The lucid dream, in its vastness, has the potential to be the mental space that will allow us to imagine solutions to the most challenging problems, from the destruction of the water sources to the dichotomy between mind and brain, from the accumulation of microplastics to the devastation of Amerindian peoples and black populations by COVID-19, from persistent police brutality to persistent male supremacy, from an epidemic of suicide to the accelerating deforestation of those unspoiled lands that remain, from extreme inequality to widespread corruption, from the most destructive addiction of all—money—to the carnage of the breeding and cruel slaughter of animals, from predatory capitalism to the end of almost all jobs, very shortly, when the robots conclude their triumphant arrival.

Finally, if we do avoid cataclysm, perhaps it will be precisely in the field of active imagination, in lucid dreams themselves, that just the right mental space will be found for us to ask the biggest question of all: Why does reality exist? Are we living in a dream, in a simulation? On the subject of what happened before the Big Bang, the pope knows just as much as the best astrophysicist: nothing. Most physicists would insist the question doesn't even make sense, because before the Big Bang, time would not have existed. So how did everything come from nothing? Non-duality stares us in the eyes, indecipherable. We are born, we live, and we die in utter metaphysical bewilderment, as we simply do not have the answers. Probably, almost certainly, we never will, but maybe, just maybe . . .

It is possible that the understanding of a phenomenon as mysterious and arbitrary as the very existence of space-time and of the objects of the universe demands—in addition to some intergalactic travel—an interior journey that is much deeper. Looking inward, in fearless abduction, toward the dizzying abyss of consciousness, might perhaps be as revealing as looking outward through the lenses of microscopes and telescopes. In the future, dreaming will be more and more like a dazzling revelation.

Epilogue

My recurring dream about the witches in the concentration camp disappeared, and life went on. I started having occasional dreams about my father. Over the decades, these dreams have explored many possibilities of return: a dead body that walks, a healthy man reborn, or simply a deserter who has gone to live elsewhere. Ever since we had Ernesto, I have not dreamed about him. Our second son, Sergio, took his name. He had been alive in me all this time, a creature of my mind. And now my mother lives there, too. I still haven't dreamed about her—not yet. I hope that one night or day I will have a full dream about them both, rediscovering in the kingdom of Aruanda the best of what I have kept and that will be passed on, in the name of the seventh generation after us.

There, he dives like Cousteau alongside a giant shark and races across the prairies of the Alcheringa like Crazy Horse riding atop a Bengal tiger. Where she reads all the books in the Babel library of my head, which are guarded at a distance by Queen Nzinga's lancers silhouetted against the horizon, protected by Apoena's warriors carrying the buriti palm logs of the ancestors across the endless future, freely exploring the oceans, fields, and mountains that I sculpted throughout my whole childhood while playing on the ground, up a tree, in the sea, and in my imagination, in books, records, and comic books, TV and the cinema, and on the Internet, traveling intensely along internal paths long before I was ready to try to explore the vast world out there.

A fine place to live, made up of luminous experiences, a shelter for the whole family for the future it allows us to pursue. My own home in which there is a place for the Yanomami and the Lakota, for aliens and souls, robots and artificial minds, which won't be long in arriving. *Imagine all the people* living inside your head. The fauna of characters and plots. The zoo of the mind. It will come.

Acknowledgments

The origins of this book go back to 1992, when Varela and Maturana blew my mind on the island of Chiloé. The decision to research dreams followed in 1995, in New York, at the start of my doctorate at Rockefeller University. The text itself began to be planned in 2001, during my postdoc at Duke University in Durham, North Carolina. In 2007, I promised publishers Globo a book about dreams, but I did not write it and finally was kindly released from the commitment. The ideas continued to mature until, in 2015, an irresistible invitation from publishers Companhia das Letras made the project take off.

Since then, the text has existed in many versions, written in Brazil (Natal, Brasília, Pirenópolis, Rio de Janeiro, São Paulo, Cotovelo, Camurutaba, Tamandaré, and Taíba), Austria (Salzburg), Chile (San Pedro de Atacama, Santiago, Antofagasta, Punta Arenas), Argentina (Buenos Aires, El Calafate, Luján), Colombia (Cartagena de Índias), Sicily (Erice), France (Paris), the United States (New York, Oakland, and Santa Barbara), and Japan (Kyoto and Tsukuba), as well as on many planes and trains. Over the years, I was able to count on all the support I could have wished for from the Brain Institute of the Federal University of Rio Grande do Norte, where I teach and research. To all these places and institutions that have inspired me and given me peaceful refuge—in silence or otherwise—my most profound thanks.

Many people have contributed to making this book a reality. The most decisive influence came from the exceptional team at Companhia das Letras. I was able first to count on the assistance of Otávio Marques da Costa and Rita Mattar, who generously helped me to organize my ideas. Then I came to rely on the attentive editing of Ricardo Teperman, whose observations and suggestions substantially improved the text as regards the structure of the chapters, as well as eliminating inconsistencies. The book then had the privilege of being read closely by Luiz Schwarcz, who patiently guided me to overcome biological hermeticism in favor of more fluent writing. I was also nourished by the encouraging words of Lilia Schwarcz, and by the esthetics of them both. To conclude things with a flourish, I benefited from the careful picture research of Paula Souza and Erica Fujito, and with the exquisite preparation of the text by Joaquim Toledo Júnior and Lucila Lombardi. Finally, as the book was translated to English, I had the excellent fortune to be translated by the scholarly yet light-handed Daniel Hahn. From start to finish, it has been a pleasure and an honor to work with such a crack team.

In addition to the meticulous editorial support, I benefited from attentive and critical reading by colleagues, friends and relatives. Since it is impossible to rank them, I am listing their names in alphabetical order, to thank them gratefully for their time, knowledge, and sensitivity: Alexandre Pontual, Cecília Hedin-Pereira, Dráulio Barros de Araújo, Fernando Arthur Tollendal Pacheco, Joaci Pereira Furtado, Joshua Martin, Leonardo Costa Braga, Luciana de Barros Jaccoud, Luís Fernando Tófoli, Pedro Roitman, Sergio Arthuro Mota-Rolim, Stevens Rehen, and Vera Lúcia Tollendal Gomes Ribeiro.

The following contributed invaluable accounts of dreams and waking, conversations, sources, doubts, and hints: Ana Lúcia Mello, Carolina Damasio dos Santos, Caterina Strambio de Castillia, Celina Roitman, Claudio Maya, Claudio Mello, Constantine Pavlides, Criméia Almeida Schmidt, Edson Sarques Prudente, Eduardo Barreira Gomes Ribeiro, Ernesto Mota Ribeiro, Fernando Antonio Bezerra Tollendal, Flavio Lobo, Gina Poe, Guilherme Brockington, Guillermo Cecchi Isaac Roitman, Jan Born, Janaina Pantoja, Jeremy Luban, Julio Tollendal Gomes Ribeiro, Luisa Tollendal Prudente, Luiz Fernando Gouvêa Labouriau, Marco Marcondes de Moura, Mariano Sigman, Mário Lisbôa Theodoro, Mauro Copelli, Mireya Suárez, Natália Bezerra Mota, Paulo Câmara, Pedro Barreira Gomes Ribeiro, Robert Stickgold, Roy Crist, Ronaldo Santos, Samuel Telles dos Santos, Sérgio Barreira Gomes Ribeiro, and Sergio Mota Ribeiro.

In a more diffuse but no less influential way, this book benefited from the words, actions, examples, and counter-examples I exchanged with Adalgisa de Rosário, Adrián Ocampo, Adriana de Barros Jaccoud, Adriana Ragoni, Adriano Tort, Albert Libchaber, Aldo Paviani, Alejandra Carboni, Alejandro Maiche, Alex Filadelfo, Alexander Henny, Alexandra Dimitri, Alexei Suárez Soares, Alice Mallet, Alírio Barreira, Allan Kardec de Barros, Alvamar Medeiros, Álvaro Cabaña, Álvaro Monteiro, Alyane Almeida de Araújo, Amanda Feilding, Amy Loesch, Ana Beatriz Presgrave, Ana Claudia Ferrari, Ana Cláudia Silva, Ana Elvira Oliveira, Ana Lucia Amaral, Ana Maria Bonetti, Ana Maria Olivera Fuentes, Ana Palmeira, Ana Palmira, Ana Paola Amaral, Ana Paola Ottoni, Ana Paula Wasilewska-Sampaio, Ana Raquel Torres, Ana Sofia Mello, André Luis Lacé Lopes, André Maya, André Pantoja, André Sant'anna, Andréa Araújo, Andréa Deslandes, Andrea Galassi, Andrea Goldin, Andrea Moro, Andrei Suárez Soares, Andrei Queiroz, Andrew Meltzoff, Ângela Maria Paiva Cruz, Angela Naschold, Angelita Araújo, Anibal Vivacqua, Aniruddha Das, Ann Kristina Hedin, Annie da Costa Souza, Antonio Battro, Antônio Celso Rodrigues, Antonio Fortes, Antonio Galves, Antonio Lopes de Alencar Junior, Antonio Pereira, Antônio Prata, Antonio Roberto Guerreiro Júnior, Antonio Roque da Silva, Antonio Sebben, Antonio Teixeira, Aparecida Vilaça, Ariadne Paixão, Armando Santos, Armenio Aguiar, Arthur Johnson, Arthur Omar, Artur França, Artur Jaccoud Theodoro, Artur Tollendal, Arturo Alvarez-Buylla, Arturo Zychlinsky, Ary Pararraios, Asif Ghazanfar, Augusto Buchweitz, Augusto Schrank, Áureo Miranda, Ava LaVonne Vinesett, Bárbara Mendes, Beatrice Crist, Beatriz Labate, Beatriz Longo, Beatriz Stransky, Beatriz Vargas, Benilton Bezerra, Benjamín Alvarez-Borda, Belinha, Beto Almeida, Bira Almeida, Bonfim Abrahão Tobias, Bori, Bradley Simmons, Brian Anderson, Bruna Koike, Bruno Caramelli, Bruno Gomes, Bruno Lobão, Bruno Torturra, Bryan Souza, Caio Mota Marinho, Cajal@babel, Carl Ebers, Carlos Alberto Guedes Corá, Carlos Fausto, Carlos Medeiros, Carlos Morel, Carlos Roberto Jamil Cury, Carlos Schwartz, Caroline Ang, Caroline Barreto, Cássio Yumatã Braz, Catia Pereira, Cecilia Inés Calero, Ceiça Almeida, Célia Maria Costa Braga, Celio Chaves, Cesar Ades, Cesar Rennó-Costa, Charbel El-Hani, Charles Gilbert, Christiane Barros, Christiane Brasileiro do Valle, Cícero Alves do Nascimento, Cilene Vieira, Cilene Rodrigues, Cíntia Barros, Claire Landmann, Clancy Cavnar, Clara Suassuna, Clarissa Maya, Claudia Domingues Vargas, Claudia Kober, Claudia Masini d'Avila-Levy, Claudia Tollendal, Claudine Veronezi Ferrão, Cláudio Almeida, Claudio Angelo, Cláudio Bellini, Claudio Cabezas, Claudio Daniel-Ribeiro, Claudio Maya, Cláu-

384 Acknowledgments

dio Queiroz, Cláudio Serfaty, Claudio Tollendal, Clausius Lima, Clecio Dias, Constance Scharff, Christian Dunker, Cristiana Schettini, Cristiano Maronna, Cristiano Porfírio, Cristiano Simões, Cristine Barreto, Cristoph Glock, D'Alembert de Barros Jaccoud, Daiane Ferreira Golbert, Dalva Alencar, Dalva Gomes Ribeiro, Damien Gervasoni, Daniel Brandão, Daniel Gomes de Almeida Filho, Daniel Herrera, Daniel Martins-de-Souza, Daniel Shulz, Daniel Takahashi, Daniela Uziel, Danilo Silva, Dante Chialvo, Dario Zamboni, Dartiu Xavier, David Bryson, David Klahr, David Vicario, Débora Costa Araújo, Denis Russo Burgierman, Derek Lomas, Desider Kremling Gomez, Desmond Dorsett, Diana Bezerra, Diego Fernández-Slezak, Diego Golombek, Diego Laplagne, Diego Mauricio Canencio, Dilene Almeida, Dimitri Daldegan, Donald Katz, Dora Ventura, Dr. Maurício, Dráulio Barros de Araujo, Edgar Morya, Edgard Altszyler, Edileuza Rufino de Melo, Edilson Silva, Edsart Besier, Edu Martins, Eduarda Alves Ribeiro, Eduardo Bouth Sequerra, Eduardo Faveret, Eduardo Martins Verticinque, Eduardo Schenberg, Edward de Robertis, Edward MacRae, Ehud Kaplan, Elena Pasquinelli, Eli Guimarães, Eliane Volchan, Elida Ojopi, Elisa Dias, Elisa Elsie, Elisabeth Ferroni, Elisaldo Carlini, Elisangela Xavier Sousa, Eliza Nobre, Elizabeth Spelke, Ellen Werther, Elta Dourado, Emilio Figueiredo, Ennio Candotti, Enzo Tagliazucchi, Erich Jarvis, Erico dos Santos Júnior, Erivan Melo, Ernesto Soares, Ernesto van Peborgh, Estrela Santos, Éverton Dantas, Fabian Borghetti, Fabiana Alvarenga, Fabio Presgrave, Fabricio Pamplona, Facundo Carrillo, Felipe Cini, Felipe Farias, Felipe Pegado, Fernanda Camargo, Fernanda Diamant, Fernando Gonzalez, Fernando Louzada, Fernando Moraes, Fernando Nottebohm, Fernando Tollendal Pacheco, Fidélis Guimarães, Fiona Doetsch, Flávia Ribeiro, Flávia Soares, Flavia Vivacqua, Flávio Torres, Francis Clifton, Francisco Alves, Francisco Inácio Bastos, Frank Wall, Frederico Horie, Frederico Prudente, Gabriel Crist, Gabriel Elias, Gabriel Lacombe, Gabriel Marini, Gabriel Mindlin, Gabriel Silva, Gabriel Vidiella Salaberry, Gabriela Costa Braga, Gabriela Simabucuru, Gabriela Tunes, Gaetano Luban, Gandhi Viswanathan, Gary Lehew, George Nascimento, Ghislaine Dehaene-Lambertz, Gildo Lemos Couto, Giles Harrison-Conwill, Gilson Dantas, Glacia Marillac, Glaucia Leal, Gláucio Ary Dillon Soares, Glaucione Gomes de Barros, Glauco Barros, Glenis Clarke, Glória Accioly, Grace Moraes, Grace Santana, Gregorio Duvivier, Guadalupe Marcondes, Guilherme Brockington, Guillermo Cecchi, Gustavo Stolovitzky, Hallison Kauan,

Harumi Visconti, Heather Jennings, Helena Bonciani Nader, Helena Borges, Hélio Barreira, Henrique Carneiro, Henrique Pacheco, Hernando Santamaría García, Herton Escobar, Hindiael Belchior, Hiroshi Asanuma, Hynek Wichterle, Ichiro Takahashi, Ignacio Sánchez Gendriz, Ildeu de Castro Moreira, Irani Martins Dantas, Íris Roitman, Isabel Prudente, Isabelle Cabral, Ismael Pereira, Ivan de Araújo, Ivan Izquierdo, Ivana Bentes, Izabel Hazin, Jacobo Sitt, Jacques Mehler, Jáder Marinho-Filho, Jaime Cirne, James Hudspeth, James Shaffery, Jan Nora Hokoç, Janaina Weissheimer, Jaques Andrade, Jeffrey Hirsch-Pasek, Jeni Vaitsman, Jessica Payne, Joana Prudente, João Alchieri, João Bosco Alves da Silva, João Emanuel Evangelista, João Felipe Souza Pegado, João Fontes, João Franca, João Maria Figueiredo da Silva, João Oliveira dos Santos, João Paulo Costa Braga, João Queiroz, João Ricardo Lacerda de Menezes, John Bruer, John Fontenele Araújo, Jonathan Winson, Jordi Riba, Jorge Macarrão, Jorge Martinez Cotrina, Jorge Medina, Jorge Muñoz, Jorge Quillfeldt, José Accioly, José Ballestrini, José Carmena, José Daniel Diniz Melo, José de Paiva Rebouças, José Eduardo Agualusa, José França, José Geraldo de Sousa Júnior, José Henrique Targino, José Ivonildo do Rego, José Luis Reyes, José Luiz Ramos, José Morais, Joselo Zambelli, Joshua White Carlstrom, Josione Batista, Josy Pontes, Joyse Medeiros, Juan Manuel Rico, Juan Valle Lisboa, Julia Todorov, Juliana Barreto, Juliana Guerra, Juliana Pimenta, Juliana Rossi, Julien Calais, Julija Filipovska, Julio Delmanto, Julio Gomes Ribeiro, Julita Lemgruber, Jurandir Accioly, Justin Halberda, Kafui Dzirasa, Karin Moreira, Karla Rocha, Katarina Leão, Katherine Hirsch-Pasek, Katie Almondes, Kerstin Schmidt, Koichi Sameshima, Larissa Queiroz, Laura Greenhalgh, Laura Oliveira, Laurent Dardenne, Lauro Morhy, Leilane Assunção, Lena Palaniyappan, Leni Almeida, Leonardo Mota, Leopoldo Petreanu, Letícia Tollendal Barros, Lia Luz, Lili Bruer, Linda Wilbrecht, Loreny Gimenes Giugliano, Lourenço Bustani, Luana Malheiros, Lucas Centeno Cecchi, Lúcia Barreira Accioly, Lúcia Santaella, Luciana Boiteux, Luciana Zaffalon, Luciano Roitman, Luciano Ribeiro Pinto Júnior, Lucile Maria Floeter Winter, Ludmila Queiroz, Luís Carlos Lisbôa Theodoro, Luís Otávio Teles Assumpção, Luís Roberto Ribeiro, Luiz Alberto Simas, Luiz Carlos Silveira, Luiz Eduardo Soares, Luiz Fernando Veríssimo, Luiz Grande, Luiz Paulo Ferreira Nogueról, Luziania Medeiros, Mailce Mota, Maite Greguol, Manuel Carreiras, Manuel Muñoz, Manuel Schabus, Marcela Peña, Marcello Dantas, Marcelo Almeida, Marcelo Tollendal Alvarenga,

Marcelo Barcinsky, Marcelo Bizerril, Marcelo Gonçalves Lima, Marcelo Lasneaux, Marcelo Leite, Marcelo Magnasco, Marcelo Roitman, Marcelo Spock, Márcio Flávio Moraes Dutra, Marco Antonio Raupp, Marco Freire, Marco Marcondes de Moura, Marcos Antonio Gomes de Carvalho, Marcos Didonet, Marcos Romualdo Costa, Marcos Frank, Marcos Trevisan, Marcus Vinicius Goulart Gonzaga, Maria Augusta Mota, Maria Bernardete Cordeiro de Sousa, Maria Brígida de Miranda, Maria Ceiça da Silva, Maria Cerise do Amaral, Maria Cristina Dal Pian, Maria Digessila Dantas Beserra, Maria do Carmo Miranda, Maria Elizabeth Mori, Maria Emilia Yamamoto, Maria Helena Bezerra, Maria Helena da Silva Oliveira, Maria Isa, Maria José da Silva, Maria Josefina Porto Goulart, Maria Léa Salgado Labouriau, Maria Luban, Maria Rita Kehl, Maria Sílvia Rossi, Maria Sonia de Oliveira Morais, Maria Stein, Mariana Medeiros, Mariana Muniz, Mariana Alves Ribeiro, Marilene Vainstein, Marília Zaluar Guimarães, Marilia Marini, Mariluce Moura, Marina Antongiovanni da Fonseca, Marina Farias, Marina Jaccoud Theodoro, Marina Nespor, Marina Ribeiro, Mário Fiorani, Mario Nelson & Cilene & Fefeu & Denis, Marisa Mamede, Marisa von Bullow, Marise Tollendal Alvarenga, Mark A. McDaniel, Marlene Queiroz, Martín Cammarota, Martin Hilbert, Martin Hopenhayn, Martín Correa, Matias López, Matteo Luban, Matthew Walker, Mauricio Dantas, Maurício Fiore, Maurício Guimarães, Mauro Copelli, Mauro Pires Salgado Moraes, Mauro Refosco, Mércia Greguol, Mércio Gomes, Mia Couto, Michael Lavine, Michael Posner, Michael Wiest, Michel Laub, Michel Rabinovitch, Miguel Angelo Laporta Nicolelis, Milon Barros, Mirinha & Larissa, Mitchel Nathan, Mizziara de Paiva, Mohammad Torabi-Nami, Monique Floer, Mrs. Taylor, Nair Bicalho, Naomar Almeida, Natal Tollendal Pacheco, Nathalia Lemos, Nathália Oliveira, Nelson Lemos, Nelson Vaz, Nelson Pretto, Nestor Capoeira, Neuza Barreira, Ney Dentes Perdigueiro, Nivaldo Antonio Portela de Vasconcelos, Nivanio Bezerra, Norma Santinoni Veras, Nuno Sousa, Ofer Tchernichovski, Onildo Marini Filho, Orlando Bueno, Orlando Jimenez, Osame Kinouchi, Otávio Velho, Otom Anselmo de Oliveira, Pablo Fuentealba, Pablo Meyer Rojas, Pablo Torterolo, Patricia Kuhl, Patricia Schaeffer, Patrícia Tollendal Pacheco, Patrick Cocquerel, Paula Marcela Herrera Gomez, Paula Tiba, Paulo Abrantes, Paulo Amarante, Paulo Cesar Silva Souza, Paulo Fontes, Paulo Lima, Paulo Mello, Paulo Roberto Petersen Hoffman, Paulo Saraiva, Pearl Hutchins, Pedro Bekinschtein, Pedro Bial,

Pedro Celestino, Pedro Maldonado, Pedro Melo, Pedro Petrovitch Maia, Pedro Roitman, Pelicano Vilas Bôas, Perla Gonzalez, Pertteson Silva, Philippe Peigneux, Phillippe Rousselot, Pierre Hervé-Luppi, Pierre Pica, Pietra Rossi, Porangui, Priscila Matos, Professor Queijo Formággio, Rafael Linden, Rafael Scott, Raimundo Alvarenga, Raimundo Furtado, Raíssa Ebert, Raquel Nunes, Raphael Bender, Rebeka Nogueira da Silva, Regina Helena Silva, Reginaldo Freitas, Régine Kolinsky, Reinaldo Lopes, Reinaldo Moraes, Renata Santinoni Veras, Renata Veras, Renato de Mendonca Lopes, Renato Filev, Renato Lopes, Renato Malcher Lopes, Renato Rozental, Renzo Torrecuso, Ricardina Almeida, Ricardo Cambeta, Ricardo Chaves, Ricardo Ferreira, Ricardo Gattass, Ricardo Lagreca, Ricardo Paixão, Ricardo Reis, Ricardo Sampaio, Richard Mooney, Richard Vinesett, Richardson Leão, Rick Doblin, Rita Mattar, Rivane Neuenschwander, Robert Desimone, Robert Stickgold, Roberto Cavalcanti, Roberto Etchenique, Roberto Lent, Roberto Viana Batista Júnior, Robson Nunes, Rodolfo Llinás, Rodrigo Cavalcanti, Rodrigo McNiven, Rodrigo Pereira, Rodrigo Portugal, Rodrigo Quiroga, Rogério Lopes de Souza, Rogerio Mesquita, Rogério Mesquita, Rogério Panizzutti, Rogério Rondon, Ronaldo Cérebro, Ronaldo Bressane, Roque Tadeu Gui, Roseli de Deus Lopes, Rossella Fabbri, Rowan Abbensetts, Rubens Naves, Rui Costa, Rute Barreira, Rute Oliveira, S. Rasika, Samuel Goldenberg, Sandro de Souza, Sara Mednick, Sebastián Lipina, Selma Jeronimo, Sergei Suárez Soares, Sergio Alves Gomes Ribeiro, Sergio Cezar, Sérgio Guerra, Sérgio Mascarenhas, Sergio Neuenschwander, Sérgio Rezende, Sergio Ricardo, Sérgio Ruschi, Shih-Chieh Lin, Sidney Simon, Sidney Strauss, Silene Lima, Silvana Benítez, Silvia Bunge, Silvia Centeno, Silvia Thomé, Silvio de Albuquerque Mota, Simone Leal, Simone Lima, Sofia Roitman Ribeiro, Solange Sato Simões, Sonia Barreira Nunes, Sonoko Ogawa, Stanislas Dehaene, Susan Fitzpatrick, Susan Sara, Sylvia Lima de Sousa Medeiros, Sylvia Pinheiro, Tainá Rossi, Takeshi Miura, Tales Tollendal Alvarenga, Tarciso Velho, Tatiana Ferreira, Tatiana Leite, Tatiana Lima Ferreira, Tersio Greguol, Thiago Cabral, Thiago Centeno Cecchi, Thiago Maya, Thiago Ribeiro, Tia Jô, Timothy Gardner, Tomas Ossandon, Torsten Wiesel, Tristán Bekinshtein, Ulisses Riedel, Valdir Pessoa, Valeska Amaral, Valfrânio Queiroz, Valquíria Michalczechen, Valter Fernandes, Vanderlan da Silva Bolzani, Vanja Dakic, Vera Graúna, Vera Santana, Veronica Nunes, Veronica Palma, Victor Albuquerque, Victor Nussenzweig, Victor Leonardi, Victor Tollendal Pacheco, Victoria Andino-Pavlovsky, Vikas Goyal, Vilma Alves Ribeiro, Vincent Brown, Vinícius Rosa Cota, Virginia Alonso, Vítor Lopes dos Santos, Vylneide Lima, Waldenor Cruz, Waldo Vieira, Wandenkolk Manoel de Oliveira, Wanderley de Souza, Wilfredo Blanco, Wilfredo Garcia, William Fishbein, Wilson Savino, Yara Barreira, Yasha Emerenciano Barros, Yogi Pacheco Filho, Yuri Suárez Soares, Yves Fregnac, Zachary Mainen, Zeca Marcondes, and Zuleica Porto.

Throughout this whole dream, the dear mother of my sons, Natália: *sine qua non.*

In the eternal light of the present, Luiza Mugnol Ugarte.

Notes

1. Why Do We Dream? (pages 3–29)

1. J. K. Boehnlein, J. D. Kinzie, R. Ben, and J. Fleck, "One-Year Follow-Up Study of Posttraumatic Stress Disorder among Survivors of Cambodian Concentration Camps," *American Journal of Psychiatry* 142 (1985): 956–59; A. Aron, "The Collective Nightmare of Central American Refugees," in *Trauma and Dreams,* ed. Deirdre Barrett (Cambridge: Harvard University Press, 1996), 140–47; E. M. Menke and J. D. Wagner, "The Experience of Homeless Female-Headed Families," *Issues in Mental Health Nursing* 18 (1997): 315–30; T. C. Neylan et al., "Sleep Disturbances in the Vietnam Generation: Findings from a Nationally Representative Sample of Male Vietnam Veterans," *American Journal of Psychiatry* 155 (1998): 929–33; K. Esposito, A. Benitez, L. Barza, T. Mellman, "Evaluation of Dream Content in Combat-Related PTSD," *Journal of Traumatic Stress* 12 (1999): 681–87; L. Wittmann, M. Schredl, and M. Kramer, "Dreaming in Posttraumatic Stress Disorder: A Critical Review of Phenomenology, Psychophysiology and Treatment," *Psychotherapy and Psychosomatics* 76 (2007): 25–39; J. Davis-Berman, "Older Women in the Homeless Shelter: Personal Perspectives and Practice Ideas," *Journal of Women and Aging* 23 (2011): 360–74; J. Davis-Berman, "Older Men in the Homeless Shelter: In-Depth Conversations Lead to Practice Implications," *Journal of Gerontological Social Work* 54 (2011): 456–74; K. E. Miller, J. A. Brownlow, S. Woodward, and P. R. Gehrman, "Sleep and Dreaming in Posttraumatic Stress Disorder," *Current Psychiatry Reports* 19 (2017): 71.
2. P. Levi, *The Truce,* trans. S. Woolf (London: The Orion Press, 1969), chap. 17.
3. D. Goldman, "Investing in the Growing Sleep-Health Economy," McKinsey & Company, 2017.
4. W. Shakespeare, *The Tempest* (London: Penguin, 2015), Act Four, scene 1.

5. P. Calderón de la Barca, *Life Is a Dream,* bilingual edition, trans. S. Appelbaum (New York: Dover, 2002).

6. B. R. Foster,"Kings of Assyria and Their Times," in *Before the Muses: An Anthology of Akkadian Literature* (Bethesda, MD: CDL Press, 2005), 308.

7. P. Clayton, *Chronicle of the Pharaohs* (London: Thames & Hudson, 1994).

8. A. F. Herold and P. C. Blum, *The Life of Buddha According to the Legends of Ancient India* (New York: A. & C. Boni, 1927), 9.

9. P. R. Goldin, *A Concise Companion to Confucius* (Hoboken: Wiley, 2017); M. Choi, *Death Rituals and Politics in Northern Song China* (Oxford: Oxford University Press, 2017).

10. Artemidorus, *The Interpretation of Dreams,* trans. M. Hammond (Oxford: Oxford University Press, 2020).

11. A. A. T. Macrobius, *Commentary on the Dream of Scipio,* trans. W. H. Stahl. (New York: Columbia University Press, 1990).

12. Artemidorus, *The Interpretation of Dreams.*

13. Ibid., 4–6.

14. Ibid., 228.

15. Macrobius, *Commentary on the Dream of Scipio.*

16. J. S. Lincoln, *The Dream in Native American and Other Primitive Cultures* (Hoboken: Dover, 2003); M. C. Jedrej et al., *Dreaming, Religion and Society in Africa* (Brill, 1997); R. K. Ong, *The Interpretation of Dreams in Ancient China* (master's thesis, University of British Columbia, 1981).

17. S. C. Gwynne, *Empire of the Summer Moon: Quanah Parker and the Rise and Fall of the Comanches, the Most Powerful Indian Tribe in American History* (New York: Scribner, 2011).

18. J. L. D. Schilz and T. F. Schilz, *Buffalo Hump and the Penateka Comanches* (El Paso: Texas Western Press, 1989).

19. F. A. Azevedo et al., "Equal Numbers of Neuronal and Nonneuronal Cells Make the Human Brain an Isometrically Scaled-Up Primate Brain," *Journal of Comparative Neurology* 513 (2009): 532–41.

20. S. Freud, *Project for a Scientific Psychology,* in *The Standard Edition of the Complete Psychological Works of Sigmund Freud,* eds. J. Strachey et al., vol. 1 (London: Hogarth Press, 1953).

21. T. V. Bliss and T. Lomo, "Long-Lasting Potentiation of Synaptic Transmission in the Dentate Area of the Anaesthetized Rabbit Following Stimulation of the Perforant Path," *Journal of Physiology* 232 (1973): 331–56.

22. M. Minsky,"Why Freud Was the First Good AI Theorist," in *The Transhumanist Reader: Classical and Contemporary Essays on the Science, Technology, and Philosophy of the Human Future,* eds. M. More and N. Vita-More (Hoboken: John Wiley and Sons, 2013).

23. S. Freud, *Beyond the Pleasure Principle; Group Psychology and the Analysis of the Ego; The Ego and the Id,* in *The Standard Edition of the Complete Psycho-*

logical Works of Sigmund Freud, eds. J. Strachey et al., vols. 18, 19 (London: Hogarth Press, 1953).

24. M. L. Andermann and B. B. Lowell, "Toward a Wiring Diagram Understanding of Appetite Control," *Neuron* 95 (2017): 757–78; W. Han et al., "A Neural Circuit for Gut-Induced Reward," *Cell* 175 (2018): 887–88; J. Panksepp, *Affective Neuroscience: The Foundations of Human and Animal Emotions* (Oxford: Oxford University Press, 1998).

25. B. Levine et al. "The Functional Neuroanatomy of Episodic and Semantic Autobiographical Remembering: A Prospective Functional MRI Study," *Journal of Cognitive Neuroscience* 16 (2004): 1633–46; R. Q. Quiroga, "Concept Cells: The Building Blocks of Declarative Memory Functions," *Nature Reviews Neuroscience* 13 (2012): 587–97; P. Martinelli, M. Sperduti, and P. Piolino, "Neural Substrates of the Self-Memory System: New Insights from a Meta-Analysis," *Human Brain Mapping* 34 (2013): 1515–29.

26. P. S. Goldman-Rakic, "The Prefrontal Landscape: Implications of Functional Architecture for Understanding Human Mentation and the Central Executive," *Philosophical Transactions of the Royal Society of London B: Biological Sciences* 351 (1996): 1445–53; F. Barcelo, S. Suwazono, and R. T. Knight, "Prefrontal Modulation of Visual Processing in Humans," *Nature Neuroscience* 3 (2000): 399–403.

27. A. Hoche et al. *Gegen Psycho-Analyse* (Munique: Verlag der Süddeutsche Monatshefte, 1931).

28. K. R. Popper, *Conjectures and Refutations: The Growth of Scientific Knowledge* (New York: Basic Books, 1962), 37.

29. F. C. Crews, ed., *Unauthorized Freud: Doubters Confront a Legend* (New York: Viking, 1998); C. Meyer and Borch-Jacobsen, *Le Livre noir de la psychanalyse: vivre, penser et aller mieux sans Freud* (Paris: Les Arènes, 2005); T. Dufresne, ed., *Against Freud: Critics Talk Back* (Stanford: Stanford University Press, 2007).

30. C. K. Morewedge and M. I. Norton, "When Dreaming is Believing: The (Motivated) Interpretation of Dreams," *Journal of Personality and Social Psychology* 96 (2009): 249–64.

31. M. C. Anderson et al., "Neural Systems Underlying the Suppression of Unwanted Memories," *Science* 303 (2004): 232–35; B. E. Depue, T. Curran, and M. T. Banich, "Prefrontal Regions Orchestrate Suppression of Emotional Memories via a Two-Phase Process," *Science* 317 (2007): 215–19.

32. K. Lorenz, *The Natural Science of the Human Species: An Introduction to Comparative Behavioral Research (The "Russian Manuscript" 1944–1948)* (Cambridge: MIT Press, 1997): 47–48.

33. F. Crick and G. Mitchison, "The Function of Dream Sleep," *Nature* 304 (1983): 111–14; F. Crick and G. Mitchison, "REM Sleep and Neural Nets," *Behavioural Brain Research* 69 (1995): 147–55.

34. Wittmann, Schredl, and Kramer, "Dreaming in Posttraumatic Stress Disorder"; Miller, Brownlow, Woodward, and Gehrman, "Sleep and Dreaming in Posttraumatic Stress Disorder"; B. A. Vanderkolk and R. Fisler, "Dissociation and the Fragmentary Nature of Traumatic Memories: Overview and Exploratory Study," *Journal of Trauma Stress* 8 (1995); H. A. Wilmer, "The Healing Nightmare: War Dreams of Vietnam Veterans," in *Trauma and Dreams*, ed. D. Barrett (Cambridge: Harvard University Press, 1996), 85–99; B. J. N. Schreuder, V. Igreja, J. van Dijk, and W. Kleijn, "Intrusive Re-Experiencing of Chronic Strife or War," *Advances in Psychiatric Treatment* 7 (2001): 102–8.

35. C. G. Jung, "General Aspects of Dream Psychology," in *Collected Works of C. G. Jung: The Structure and Dynamics of the Psyche* (Princeton: Princeton University Press, 1916), 493.

36. C. G. Jung, "The Unconscious," in *The Collected Works of C. G. Jung*, vol. 5 (London: Routledge and K. Paul, 1966).

2. The Ancestral Dream (pages 30–58)

1. J. J. Hublin et al., "New Fossils from Jebel Irhoud, Morocco and the Pan-African Origin of *Homo sapiens*," *Nature* 546 (2017): 289–92; D. Richter et al., "The Age of the Hominin Fossils from Jebel Irhoud, Morocco, and the Origins of the Middle Stone Age," *Nature* 546 (2017): 293–96.

2. A. W. Pike et al., "U-Series Dating of Paleolithic Art in 11 Caves in Spain," *Science* 336 (2012): 1409–13; M. Aubert et al., "Pleistocene Cave Art from Sulawesi, Indonesia," *Nature* 514 (2014): 223–27; D. L. Hoffmann et al., "U-Th Dating of Carbonate Crusts Reveals Neanderthal Origin of Iberian Cave Art," *Science* 359 (2018): 912–15.

3. K. Lohse and L. A. Frantz, "Neandertal Admixture in Eurasia Confirmed by Maximum-Likelihood Analysis of Three Genomes," *Genetics* 196 (2014): 1241–51; S. Sankararaman et al., "The Genomic Landscape of Neanderthal Ancestry in Present-Day Humans," *Nature* 507 (2014): 354–57; S. R. Browning et al., "Analysis of Human Sequence Data Reveals Two Pulses of Archaic Denisovan Admixture," *Cell* 173 (2018): 53–61; V. Slon et al., "The Genome of the Offspring of a Neanderthal Mother and a Denisovan Father," *Nature* 561 (2018).

4. A. Sieveking, *The Cave Artists: Ancient Peoples and Places* (London: Thames and Hudson, 1979), 93.

5. A. Leroi-Gourhan, *L'Art des cavernes: atlas des grottes ornées paléolithiques françaises*, Atlas Archéologiques de la France (Paris: Ministère de la culture, Direction du patrimoine, Impr. Nationale, 1984).

6. H. Bégouën, "Un Dessin relevé dans la caverne des Trois-frères, à Montesquieu-Avantès (Ariège)," *Comptes rendus des séances de l'Académie des Inscriptions et Belles-Lettres* 64 (1920): 303–10.

7. O. Grøn, "A Siberian Perspective on the North European Hamburgian Culture: A Study in Applied Hunter-Gatherer Ethnoarchaeology," *Before Farming* 1 (2005).

8. O. Soffer, *Upper Paleolithic of the Central Russian Plain* (Cambridge: Academic Press, 1985).

9. M. Germonpré and R. Hämäläinen, "Fossil Bear Bones in the Belgian Upper Paleolithic: The Possibility of a Proto Bear-Ceremonialism," *Arctic Anthropology* 44 (2007): 1–30.

10. E. Hill, "Animals as Agents: Hunting Ritual and Relational Ontologies in Prehistoric Alaska and Chukotka," *Cambridge Archaeological Journal* 21 (2011): 407–26.

11. W. Roebroeks and P. Villa, "On the Earliest Evidence for Habitual Use of Fire in Europe," *Proceedings of the National Academy of Sciences of the USA* 108 (2011): 5209–14; R. Shimelmitz et al., " 'Fire at Will': The Emergence of Habitual Fire Use 350,000 Years Ago," *Journal of Human Evolution* 77 (2014): 196–203.

12. C. Lévi-Strauss, *The Raw and the Cooked* (New York: Harper & Row, 1969).

13. F. W. Nietzsche, *Human, All Too Human,* trans. M. Faber and S. Lehmann (London: Penguin Classics, 2004), 16.

14. E. Durkheim, *The Elementary Forms of Religious Life,* trans. C. Cosman (Oxford: Oxford University Press, 2001), 49.

15. B. Vandermeersch, *Les Hommes fossiles de Qafzeh, Israël,* Cahiers de paléontologie Paléoanthropologie (Paris: Éditions du Centre National de la Recherche Scientifique, 1981); I. Wunn, "Beginning of Religion," *Numen* 47 (2000): 417–52.

16. M. P. Cabral and J. D. d. M. Saldanha, "Paisagens megalíticas na costa norte do Amapá," *Revista de Arqueologia da Sociedade de Arqueologia Brasileira* 21 (2008).

17. Ibid.

18. J. S. Lincoln, *The Dream in Native American and Other Primitive Cultures* (Hoboken: Dover, 2003).

19. Ibid.; J. O. Santos, *Vagares da alma: elaborações ameríndias acerca do sonhar* (master's thesis, Departamento de Antropologia, Universidade de Brasília, 2010); K. G. Shiratori, *O acontecimento onírico ameríndio: o tempo desarticulado e as veredas dos possíveis* (master's thesis, Museu Nacional, Universidade Federal do Rio de Janeiro, 2013).

20. D. Q. Fuller et al., "Convergent Evolution and Parallelism in Plant Domestication Revealed by an Expanding Archaeological Record," *Proceedings of the National Academy of Sciences of the USA* 111 (2014): 6147–52.

21. G. Larson et al., "Rethinking Dog Domestication by Integrating Genetics, Archeology, and Biogeography," *Proceedings of the National Academy of Sciences of the USA* 109 (2012): 8878–83; A. Perri, "A Wolf in Dog's Clothing:

Initial Dog Domestication and Pleistocene Wolf Variation," *Journal of Archaeological Science* 68 (2016): 1–4.

22. D. R. Piperno, "The Origins of Plant Cultivation and Domestication in the New World Tropics: Patterns, Process, and New Developments," *Current Anthropology* 52 (2011): S453–70.

23. K. Schmidt, "Göbekli Tepe: A Neolithic Site in Southwestern Anatolia," in *The Oxford Handbook of Ancient Anatolia*, eds. S. R. Steadman and G. McMahon (Oxford: Oxford University Press, 2011), 917.

24. M. Gaspar, *Sambaqui: Arquelogia do litoral brasileiro* (Rio de Janeiro: Zahar, 2000); S. K. Fish, P. De Blasis, M. D. Gaspar, and P. R. Fish, "Eventos Incrementais na Construção de Sambaquis, Litoral Sul do Estado de Santa Catarina," *Revista do Museu de Arqueologia e Etnologia* 10 (2000): 69–87; D. M. Klokler, *Food for Body and Soul: Mortuary Ritual in Shell Mounds (Laguna—Brazil)* (master's thesis in anthropology, University of Arizona, 2008).

25. M. M. Okumura and S. Eggers, "The People of Jabuticabeira II: Reconstruction of the Way of Life in a Brazilian Shellmound," *Homo* 55 (2005): 263–81.

26. D. Tedlock, trans., *Popol Vuh* (New York: Touchstone, 1996).

27. V. Brown, *The Reaper's Garden: Death and Power in the World of Atlantic Slavery* (Cambridge: Harvard University Press, 2010).

28. F. D. Goodman, J. H. Henney, and E. Pressel, *Trance, Healing, and Hallucination; Three Field Studies in Religious Experience* (Hoboken: J. Wiley, 1974); L. F. S. Leite, *Relacionando Territórios: O "sonho" como objeto antropológico* (master's thesis in social anthropology, Museu Nacional, Universidade Federal do Rio de Janeiro, 2003); W. Zangari, "Experiências anômalas em médiuns de Umbanda: Uma avaliação fenomenológica e ontológica," *Boletim da Academia Paulista de Psicologia* 27 (2007): 67–86; L. F. Q. A. Leite, "Algumas categorias para análise dos sonhos no candomblé," *Prelúdios* 1 (2013): 73–99.

29. J. K. Thornton, "Religion and Ceremonial Life in the Kongo and Mbundu Areas, 1500–1700," in *Central Africans and Cultural Transformations in the American Diaspora*, ed. L. Heywood (Cambridge: Cambridge University Press, 2001).

30. A. Battell, *The Strange Adventures of Andrew Battell of Leigh, in Angola and the Adjoining Regions* (London: The Hakluyt Society, 1901).

31. M. H. Kingsley, *West African Studies* (New York: Macmillan, 1899).

32. J. Binet, "Drugs and Mysticism: The Bwiti Cult of the Fang," *Diogenes* 86 (1974): 31–54; J. W. Fernandez, *Bwiti: An Ethnography of the Religious Imagination in Africa* (Princeton: Princeton University Press, 1982).

33. P. Ariès, *Western Attitudes Toward Death from the Middle Ages to the Present* (Baltimore: Johns Hopkins University Press, 1974); P. Metcalf and

R. Huntington, *Celebrations of Death: The Anthropology of Mortuary Ritual* (Cambridge: Cambridge University Press, 1991); M. Parker Pearson, *The Archaeology of Death and Burial*, Texas A&M University anthropology series (College Station: Texas A&M University Press, 1999); A. C. G. M. Robben, *Death, Mourning, and Burial: A Cross-Cultural Reader* (Malden: Wiley Blackwell, 2018).

34. J. R. Anderson, A. Gillies, and L. C. Lock, "Pan Thanatology," *Current Biology* 20 (2010): R349–51.

35. D. Biro et al., "Chimpanzee Mothers at Bossou, Guinea, Carry the Mummified Remains of Their Dead Infants," *Current Biology* 20 (2010): R351–52.

36. F. G. P. De Ayala, *El primer nueva corónica y buen gobierno 1615/1616*, v. GkS 2232 4to Quires, Sheets, and Watermarks, Royal Library, 1615; S. MacCormack, *Religion in the Andes: Vision and Imagination in Early Colonial Peru* (Princeton: Princeton University Press, 1993).

37. D. Tedlock, trans., *Popol Vuh* (New York: Touchstone, 1996).

38. J. Jaynes, *The Origin of Consciousness in the Breakdown of the Bicameral Mind* (New York: Mariner Books, 2000), chap. 2.

39. S. Freud, *Group Psychology and the Analysis of the Ego*, in *The Standard Edition of the Complete Psychological Works of Sigmund Freud*, eds. J. Strachey et al., vol. 18 (London: Hogarth Press, 1953), 124.

40. G. Turville-Petre, *Nine Norse Studies*, text series: Viking Society for Northern Research, vol. 5 (London: Viking Society for Northern Research, University College London, 1972).

41. G. D. Kelchner, *Dreams in Old Norse Literature and Their Affinities in Folklore: With an Appendix Containing the Icelandic Texts and Translations* (Norwood, UK: Norwood Editions, 1978).

42. S. Sturluson, *Halfdan the Black Saga*, in *Heimskringla or The Chronicle of the Kings of Norway* (London: Longman, Brown, Green and Longmans, 1844).

43. G. Jones, *A History of the Vikings* (Oxford: Oxford University Press, 2001).

44. R. K. Ong, *The Interpretation of Dreams in Ancient China* (master's thesis, Vancouver, University of British Columbia, 1981).

45. I. Edgar, *The Dream in Islam: From Qur'anic Tradition to Jihadist Inspiration* (New York: Berghahn, 2011), 178; I. R. Edgar and D. Henig, "Istikhara: The Guidance and Practice of Islamic Dream Incubation Through Ethnographic Comparison," *History and Anthropology* 21 (2010): 251–62.

46. S. N. Kramer, *The Sumerians: Their History, Culture, and Character* (Chicago: The University of Chicago Press, 1963).

47. B. Eranimos and A. Funkhouser, "The Concept of Dreams and Dreaming: A Hindu Perspective," *The International Journal of Indian Psychology* 4 (2017): 108–16.

48. B. R. Foster, *The Epic of Gilgamesh* (New York: W. W. Norton & Company, 2018).

49. Homer, *The Iliad,* trans. Robert Fagles (London: Penguin, 1990).
50. P. Kriwaczek, *Babylon: Mesopotamia and the Birth of Civilization* (New York: Thomas Dunne / St. Martin's, 2012).
51. Enheduanna and B. D. S. Meador, *Inanna, Lady of Largest Heart: Poems of the Sumerian High Priestess Enheduanna* (Austin: University of Texas Press, 2000); The Electronic Text Corpus of Sumerian Literature, http://etcsl .orinst.ox.ac.uk/section4/tr4073.htm.
52. Anon., *Gudea and his Dynasty,* vol. 3:1, The Royal Inscriptions of Mesopotamia, Early Periods (Toronto: University of Toronto Press, 1997), 71–72.
53. S. N. Kramer, *The Sumerians.*
54. S. Bar, *A Letter That Has Not Been Read: Dreams in the Hebrew Bible,* New Century Edition of the Works of Emanuel Swedenborg (Cincinnati: Hebrew Union College Press, 2001).
55. Herodotus, *Histories,* eds. P. Mensch and J. S. Romm (Indianapolis: Hackett Publishing, 2014).
56. Artemidorus, *The Interpretation of Dreams,* trans. M. Hammond (Oxford: Oxford University Press, 2020).
57. C. Roebuck, *Corinth: The Asklepieion and Lerna,* vol. 14 (Princeton: American School of Classical Studies at Athens, 1951); S. B. Aleshire, *The Athenian Asclepeion: Their People, Their Dedications, and Their Inventories* (Amsterdam: J. C. Gieben, 1989).
58. S. M. Oberhelman, ed., *Dreams, Healing, and Medicine in Greece: From Antiquity to the Present* (Farnham: Ashgate, 2013).
59. W. Rouse, *Greek Votive Offerings: An Essay in the History of Greek Religion* (Cambridge: Cambridge University Press, 1902); S. M. Oberhelman, "Anatomical Votive Reliefs as Evidence for Specialization at Healing Sanctuaries in the Ancient Mediterranean World," *Athens Journal of Health* 1 (2014): 47–62.
60. Suetonius, *Life of Augustus (Vita divi Augusti),* ed. D. Wardle (Oxford: Oxford University Press, 2014).
61. Suetonius, *The Twelve Caesars,* eds. R. Graves and M. Grant (London: Penguin, 2003).

3. From Living Gods to Psychoanalysis (pages 59–76)

1. R. Drews, *The End of The Bronze Age: Changes in Warfare and the Catastrophe ca. 1200 B.C.* (Princeton: Princeton University Press, 1993); P. B. DeMenocal, "Cultural Responses to Climate Change during the Late Holocene," *Science* 292 (2001): 667–73; J. M. Diamond, *Collapse: How Societies Choose to Fail or Succeed* (London: Penguin Books, 2011).
2. C. G. Diuk et al., "A Quantitative Philology of Introspection," *Frontiers in Integrative Neuroscience* 6 (2012): 80.

3. A. F. Herold and P. C. Blum, *The Life of Buddha According to the Legends of Ancient India* (New York: A. & C. Boni, 1927).

4. Ibid., 21.

5. Ibid., 31.

6. R. K. Ong, *The Interpretation of Dreams in Ancient China* (master's thesis, University of British Columbia, 1981).

7. W. E. Soothill, *The Three Religions of China; Lectures Delivered at Oxford* (New York: Hyperion, 1973), 75.

8. Plato, *Theaetetus* 158, *Laws* 461, in *Complete Works*, ed. J. Cooper (London: Hackett Publishing, 1997).

9. Aristotle, *On Sleep and Dreams*, ed. and trans. D. Gallop (Liverpool: Liverpool University Press, 1996).

10. Matthew 1:20–2:22 (King James Version).

11. Matthew 27:19 (King James Version).

12. Acts 16:9–10 (King James Version).

13. I. Edgar, *The Dream in Islam: From Qur'anic Tradition to Jihadist Inspiration* (New York: Berghahn Books, 2011), 178; C. M. Naim, "'Prophecies' in South Asian Muslim Political Discourse: The Poems of Shah Ni'matullah Wali," *Economic and Political Weekly* 46 (2011): 49–58.

14. Augustine, *Confessions*, trans. H. Chadwick (Oxford: Oxford University Press, 1998), 203.

15. J. Verdon, *Night in the Middle Ages* (Notre Dame: University of Notre Dame Press, 2002); A. R. Ekirch, *At Day's Close: Night in Times Past* (New York: W. W. Norton, 2005).

16. C. Vogel, *Le Pécheur et la pénitence dans l'Église ancienne, textes choisis* (Paris: Éditions du Cerf, 1966).

17. T. Aquinas, trans. Fathers of the English Dominican Province, *The Summa Theologica* (New York: Catholic Way Publishing, 2014), 2–2, 94, 6.

18. J. Passavanti and G. Auzzas, *Lo Specchio della Vera Penitenzia*, Scrittori Italiani e Testi Antichi (Florença: Accademia della Crusca, 2014).

19. C. Speroni, "Dante's Prophetic Morning-Dreams," *Studies in Philology* 45 (1948): 50–59.

20. O. Kraut, *Ninety-Five Theses* (New York: Pioneer, 1975), 150.

21. J. A. Wylie, *The History of Protestantism* (Neerlandia, AB: Inheritance, 2018), chap. 9.

22. R. Descartes, *Discourse on Method; And, Meditations on First Philosophy*, trans. D. A. R. Cress (Indianapolis: Hackett, 1998).

23. S. Freud, *The Interpretation of Dreams*, in *The Standard Edition of the Complete Psychological Works of Sigmund Freud*, eds. J. Strachey et al., vols. 4, 5 (London: Hogarth Press, 1953).

24. S. Bar, *A Letter That Has Not Been Read: Dreams in the Hebrew Bible*, vol. 25, New Century Edition of the Works of Emanuel Swedenborg (Cincin-

nati: Hebrew Union College Press, 2001), 6; see also Babylonian Talmud, Berakhot, 55b.

4. Unique Dreams and Typical Dreams (pages 77–98)

1. W. B. Webb and H. W. Agnew, "Are We Chronically Sleep Deprived?" *Bulletin of the Psychonomic Society* 6 (1975): 47–48.
2. G. W. Domhoff and A. Schneider, "Studying Dream Content Using the Archive and Search Engine on DreamBank.net," *Consciousness and Cognition* 17 (2008): 1238–47.
3. D. Foulkes, *Dreaming: A Cognitive-Psychological Analysis* (New Jersey: Lawrence Erlbaum Associates, 1985); G. Domhoff, *Finding Meaning in Dreams: A Quantitative Approach* (New York: Plenum Press, 1996).
4. P. McNamara, "Counterfactual Thought in Dreams," *Dreaming* 10 (2000): 232–45; P. McNamara et al. "Counterfactual Cognitive Operations in Dreams," *Dreaming* 12 (2002): 121–33.
5. D. Kahneman, "Varieties of Counterfactual Thinking" and C. G. Davis and D. R. Lehman, "Counterfactual Thinking and Coping with Traumatic Life Events," in *What Might Have Been: The Social Psychology of Counterfactual Thinking*, eds. J. M. Olson and N. J. Roese (New Jersey: Lawrence Erlbaum Associates, 1995), 375–96.
6. A. Nwoye, "The Psychology and Content of Dreaming in Africa," *Journal of Black Psychology* 43 (2015): 3–26.
7. D. F. Perry, J. DiPietro, and K. Costigan, "Are Women Carrying 'Basketballs' Really Having Boys? Testing Pregnancy Folklore," *Birth Defects Research B: Developmental and Reproductive Toxicology* 26 (1999): 172–77.
8. W. Shakespeare, *Hamlet* (London: Penguin, 2015), Act 2, scene 2.
9. J. L. Borges, "The Library of Babel," in *Labyrinths: Selected Stories and Other Writings* (London: Penguin Books, 1970).
10. W. C. Dement, with C. Vaughan, *The Promise of Sleep: A Pioneer in Sleep Medicine Explores the Vital Connection Between Health, Happiness, and a Good Night's Sleep* (New York: Dell, 1999).
11. F. Boas, *Contributions to the Ethnology of the Kwakiutl*, vol. 3 (New York: Columbia University Contributions to Anthropology, 1925).

5. First Images (pages 99–113)

1. W. O'Grady and S. W. Cho, "First Language Acquisition," in *Contemporary Linguistics: An Introduction* (Boston: Bedford St. Martin's, 2001), 326–62.
2. A. Machado, "Parábolas," in *Poesías Completas* (Barcelona: Austral, 2015).
3. S. Freud, *Three Essays on the Theory of Sexuality* and *Introductory Lectures on Psychoanalysis*, in *The Standard Edition of the Complete Psychological Works of Sigmund Freud*, eds. J. Strachey et al., vols. 7, 15, 16 (London: Hogarth Press, 1953); M. Klein, *The Psychoanalysis of Children; Authorized Translation by Alix*

Strachey (New York: Grove Press, 1960); P. King, R. Steiner, and British Psycho-Analytical Society, *The Freud-Klein Controversies, 1941–45* (London: Tavistock/Routledge, 1991).

4. D. Foulkes, *Children's Dreams: Longitudinal Studies* (New York: Wiley, 1982).
5. Ibid., 66.
6. Ibid., 68.
7. C. Hall and B. Domhoff, "A Ubiquitous Sex Difference in Dreams," *Journal of Abnormal and Social Psychology* 66 (1963): 278–80; C. S. Hall et al., "The Dreams of College Men and Women in 1959 and 1980: A Comparison of Dream Contents and Sex Differences," *Sleep* 5 (1982): 188–94.
8. M. Lortie-Lussier, C. Schwab, and J. De Koninck, "Working Mothers Versus Homemakers: Do Dreams Reflect the Changing Roles of Women?" *Sex Roles* 12 (1985): 1009–21; J. Mathes, and M. Schredl, "Gender Differences in Dream Content: Are They Related to Personality?" *International Journal of Dream Research* 6 (2013): 104–9.
9. D. Foulkes, *Children's Dreams*, 137.
10. P. Sandor, S. Szakadat, and R. Bodizs, "Ontogeny of Dreaming: A Review of Empirical Studies," *Sleep Medicine Reviews* 18 (2014): 435–49; P. Sandor, S. Szakadat, K. Kertesz, and R. Bodizs, "Content Analysis of 4 to 8 Year-Old Children's Dream Reports," *Frontiers in Psychology* 6 (2015): 534.
11. K. Valli and A. Revonsuo, "The Threat Simulation Theory in Light of Recent Empirical Evidence: A Review," *American Journal of Psychology* 122 (2009): 17–38.
12. M. G. Umlauf et al., "The Effects of Age, Gender, Hopelessness, and Exposure to Violence on Sleep Disorder Symptoms and Daytime Sleepiness among Adolescents in Impoverished Neighborhoods," *Journal of Youth Adolescence* 44 (2015): 518–42.
13. L. Hale, L. M. Berger, M. K. LeBourgeois, and J. Brooks-Gunn, "Social and Demographic Predictors of Preschoolers' Bedtime Routines," *Journal of Developmental and Behavior Pediatrics* 30 (2009): 394–402.
14. M. T. Hyyppa, E. Kronholm, E. Alanen, "Quality of Sleep during Economic Recession in Finland: A Longitudinal Cohort Study," *Social Science and Medicine* 45 (1997): 731–38.
15. D. L. Bliwise, "Historical Change in the Report of Daytime Fatigue," *Sleep* 19 (1996): 462–64; J. E. Broman, L. G. Lundh, and J. Hetta, "Insufficient Sleep in the General Population," *Neurophysiology Clinic* 26 (1996): 30–39; M. M. Mitler et al., "The Sleep of Long-Haul Truck Drivers," *The New England Journal of Medicine* 337 (1997): 755–61.
16. S. Stranges et al., "Sleep Problems: An Emerging Global Epidemic? Findings from the INDEPTH WHO-SAGE Study Among More Than 40,000 Older Adults from 8 Countries Across Africa and Asia," *Sleep* 35 (2012): 1173–81.

17. L. R. Teixeira et al., "Sleep Patterns of Day-Working, Evening High-Schooled Adolescents of São Paulo, Brazil," *Chronobiology International* 21 (2004): 239–52.
18. A. L. D. Medeiros, D. B. F. Mendes, P. F. Lima, and J. R. Araujo, "The Relationships Between Sleep-Wake Cycle and Academic Performance in Medical Students," *Biological Rhythm Research* 32 (2001): 263–70.
19. M. E. Hartmann and J. R. Prichard, "Calculating the Contribution of Sleep Problems to Undergraduates' Academic Success," *Sleep Health* 4 (2018): 463–71.
20. A. K. Leung and W. L. Robson, "Nightmares," *Journal of the National Medical Association* 85 (1993): 233–35; A. Gauchat, J. R. Seguin, and A. Zadra, "Prevalence and Correlates of Disturbed Dreaming in Children," *Pathologie Biologie (Paris)* 62 (2014): 311–18.
21. J. Borjigin, et al., "Surge of Neurophysiological Coherence and Connectivity in the Dying Brain," *Proceedings of the National Academy of Sciences of the USA* 110 (2013): 14432–37.

6. The Evolution of Dreaming (pages 114–128)

1. M. S. Dodd et al., "Evidence for Early Life in Earth's Oldest Hydrothermal Vent Precipitates," *Nature* 543 (2017): 60–64.
2. D. R. Mitchell, "Evolution of Cilia," *Cold Spring Harbor Perspectives in Biology* 9 (2017).
3. H. Wijnen and M. W. Young, "Interplay of Circadian Clocks and Metabolic Rhythms," *Annual Review of Genetics* 40 (2006): 409–48.
4. R. D. Nath et al., "The Jellyfish Cassiopea Exhibits a Sleep-like State," *Current Biology* 27 (2017): 2983–90.
5. M. A. Tosches, D. Bucher, P. Vopalensky, and D. Arendt, "Melatonin Signaling Controls Circadian Swimming Behavior in Marine Zooplankton," *Cell* 159 (2014): 46–57.
6. C. A. Czeisler et al., "Stability, Precision, and Near-24-Hour Period of the Human Circadian Pacemaker," *Science* 284 (1999): 2177–81.
7. J. J. Hublin et al., "New Fossils from Jebel Irhoud, Morocco and the Pan-African Origin of *Homo sapiens*," *Nature* 546 (2017): 289–92; D. Richter et al., "The Age of the Hominin Fossils from Jebel Irhoud, Morocco, and the Origins of the Middle Stone Age," *Nature* 546 (2017): 293–96.
8. W. Kaiser and J. Steiner-Kaiser, "Neuronal Correlates of Sleep, Wakefulness and Arousal in a Diurnal Insect," *Nature* 301 (1983): 707–79; K. M. Hartse, *Sleep in Insects and Nonmammalian Vertebrates,* Principles and Practice of Sleep Medicine (Philadelphia: W. B. Saunder, 1989); I. I. Tobler and M. Neuner-Jehle, "24-H Variation of Vigilance in the Cockroach Blaberus Giganteus," *Journal of Sleep Research* 1 (1992): 231–39; S. Sauer, E. Herr-

mann, and W. Kaiser, "Sleep Deprivation in Honey Bees," *Journal of Sleep Research* 13 (2004): 145–52.

9. J. C. Hendricks et al., "Rest in Drosophila Is a Sleep-Like State," *Neuron* 25 (2000): 129–38; P. J. Shaw, C. Cirelli, R. J. Greenspan, and G. Tononi, "Correlates of Sleep and Waking in *Drosophila Melanogaster*," *Science* 287 (2000): 1834–37.

10. J. M. Siegel, "Do All Animals Sleep?" *Trends in Neuroscience* 31 (2008): 208–13.

11. I. Tobler and A. A. Borbely, "Effect of Rest Deprivation on Motor Activity of Fish," *Journal of Comparative Physiology A* 157 (1985): 817–22; I. V. Zhdanova, S. Y. Wang, O. U. Leclair, and N. P. Danilova, "Melatonin Promotes Sleep-Like State in Zebrafish," *Brain Research* 903 (2001): 263–68; T. Yokogawa et al., "Characterization of Sleep in Zebrafish and Insomnia in Hypocretin Receptor Mutants," *PLOS Biology* 5 (2007): e277; B. B. Arnason, H. Thornorsteinsson, and K. A. E. Karlsson, "Absence of Rapid Eye Movements during Sleep in Adult Zebrafish," *Behavioural Brain Research* 291 (2015): 189–94.

12. J. A. Hobson, "Electrographic Correlates of Behavior in the Frog with Special Reference to Sleep," *Electroencephalography Clinical Neurophysiology* 22 (1967): 113–21; J. A. Hobson, O. B. Goin, and C. J. Goin, "Electrographic Correlates of Behaviour in Tree Frogs," *Nature* 220 (1968): 386–87.

13. A. W. Crompton, C. R. Taylor, and J. A. Jagger, "Evolution of Homeothermy in Mammals," *Nature* 272 (1978): 333–36.

14. M. Shein-Idelson et al., "Slow Waves, Sharp Waves, Ripples, and REM in Sleeping Dragons," *Science* 352 (2016): 590–95.

15. S. C. Nicol, N. A. Andersen, N. H. Phillips, and R. J. Berger, "The Echidna Manifests Typical Characteristics of Rapid Eye Movement Sleep," *Neuroscience Letters* 283 (2000): 49–52.

16. J. M. Siegel et al., "Sleep in the Platypus," *Neuroscience* 91 (1999): 391–400.

17. J. A. Lesku et al., "Ostriches Sleep like Platypuses," *PLOS One* 6 (2011): e23203.

18. R. N. Martinez et al., "A Basal Dinosaur from the Dawn of the Dinosaur Era in Southwestern Pangaea," *Science* 331 (2011): 206–10; S. J. Nesbitt et al., "The Oldest Dinosaur? A Middle Triassic Dinosauriform from Tanzania," *Biology Letters* 9 (2013).

19. X. Xu and M. A. Norell, "A New Troodontid Dinosaur from China with Avian-Like Sleeping Posture," *Nature* 431 (2004): 838–41; C. Gao et al., "A Second Soundly Sleeping Dragon: New Anatomical Details of the Chinese Troodontid Mei long with Implications for Phylogeny and Taphonomy," *PLOS One* 7 (2012).

20. A. Tiriac, G. Sokoloff, and M. S. Blumberg, "Myoclonic Twitching and

Sleep-Dependent Plasticity in the Developing Sensorimotor System," *Current Sleep Medicine Reports* 1 (2015): 74–79; M. S. Blumberg et al., "Development of Twitching in Sleeping Infant Mice Depends on Sensory Experience," *Current Biology* 25 (2015): 656–62.

21. P. R. Renne et al., "Time Scales of Critical Events around the Cretaceous-Paleogene Boundary," *Science* 339 (2013): 684–87.

22. K. O. Pope, K. H. Baines, A. C. Ocampo, and B. A. Ivanov, "Impact Winter and the Cretaceous/Tertiary Extinctions: Results of a Chicxulub Asteroid Impact Model," *Earth and Planetary Science Letters* 128 (1994): 719–25; J. Vellekoop et al., "Rapid Short-Term Cooling Following the Chicxulub Impact at the Cretaceous-Paleogene Boundary," *Proceedings of the National Academy of Sciences of the USA* 111 (2014): 7537–41.

23. R. Maor, T. Dayan, H. Ferguson-Gow, and K. E. Jones, "Temporal Niche Expansion in Mammals from a Nocturnal Ancestor after Dinosaur Extinction," *Nature Ecology and Evolution* 1 (2017): 1889–95.

24. Nicol et al., "The Echidna Manifests Typical Characteristics."

25. S. T. Piantadosi and C. Kidd, "Extraordinary Intelligence and the Care of Infants," *Proceedings of the National Academy of Sciences of the USA* 113 (2016): 6874–79.

26. Y. Mitani et al., "Three-Dimensional Resting Behaviour of Northern Elephant Seals: Drifting like a Falling Leaf," *Biology Letters* 6 (2010): 163–66.

27. J. D. R. Houghton et al., "Measuring the State of Consciousness in a Free-Living Diving Sea Turtle," *Journal of Experimental Marine Biology and Ecology* 356 (2008): 115–20.

28. A. I. Oleksenko et al., "Unihemispheric Sleep Deprivation in Bottlenose Dolphins," *Journal of Sleep Research* 1 (1992): 40–44; O. I. Lyaminet et al., "Unihemispheric Slow Wave Sleep and the State of the Eyes in a White Whale," *Behavioural Brain Research* 129 (2002): 125–29; O. Lyamin, J. Pryaslova, V. Lance, and J. Siegel, "Animal Behaviour: Continuous Activity in Cetaceans after Birth," *Nature* 435 (2005): 1177; L. M. Mukhametov, "Sleep in Marine Mammals," *Experimental Brain Research* 8 (2007): 227–38.

29. G. G. Mascetti, "Unihemispheric Sleep and Asymmetrical Sleep: Behavioral, Neurophysiological, and Functional Perspectives," *Nature and Science of Sleep* 8 (2016): 221–38.

30. N. C. Rattenborg et al., "Migratory Sleeplessness in the White-Crowned Sparrow *(Zonotrichia leucophrys gambelii)*," *PLOS Biology* 2 (2004): e212.

31. N. C. Rattenborg et al., "Evidence that Birds Sleep in Mid-Flight," *Nature Communications* 7 (2016): 12468.

32. N. C. Rattenborg, S. L. Lima, and C. J. Amlaner, "Half-Awake to the Risk of Predation," *Nature* 397 (1999): 397–98; N. C. Rattenborg, S. L. Lima, and C. J. Amlaner, "Facultative Control of Avian Unihemispheric Sleep under the Risk of Predation," *Behavioural Brain Research* 105 (1999): 163–72.

33. N. Gravett et al., "Inactivity/Sleep in Two Wild Free-Roaming African Elephant Matriarchs: Does Large Body Size Make Elephants the Shortest Mammalian Sleepers?" *PLOS One* 12 (2017): e0171903.
34. R. Noser, L. Gygax, and I. Tobler, "Sleep and Social Status in Captive Gelada Baboons (*Theropithecus Gelada*)," *Behavioural Brain Research* 147 (2003): 9–15.
35. D. R. Samson et al., "Segmented Sleep in a Nonelectric, Small-Scale Agricultural Society in Madagascar," *American Journal of Human Biology* 29 (2017).
36. G. Yetish et al., "Natural Sleep and Its Seasonal Variations in Three Pre-Industrial Societies," *Current Biology* 25 (2015): 2862–68.
37. D. R. Samson et al., "Chronotype Variation Drives Night-Time Sentinel-Like Behaviour in Hunter-Gatherers," *Proceedings of the Royal Society: Biological Sciences* 284 (2017).
38. L. A. Zhivotovsky, N. A. Rosenberg, and M. W. Feldman, "Features of Evolution and Expansion of Modern Humans, Inferred from Genomewide Microsatellite Markers," *The American Journal of Human Genetics* 72 (2003): 1171–86.
39. H. O. De la Iglesia et al., "Ancestral Sleep," *Current Biology* 26 (2016): R271–72.

7. The Biochemistry of Dreams (pages 129–144)

1. E. Aserinsky and N. Kleitman, "Regularly Occurring Periods of Eye Motility, and Concomitant Phenomena, during Sleep," *Science* 118 (1953): 273–74.
2. M. Roth, J. Shaw, and J. Green, "The Form Voltage Distribution and Physiological Significance of the K-Complex," *Electroencephalography and Clinical Neurophysiology* 8 (1956): 385–402; M. Steriade and F. Amzica, "Slow Sleep Oscillation, Rhythmic K-Complexes, and Their Paroxysmal Developments," *Journal of Sleep Research* 7 (1998): 30–35; A. G. Siapas and M. A. Wilson, "Coordinated Interactions between Hippocampal Ripples and Cortical Spindles during Slow-Wave Sleep," *Neuron* 21 (1998): 1123–28; N. K. Logothetis et al., "Hippocampal-Cortical Interaction during Periods of Subcortical Silence," *Nature* 491 (2012): 547–53.
3. W. Dement and N. Kleitman, "Cyclic Variations in EEG during Sleep and Their Relation to Eye Movements, Body Motility, and Dreaming," *Electroencephalography and Clinical Neurophysiology* 9 (1957): 673–90; W. Dement and N. Kleitman, "The Relation of Eye Movements during Sleep to Dream Activity: An Objective Method for the Study of Dreaming," *Journal of Experimental Psychology* 53 (1957): 339–46.
4. W. Dement and N. Kleitman, "The Relation of Eye Movements during Sleep to Dream Activity: An Objective Method for the Study of Dreaming," *Journal of Experimental Psychology* 53 (1957): 339–46; M. Jouvet and

D. Jouvet, "A Study of the Neurophysiological Mechanisms of Dreaming," *Electroencephalography and Clinical Neurophysiology*, Suppl. 24 (1963): 133–157.

5. F. D. Foulkes, "Dream Reports from Different Stages of Sleep," *Journal of Abnormal Psychology* 65 (1962): 14–25.

6. G. G. Abel, W. D. Murphy, J. V. Becker, and A. Bitar, "Women's Vaginal Responses during REM Sleep," *Journal of Sex and Marital Therapy* 5 (1979): 5–14; G. S. Rogers, R. L. Van de Castle, W. S. Evans, and J. W. Critelli, "Vaginal Pulse Amplitude Response Patterns during Erotic Conditions and Sleep," *Archives of Sexual Behaviour* 14 (1985): 327–42.

7. C. Fisher, J. Gorss, and J. Zuch, "Cycle of Penile Erection Synchronous with Dreaming (REM) Sleep," Preliminary Report, *Archives of General Psychiatry* 12 (1965): 29–45.

8. T. A. Wehr, "A Brain-Warming Function for REM Sleep," *Neuroscience and Biobehavioral Reviews* 16 (1992): 379–97.

9. L. Xie et al., "Sleep Drives Metabolite Clearance from the Adult Brain," *Science* 342 (2013): 373–77.

10. H. Lee et al., "The Effect of Body Posture on Brain Glymphatic Transport," *Journal of Neuroscience* 35 (2015): 11034–44.

11. A. S. Urrila et al., "Sleep Habits, Academic Performance, and the Adolescent Brain Structure," *Scientific Reports* 7 (2017): 41678.

12. R. L. Weinmann, "Levodopa and Hallucination," *Journal of the American Medical Association* 221 (1972): 1054; K. Kamakura et al., "Therapeutic Factors Causing Hallucination in Parkinson's Disease Patients, Especially Those Given Selegiline," *Parkinsonism and Related Disorders* 10 (2004): 235–42.

13. M. Taheri and E. Arabameri, "The Effect of Sleep Deprivation on Choice Reaction Time and Anaerobic Power of College Student Athletes," *Asian Journal of Sports Medicine* 3 (2012): 15–20; K. Tokizawa et al., "Effects of Partial Sleep Restriction and Subsequent Daytime Napping on Prolonged Exertional Heat Strain," *Occupational and Environmental Medicine* 72 (2015): 521–28; A. Sufrinko, E. W. Johnson, and L. C. Henry, "The Influence of Sleep Duration and Sleep-Related Symptoms on Baseline Neurocognitive Performance among Male and Female High School Athletes," *Neuropsychology* 30 (2016): 484–91; R. Ben Cheikh, I. Latiri, M. Dogui, and H. Ben Saad, "Effects of One-Night Sleep Deprivation on Selective Attention and Isometric Force in Adolescent Karate Athletes," *The Journal of Sports Medicine and Physical Fitness* 57 (2017): 752–59.

14. R. Leproult and E. Van Cauter, "Effect of 1 Week of Sleep Restriction on Testosterone Levels in Young Healthy Men," *Journal of the American Medical Association* 305 (2011): 2173–74.

15. C. Cajochen et al., "EEG and Ocular Correlates of Circadian Melatonin

Phase and Human Performance Decrements during Sleep Loss," *American Journal of Physiology* 277 (1999): R640–49.

16. S. F. Sorrells et al., "Human Hippocampal Neurogenesis Drops Sharply in Children to Undetectable Levels in Adults," *Nature* 555 (2018): 377–81.

17. C. Liston et al., "Circadian Glucocorticoid Oscillations Promote Learning-Dependent Synapse Formation and Maintenance," *Nature Neuroscience* 16 (2013); 698–705.

18. C. Pavlides, L. G. Nivon, and B. S. McEwen, "Effects of Chronic Stress on Hippocampal Long-Term Potentiation," *Hippocampus* 12 (2002): 245–57.

19. R. Legendre and H. Piéron, "De la Propriété hypnotoxique des humeurs développée au cours d'une veille prolongée," *Comptes Rendus de la Société de Biologie de Paris* 70 (1912): 210–12.

20. J. M. Krueger, J. R. Pappenheimer, and M. L. Karnovsky, "Sleep-Promoting Effects of Muramyl Peptides," *Proceedings of the National Academy of Sciences of the USA* 79 (1982): 6102–6; S. Shoham and J. M. Krueger, "Muramyl Dipeptide-Induced Sleep and Fever: Effects of Ambient Temperature and Time of Injections," *American Journal of Physiology* 255 (1988): R157–65; J. M. Krueger and M. R. Opp, "Sleep and Microbes," *International Review of Neurobiology* 131 (2016): 207–25.

21. J. A. MacCulloch, "Fasting (Introductory and Non-Christian)" and G. Foucart, "Dreams and Sleep: Egyptian" in *Encyclopedia of Religion and Ethics,* ed. J. Hastings, vol. 5 (New York: Charles Scribner's Sons, 1912); J. S. Lincoln, *The Dream in Native American and Other Primitive Cultures* (Hoboken: Dover, 2003).

22. T. Nielsen and R. A. Powell, "Dreams of the Rarebit Fiend: Food and Diet as Instigators of Bizarre and Disturbing Dreams," *Frontiers in Psychology* 6 (2015): 47.

23. R. G. Pertwee, *Handbook of Cannabis* (Oxford: Oxford University Press, 2014).

24. D. E. Nichols, "Psychedelics," *Pharmacological Reviews* 68 (2016): 264–355.

25. J. G. Soares Maia and W. A. Rodrigues, "*Virola theiodora* como alucinógena e tóxica," *Acta Amazonica* 4 (1974): 21–23.

26. A. Berardi, G. Schelling, and P. Campolongo, "The Endocannabinoid System and Post Traumatic Stress Disorder (PTSD): From Preclinical Findings to Innovative Therapeutic Approaches in Clinical Settings," *Pharmacological Research* 111 (2016): 668–78.

27. E. Tagliazucchi et al., "Increased Global Functional Connectivity Correlates with LSD-Induced Ego Dissolution," *Current Biology* 26 (2016): 1043–50; R. Kraehenmann, "Dreams and Psychedelics: Neurophenomenological Comparison and Therapeutic Implications," *Current Neuropharmacology* 15 (2017): 1032–42; R. Kraehenmann et al., "Dreamlike Effects of

LSD on Waking Imagery in Humans Depend on Serotonin 2A Receptor Activation," *Psychopharmacology (Berlin)* 234 (2017): 2031–46; C. Sanz et al., "The Experience Elicited by Hallucinogens Presents the Highest Similarity to Dreaming within a Large Database of Psychoactive Substance Reports," *Frontiers in Neuroscience* 12 (2018): 7.

28. Nichols, "Psychedelics."

29. J. Riba et al., "Topographic Pharmaco-EEG Mapping of the Effects of the South American Psychoactive Beverage Ayahuasca in Healthy Volunteers," *British Journal of Clinical Pharmacology* 53 (2002): 613–28.

30. S. M. Kosslyn et al., "The Role of Area 17 in Visual Imagery: Convergent Evidence from PET and rTMS," *Science* 284 (1999): 167–70.

31. D. B. de Araújo et al., "Seeing with the Eyes Shut: Neural Basis of Enhanced Imagery Following Ayahuasca Ingestion," *Human Brain Mapping* 33 (2012): 2550–60.

32. R. L. Carhart-Harris et al., "Neural Correlates of the LSD Experience Revealed by Multimodal Neuroimaging," *Proceedings of the National Academy of Sciences of the USA* 113 (2016): 4853–58.

33. A. Viol et al., "Shannon Entropy of Brain Functional Complex Networks under the Influence of the Psychedelic Ayahuasca," *Scientific Reports* 7 (2017): 7388.

34. E. Tagliazucchi et al., "Enhanced Repertoire of Brain Dynamical States during the Psychedelic Experience," *Human Brain Mapping* 35 (2014): 5442–56; A. V. Lebedev et al., "LSD-Induced Entropic Brain Activity Predicts Subsequent Personality Change," *Human Brain Mapping* 37 (2016): 3203–13; M. M. Schartner et al., "Increased Spontaneous MEG Signal Diversity for Psychoactive Doses of Ketamine, LSD and Psilocybin," *Scientific Reports* 7 (2017): 46421.

35. P. Luz, "O uso ameríndio do caapi," and B. Keifenheim, "Nixi pae como participação sensível no princípio de transformação da criação primordial entre os índios kaxinawa no leste do Peru," in *O uso ritual da ayahuasca*, eds. B. C. Labate and W. S. Araujo (Campinas: Mercado de Letras, 2002), 37–68, 97–127.

8. Madness Is a Dream One Dreams Alone (pages 145–158)

1. C. Okorome Mume, "Nightmare in Schizophrenic and Depressed Patients," *The European Journal of Psychiatry* 23 (2009); 177–83; F. Michels et al., "Nightmare Frequency in Schizophrenic Patients, Healthy Relatives of Schizophrenic Patients, Patients at High Risk States for Psychosis, and Healthy Controls," *International Journal of Dream Research* 7 (2014): 9–13.

2. J. C. Skancke, I. Holsen, and M. Schredl, "Continuity between Waking Life and Dreams of Psychiatric Patients: A Review and Discussion of the

Implications for Dream Research," *International Journal of Dream Research* 7 (2014): 39–53.

3. K. Dzirasa et al., "Dopaminergic Control of Sleep-Wake States," *Journal of Neuroscience* 26 (2006): 10577–89.

4. J. Lacan, *Anxiety*, in *The Seminar of Jacques Lacan*, trans. A. R. Price, vol. 10 (Cambridge: Polity, 2016).

5. S. Beckett, *Waiting for Godot* (London: Faber & Faber, 2006), 54.

6. C. G. Jung, *Symbols of Transformation*, in *The Collected Works of C. G. Jung*, vol. 5. (London: Routledge and K. Paul, 1966).

7. S. Freud, *Totem and Taboo*, in *The Standard Edition of the Complete Psychological Works of Sigmund Freud*, eds. J. Strachey et al., vol. 13 (London: Hogarth Press, 1953).

8. Ibid., 89.

9. S. Freud, *Introductory Lectures on Psychoanalysis*, in *The Standard Edition of the Complete Psychological Works of Sigmund Freud*, eds. J. Strachey et al., vols. 15, 16 (London: Hogarth Press, 1953).

10. S. Freud, *The Future of an Illusion*, in *The Standard Edition of the Complete Psychological Works of Sigmund Freud*, eds. J. Strachey et al., vol. 21 (London: Hogarth Press, 1953), 53.

11. M. Klein, "Criminal Tendencies in Normal Children," *British Journal of Medical Psychology* 74 (1927); M. Klein, *Narrative of a Child Analysis; The Conduct of the Psychoanalysis of Children as Seen in the Treatment of a Ten Year Old Boy* (New York: Basic Books, 1961).

12. M. Klein, *The Psychoanalysis of Children; Authorized Translation by Alix Strachey* (New York: Grove Press, 1960).

13. M. Kramer, "Dream Differences in Psychiatric Patients," in *Sleep and Mental Illness*, eds. S. R. Pandi-Perumal and M. Kramer (Cambridge: Cambridge University Press, 2010): 375–382.

14. N. B. Mota et al., "Speech Graphs Provide a Quantitative Measure of Thought Disorder in Psychosis," *PLOS One* 7 (2012): e34928; N. B. Mota et al., "Graph Analysis of Dream Reports Is Especially Informative about Psychosis," *Scientific Reports* 4 (2014): 3691; N. B. Mota, M. Copelli, and S. Ribeiro, "Thought Disorder Measured as Random Speech Structure Classifies Negative Symptoms and Schizophrenia Diagnosis 6 Months in Advance," *npj Schizophrenia* 3 (2017): 1–10.

9. Sleeping and Remembering (pages 159–173)

1. J. B. Jenkins and K. M. Dallenbach, "Oblivescence during Sleep and Waking," *The American Journal of Psychology* 35 (1924): 605–12.

2. C. A. Pearlman, "Effect of Rapid Eye Movement (Dreaming) Sleep Deprivation on Retention of Avoidance Learning in Rats," *Reports of the US*

Navy Submarine Medical Center 563 (1969): 1–4; P. Leconte and V. Bloch, "Effect of Paradoxical Sleep Deprivation on the Acquisition and Retention of Conditioning in Rats," *Journal de Physiologie (Paris)* 62 (1970): 290; W. C. Stern, "Acquisition Impairments Following Rapid Eye Movement Sleep Deprivation in Rats," *Physiology and Behavior* 7 (1971): 345–52.

3. C. Smith and S. Butler, "Paradoxical Sleep at Selective Times Following Training is Necessary for Learning," *Physiology and Behavior* 29 (1982): 469–73; C. Smith and G. Kelly, "Paradoxical Sleep Deprivation Applied Two Days after End of Training Retards Learning," *Physiology and Behavior* 43 (1988): 213–16; C. Smith and G. M. Rose, "Evidence for a Paradoxical Sleep Window for Place Learning in the Morris Water Maze," *Physiology & Behavior* 59 (1996): 93–97; C. Smith and G. M. Rose, "Posttraining Paradoxical Sleep in Rats Is Increased after Spatial Learning in the Morris Water Maze," *Behavioral Neuroscience* 111 (1997): 1197–204.

4. R. Stickgold et al., "Replaying the Game: Hypnagogic Images in Normals and Amnesics," *Science* 290 (2000): 350–53.

5. R. Stickgold, L. James, and J. A. Hobson, "Visual Discrimination Learning Requires Sleep after Training," *Nature Neuroscience* 3 (2000): 1237–38.

6. S. C. Mednick et al., "The Restorative Effect of Naps on Perceptual Deterioration," *Nature Neuroscience* 5 (2002): 677–81.

7. S. Mednick, K. Nakayama, and R. Stickgold, "Sleep-Dependent Learning: A Nap Is as Good as a Night," *Nature Neuroscience* 6 (2003): 697–98.

8. S. S. Yoo et al., "A Deficit in the Ability to Form New Human Memories without Sleep," *Nature Neuroscience* 10 (2007): 385–92.

9. W. Plihal and J. Born, "Effects of Early and Late Nocturnal Sleep on Declarative and Procedural Memory," *Journal of Cognitive Neuroscience* 9 (1997): 534–47; W. Plihal and J. Born, "Effects of Early and Late Nocturnal Sleep on Priming and Spatial Memory," *Psychophysiology* 36 (1999): 571–82.

10. L. J. Batterink, C. E. Westerberg, and K. A. Paller, "Vocabulary Learning Benefits from REM after Slow-Wave Sleep," *Neurobiology of Learning and Memory* 144 (2017).

11. N. Lemos, J. Weissheimer, and S. Ribeiro, "Naps in School Can Enhance the Duration of Declarative Memories Learned by Adolescents," *Frontiers in Systems Neuroscience* 8 (2014): 103.

12. T. Cabral et al., "Post-Class Naps Boost Declarative Learning in a Naturalistic School Setting," *npj Science of Learning* 3 (2018): 14.

13. C. Beck, "Students Allowed to Nap at School With Sleep Pods," NBC News, Mar. 6, 2017, https://www.nbcnews.com/health/kids-health/students-allowed-nap-school-sleep-pods-n729881; S. Danzy, "High Schools Are Allowing Sleep-deprived Students to Take Midday Naps," *People,* Feb. 22, 2017, https://people.howstuffworks.com/high-schools-are-allowing

-sleepdeprived-students-take-midday-naps.htm; D. Willis, "N. M. Schools Roll Out High-Tech Sleep Pods for Students," *USA Today*, Mar. 1, 2017, https://www.usatoday.com/story/tech/nation-now/2017/03/01/nm -schools-roll-out-high-tech-sleep-pods-students/98619548/; N. Borges, "Tempo integral: a experiência das escolas de Santa Cruz," GAZ, Jun. 15, 2018, http://www.gaz.com.br/conteudos/educacao/2018/06/15/122501 -tempo_integral_a_experiencia_das_escolas_de_santa_cruz.html.php; G. Pin, "Quitar la Siesta al Niño cuando Llega al Colegio, ¡Un Grave Error!" Serpadres, 2018, https://www.serpadres.es/3-6-anos/educacion-desarrollo /articulo/quitar-la-siesta-al-nino-cuando-llega-al-colegio-un-grave-error.

14. D. L . Hummer and T. M. Lee, "Daily Timing of the Adolescent Sleep Phase: Insights from a Cross-Species Comparison," *Neuroscience & Biobehavioral Reviews* 70 (2016): 171–81.

15. G. P. Dunster et al., "Sleepmore in Seattle: Later School Start Times Are Associated with More Sleep and Better Performance in High School Students," *Science Advances* 4 (2018).

10. The Reverberation of Memories (pages 174-197)

1. W. Penfield, "Some Mechanisms of Consciousness Discovered during Electrical Stimulation of the Brain," *Proceedings of the National Academy of Sciences USA* 44 (1958): 51–66.

2. D. Hebb, *The Organization of Behavior* (Hoboken: Wiley, 1949).

3. Ibid., F9.

4. C. Pavlides and J. Winson, "Influences of Hippocampal Place Cell Firing in the Awake State on the Activity of These Cells during Subsequent Sleep Episodes," *Journal of Neuroscience* 9 (1989): 2907–18.

5. S. Ribeiro et al., "Long-Lasting Novelty-Induced Neuronal Reverberation during Slow-Wave Sleep in Multiple Forebrain Areas," *PLOS Biology* 2 (2004): E24; J. O'Neill, T. Senior, and J. Csicsvari, "Place-Selective Firing of CA1 Pyramidal Cells during Sharp Wave/Ripple Network Patterns in Exploratory Behavior," *Neuron* 49 (2006): 143–55.

6. F. Niemtschek, *Leben des K.K. Kapellmeisters Wolfgang Gottlieb Mozart, nach Originalquellen beschrieben* (Praga: Herrlischen Buchhandlung, 1798).

7. T. Lomo, "Potentiation of Monosynaptic EPSPs in Cortical Cells by Single and Repetitive Afferent Volleys," *Journal of Physiology* 194 (1968): 84–85P; T. V. Bliss and T. Lomo, "Long-Lasting Potentiation of Synaptic Transmission in the Dentate Area of the Anaesthetized Rabbit Following Stimulation of the Perforant Path," *Journal of Physiology* 232 (1973): 331–56.

8. J. R. Whitlock, A. J. Heynen, M. G. Shuler, and M. F. Bear, "Learning Induces Long-Term Potentiation in the Hippocampus," *Science* 313 (2006): 1093–97.

9. C. Pavlides, Y. J. Greenstein, M. Grudman, and J. Winson, "Long-Term Potentiation in the Dentate Gyrus Is Induced Preferentially on the Positive Phase of Theta-Rhythm," *Brain Research* 439 (1988): 383–87.

10. C. Holscher, R. Anwyl, and M. J. Rowan, "Stimulation on the Positive Phase of Hippocampal Theta Rhythm Induces Long-Term Potentiation That Can Be Depotentiated by Stimulation on the Negative Phase in Area CA1 in Vivo," *Journal of Neuroscience* 17 (1997): 6470–77; J. Hyman et al., "Stimulation in Hippocampal Region CA1 in Behaving Rats Yields Long-Term Potentiation when Delivered to the Peak of Theta and Long-Term Depression when Delivered to the Trough," *Journal of Neuroscience* 23 (2003): 11725–31; P. T. Huerta and J. E. Lisman, "Bidirectional Synaptic Plasticity Induced by a Single Burst During Cholinergic Theta Oscillation in CA1 in Vitro," *Neuron* 15 (1995): 1053–63.

11. J. E. Lisman and O. Jensen, "The Theta-Gamma Neural Code," *Neuron* 77 (2013): 1002–16; V. Lopes-Dos-Santos et al., "Parsing Hippocampal Theta Oscillations by Nested Spectral Components during Spatial Exploration and Memory-Guided Behavior," *Neuron* 100 (2018): 950–52.

12. H. C. Heller and S. F. Glotzbach, "Thermoregulation during Sleep and Hibernation," *International Review of Physiology* 15 (1977): 147–88.

13. G. R. Poe, D. A. Nitz, B. L. McNaughton, and C. A. Barnes, "Experience-Dependent Phase-Reversal of Hippocampal Neuron Firing during REM Sleep," *Brain Research* 855 (2000): 176–80.

14. P. Maquet et al., "Experience-Dependent Changes in Cerebral Activation during Human REM Sleep," *Nature Neuroscience* 3 (2000): 831–36; P. Peigneux et al., "Learned Material Content and Acquisition Level Modulate Cerebral Reactivation during Posttraining Rapid-Eye-Movements Sleep," *Neuroimage* 20 (2003): 125–34.

15. R. Huber, M. F. Ghilardi, M. Massimini, and G. Tononi, "Local Sleep and Learning," *Nature* 430 (2004): 78–81.

16. R. Boyce, S. D. Glasgow, S. Williams, and A. Adamantidis, "Causal Evidence for the Role of REM Sleep Theta Rhythm in Contextual Memory Consolidation," *Science* 352 (2016): 812–16.

17. L. Marshall, H. Helgadottir, M. Molle, and J. Born, "Boosting Slow Oscillations during Sleep Potentiates Memory," *Nature* 444 (2006): 610–13.

18. H. V. Ngo, T. Martinetz, J. Born, and M. Molle, "Auditory Closed-Loop Stimulation of the Sleep Slow Oscillation Enhances Memory," *Neuron* 78 (2013): 545–53.

19. J. Seibt et al., "Cortical Dendritic Activity Correlates with Spindle-Rich Oscillations during Sleep in Rodents," *Nature Communications* 8 (2017): 684.

20. B. Rasch, C. Buchel, S. Gais, and J. Born, "Odor Cues During Slow-Wave Sleep Prompt Declarative Memory Consolidation," *Science* 315 (2007): 1426–29.

21. A. Bilkei-Gorzo et al., "A Chronic Low Dose of Delta9-tetrahydrocannabinol (THC) Restores Cognitive Function in Old Mice," *Nature Medicine* 23 (2017): 782–87.
22. A. Guerreiro, *Ancestrais e suas sombras: uma etnografia da chefia kalapao e seu ritual mortuário* (Campinas: Unicamp, 2015).

11. Genes and Memes (pages 198–215)

1. J. L. Borges, "Funes the Memorious," in *Labyrinths: Selected Stories and Other Writings* (London: Penguin Books, 1970).
2. M. Pompeiano, C. Cirelli, and G. Tononi, "Effects of Sleep Deprivation on Fos-Like Immunoreactivity in the Rat Brain," *Archives Italiennes de Biologie* 130 (1992): 325–35; C. Cirelli, M. Pompeiano, and G. Tononi, "Fos-Like Immunoreactivity in the Rat Brain in Spontaneous Wakefulness and Sleep," *Archives Italiennes de Biologie* 131 (1993): 327–30; M. Pompeiano, C. Cirelli, and G. Tononi, "Immediate-Early Genes in Spontaneous Wakefulness and Sleep: Expression of C-Fos and NGFI-A mRNA and Protein," *Journal of Sleep Research* 3 (1994): 80–96.
3. C. A. Pearlman, "Effect of Rapid Eye Movement (Dreaming) Sleep Deprivation on Retention of Avoidance Learning in Rats," *Reports of the US Navy Submarine Medical Center* 563 (1969): 1–4; P. Leconte and V. Bloch, "Effect of Paradoxical Sleep Deprivation on the Acquisition and Retention of Conditioning in Rats," *Journal de Physiologie (Paris)* 62 (1970): 290; W. C. Stern, "Acquisition Impairments Following Rapid Eye Movement Sleep Deprivation in Rats," *Physiology & Behavior* 7 (1971): 345–52.
4. A. Giuditta et al., "The Sequential Hypothesis of the Function of Sleep," *Behavioural Brain Research* 69 (1995): 157–66.
5. G. Tononi and C. Cirelli, "Modulation of Brain Gene Expression during Sleep and Wakefulness: A Review of Recent Findings," *Neuropsychopharmacology* 25 (2001): S28–35.
6. V. V. Vyazovskiy et al., "Cortical Firing and Sleep Homeostasis," *Neuron* 63 (2009): 865–78; Z. W. Liu et al., "Direct Evidence for Wake-Related Increases and Sleep-Related Decreases in Synaptic Strength in Rodent Cortex," *Journal of Neuroscience* 30 (2010): 8671–75.
7. D. Bushey, G. Tononi, and C. Cirelli, "Sleep and Synaptic Homeostasis: Structural Evidence in *Drosophila*," *Science* 332 (2011): 1576–81.
8. G. G. Turrigiano et al., "Activity-Dependent Scaling of Quantal Amplitude in Neocortical Neurons," *Nature* 391 (1998): 892–96.
9. G. Tononi and C. Cirelli, "Sleep and Synaptic Homeostasis: A Hypothesis," *Brain Research Bulletin* 62 (2003): 143–150.
10. S. Ribeiro and M. A. Nicolelis, "Reverberation, Storage, and Postsynaptic Propagation of Memories during Sleep," *Learning and Memory* 11 (2004): 686–96; S. Ribeiro et al., "Downscale or Emboss Synapses during Sleep?"

Frontiers in Neuroscience 3 (2009); S. Ribeiro, "Sleep and Plasticity," *Pflugers Archiv* 463 (2012): 111–20.

11. M. G. Frank, N. P. Issa, and M. P. Stryker, "Sleep Enhances Plasticity in the Developing Visual Cortex," *Neuron* 30 (2001): 275–87; J. Ulloor and S. Datta, "Spatio-temporal Activation of Cyclic AMP Response Element-Binding Protein, Activity-Regulated Cytoskeletal-Associated Protein and Brain-Derived Nerve Growth Factor: A Mechanism for Pontine-Wave Generator Activation-Dependent Two-Way Active-Avoidance Memory Processing in the Rat," *Journal of Neurochemistry* 95 (2005): 418–28; I. Ganguly-Fitzgerald, J. Donlea, and P. J. Shaw, "Waking Experience Affects Sleep Need in *Drosophila*," *Science* 313 (2006): 1775–81; J. M. Donlea et al., "Inducing Sleep by Remote Control Facilitates Memory Consolidation in *Drosophila*," *Science* 332 (2011): 1571–76; J. B. Calais et al., "Experience-Dependent Upregulation of Multiple Plasticity Factors in the Hippocampus during early REM Sleep," *Neurobiology of Learning and Memory* 122 (2015); C. G. Vecsey et al., "Sleep Deprivation Impairs cAMP Signalling in the Hippocampus," *Nature* 461 (2009): 1122–25; P. Ravassard et al., "REM Sleep-Dependent Bidirectional Regulation of Hippocampal-Based Emotional Memory and LTP," *Cerebral Cortex* 26 (2016): 1488–500.

12. G. Tononi and C. Cirelli, "Sleep and the Price of Plasticity: From Synaptic and Cellular Homeostasis to Memory Consolidation and Integration," *Neuron* 81 (2014): 12–34.

13. G. Yang et al., "Sleep Promotes Branch-Specific Formation of Dendritic Spines after Learning," *Science* 344 (2014): 1173–78.

14. W. Li, L. Ma, G. Yang, and W. B. Gan, "REM Sleep Selectively Prunes and Maintains New Synapses in Development and Learning," *Nature Neuroscience* 20 (2017): 427–37.

15. Ibid.

12. Sleeping to Create (pages 216–246)

1. T. W.-M. Draper, *The Bemis History and Genealogy: Being an Account, in Greater Part of the Descendants of Joseph Bemis, of Watertown, Mass.* (San Francisco: Stanley-Taylor Co. Print., 1900), 160.

2. J. Essinger, *Jacquard's Web: How a Hand-Loom Led to the Birth of the Information Age* (Oxford: Oxford University Press, 2007); M. Tedre, *The Science of Computing: Shaping a Discipline* (Boca Raton: CRC Press, 2014).

3. J. J. L. F. Lalande, *Voyage en Italie, contenant l'histoire & les anecdotes les plus singulieres de l'Italie, & sa description; les usages, le gouvernement, le commerce, la littérature, les arts, l'histoire naturelle, & les antiquités* (Paris: Veuve Desaint, 1786), 293–94.

4. S. Turner, *A Hard Day's Write: The Stories behind Every Beatles Song* (New York: HarperPerennial, 1999), 83.

5. Albrecht Dürer, *Speis der maier knaben (Nourishment for Young Painters)*, "Dürer's Dream of 1525."

6. Ibid.

7. A. A. T. Macrobius, *Commentary on the Dream of Scipio*, trans. W. H. Stahl (New York: Columbia University Press, 1990).

8. A. M. Peden, "Macrobius and Mediaeval Dream Literature," *Medium Ævum* 54 (1985): 59–73.

9. A. J. Kabir, *Paradise, Death and Doomsday in Anglo-Saxon Literature* (Cambridge: Cambridge University Press, 2001).

10. M. de Cervantes, *Don Quixote*, trans. E. Grossman (London: Vintage, 2005), 21.

11. F. Pessoa, *The Book of Disquiet: The Complete Edition*, trans. M. J. Costa (London: Serpent's Tail, 2018), 230.

12. F. Pessoa, *Poesia completa de Álvaro de Campos* (São Paulo: Companhia das Letras, 2007), 287.

13. J. E. Agualusa in interview with S. Ribeiro, *Limiar: Uma década entre o cérebro e a mente* (São Paulo: Vieira Lent, 2015), 29–31.

14. L. Trotsky, *Trotsky's Diary in Exile, 1935* (Cambridge: Harvard University Press, 1976), 145–46.

15. G. Orwell, "My Country Right or Left," in *The Collected Essays, Journalism and Letters*, Vol. 1 (London: Penguin Books, 1970), 590–91.

16. A. Kekulé, "Sur la constitution des substances aromatiques," *Bulletin de la Société Chimique de Paris* 3 (1865): 98–110.

17. E. Hornung, *The Ancient Egyptian Books of the Afterlife* (Ithaca, NY: Cornell University Press, 1999).

18. S. F. Rudofsky and J. H. Wotiz, "Psychologists and the Dream Accounts of August Kekulé," *Ambix* 35 (1988): 31–38.

19. O. B. Ramsay and A. J. Rocke, "Kekulé's Dreams: Separating the Fiction from the Fact," *Chemistry in Britain* 20 (1984): 1093–94.

20. O. Loewi, "From the Workshop of Discoveries," *Perspectives in Biology and Medicine* 4 (1960): 1–25.

21. A. R. Wallace, *My Life: A Record of Events and Opinions*, vol. 1 (London: Chapman and Hall, 1905), 361.

22. J. Benton, "Descartes' Olympica," *Philosophy and Literature* 2 (1980): 163–66.

23. G. Leibniz, *Philosophical Papers and Letters*, ed. and trans. L. E. Loemker (Dordrecht: Kluwer Academic Publishers, 1989), 114.

24. H. Poincaré, "Mathematical Creation," in *The Foundations of Science: Science and Hypothesis, the Value of Science, Science and Method* (Amazon Digital Services, 2018), 389.

25. J. Hadamard, *The Psychology of Invention in the Mathematical Field* (Mineola: Dover, 1954).

26. S. Dehaene and L. Cohen, "The Unique Role of the Visual Word Form Area in Reading," *Trends in Cognitive Science* 15 (2011): 254–62.
27. P. Tholey, "Consciousness and Abilities of Dream Characters Observed during Lucid Dreaming," *Perceptual and Motor Skills* 68 (1989): 567–78; T. Stumbrys and M. Daniels, "An Exploratory Study of Creative Problem Solving in Lucid Dreams: Preliminary Findings and Methodological Considerations," *International Journal of Dream Research* 3 (2010): 121–29; T. Stumbrys, D. Erlacher, and S. Schmidt, "Lucid Dream Mathematics: An Explorative Online Study of Arithmetic Abilities of Dream Characters," *International Journal of Dream Research* 4 (2011): 35–40.
28. G. H. Hardy, *Ramanujan: Twelve Lectures on Subjects Suggested by His Life and Work* (Cambridge: AMS: Chelsea Publishing Co., 1940), 9.
29. G. H. Hardy, "Obituary, S. Ramanujan," *Nature* 105 (1920): 494–95.
30. S. Ramanujan, *Ramanujan: Letters and Reminiscences*, vol. 1, Memorial Number (Muthialpet High School, 1968); B. Krishnayya, *Ramanujan: The Man and the Mathematician* (New York: Thomas Nelson and Sons Ltd, 1967), 87.
31. B. Russell, *Human Knowledge: Its Scope and Limits* (New York: Simon & Schuster, 1948), 172.
32. A. Antunes, *Como é que chama o nome disso: Antologia* (São Paulo: Publifolha, 2006).
33. U. Wagner et al., "Sleep Inspires Insight," *Nature* 427 (2204): 352–55.
34. M. P. Walker, C. Liston, J. A. Hobson, and R. Stickgold, "Cognitive Flexibility Across the Sleep-Wake Cycle: REM-sleep Enhancement of Anagram Problem Solving," *Cognitive Brain Research* 14 (2002): 317–24.
35. D. J. Cai et al., "REM, not Incubation, Improves Creativity by Priming Associative Networks," *Proceedings of the National Academy of Sciences of the USA* 106 (2009): 10130–34.
36. S. Deregnaucourt et al., "How Sleep Affects the Developmental Learning of Bird Song," *Nature* 433 (2005): 710–16.
37. W. A. Liberti III et al., "Unstable Neurons Underlie a Stable Learned Behavior," *Nature Neuroscience* 19 (2016): 1665–71.
38. E. J. Wamsley et al., "Cognitive Replay of Visuomotor Learning at Sleep Onset: Temporal Dynamics and Relationship to Task Performance," *Sleep* 33 (2010): 59–68.
39. D. W. Singer, *Giordano Bruno: His Life and Thought. With Annotated Translation of His Work On the Infinite Universe and Worlds* (New York: Schuman, 1950).
40. A. Druyan and S. Soter, in *Cosmos: A Spacetime Odyssey*, ed. B. Braga (Santa Fe: Netflix, 2014). Unfortunately, I have been unable to find the original source for this quotation, which allows me to use a lapidary phrase that

is also attributed without evidence to Giordano: "Se non è vero, è molto ben trovato."

41. Singer, *Giordano Bruno*; G. Bruno, *On the Infinite, the Universe and the Worlds: Five Cosmological Dialogues*, vol. 2 (Scotts Valley, CA: CreateSpace Independent Publishing Platform, 2014).

42. I. A. Ahmad, "The Impact of the Qur'anic Conception of Astronomical Phenomena on Islamic Civilization," *Vistas in Astronomy* 39 (1995): 395–403.

43. J. Kepler, "Letter from Johannes Kepler to Galileo Galilei, 1610," *Johannes Kepler Gesammelte Werke*, vol. 4 (Bonn: Deutsche Forschungsgemeinschaft, 2009), 287–310.

44. D. O. Hebb, "The Effects of Early and Late Brain Injury upon Test Scores, and the Nature of Normal Adult Intelligence," *Proceedings of the American Philosophical Society* 85 (1942): 275 –92.

45. W. B. Scoville and B. Milner, "Loss of Recent Memory after Bilateral Hippocampal Lesions," *Journal of Neurology, Neurosurgery and Psychiatry* 20 (1957): 11–21.

46. S. Ribeiro et al., "Induction of Hippocampal Long-Term Potentiation during Waking Leads to Increased Extrahippocampal Zif-268 Expression during Ensuing Rapid-Eye-Movement Sleep," *Journal of Neuroscience* 22 (2002): 10914–23.

47. S. Ribeiro et al., "Novel Experience Induces Persistent Sleep-Dependent Plasticity in the Cortex but Not in the Hippocampus," *Frontiers in Neuroscience* 1 (2007): 43–55.

13. REM Sleep Isn't Dreaming (pages 247-266)

1. M. Solms, *The Neuropsychology of Dreams: A Clinico-Anatomical Study* (New Jersey, Lawrence Erlbaum Associates, 1997); M. Solms, "Dreaming and REM Sleep Are Controlled by Different Brain Mechanisms,"*Behavioral and Brain Sciences* 23 (2000): 843–50.

2. W. R. Adey, E. Bors, and R. W. Porter, "EEG Sleep Patterns after High Cervical Lesions in Man," *Archives of Neurology* 19 (1968): 377–83; T. N. Chase, L. Moretti, and A. L. Prensky, "Clinical and Electroencephalographic Manifestations of Vascular Lesions of the Pons," *Neurology* 18 (1968): 357–68; J. L. Cummings, and R. Greenberg, "Sleep Patterns in the 'Locked-In' Syndrome," *Electroencephalography and Clinical Neurophysiology* 43 (1977): 270–71; P. Lavie et al., "Localized Pontine Lesion: Nearly Total Absence of REM Sleep," *Neurology* 34 (1984): 118–20.

3. M. Solms, *The Neuropsychology of Dreams: A Clinico-Anatomical Study* (New Jersey: Lawrence Erlbaum Associates, 1997), 186.

4. J.-M. Charcot, "Un Cas de suppression brusque et isolée de la vision mentale des signes et des objets (formes et couleurs)," *Le Progrès Médical* 11 (1883).

5. M. Bischof and C. L. Bassetti, "Total Dream Loss: A Distinct Neuropsychological Dysfunction after Bilateral PCA Stroke," *Annals of Neurology* 56 (2004): 583–86.

6. H. W. Lee et al., "Mapping of Functional Organization in Human Visual Cortex: Electrical Cortical Stimulation," *Neurology* 54 (2000): 849–54; H. Kimmig et al., "fMRI Evidence for Sensorimotor Transformations in Human Cortex during Smooth Pursuit Eye Movements," *Neuropsychologia* 46 (2008): 2203–13; P. Fattori, S. Pitzalis, and C. Galletti, "The Cortical Visual Area V6 in Macaque and Human Brains," *Journal of Physiology Paris* 103 (2009): 88– 97; G. Handjaras et al., "How Concepts Are Encoded in the Human Brain: A Modality Independent, Category-Based Cortical Organization of Semantic Knowledge," *Neuroimage* 135 (2016): 232–42.

7. H. C. Tsai et al., "Phasic Firing in Dopaminergic Neurons Is Sufficient for Behavioral Conditioning," *Science* 324 (2009): 1080–84; A. H. Luo et al., "Linking Context with Reward: A Functional Circuit from Hippocampal CA3 to Ventral Tegmental Area," *Science* 333 (2011): 353–57; J. Y. Cohen et al., "Neuron-Type-Specific Signals for Reward and Punishment in the Ventral Tegmental Area," *Nature* 482 (2012): 85–88.

8. S. Fujisawa and G. Buzsaki, "A 4 Hz Oscillation Adaptively Synchronizes Prefrontal, VTA, and Hippocampal Activities," *Neuron* 72 (2011): 153–65; S. N. Gomperts, F. Kloosterman, and M. A. Wilson, "VTA Neurons Coordinate with the Hippocampal Reactivation of Spatial Experience," *eLife* 4 (2015): e05360.

9. J. L. Valdés, B. L. McNaughton, and J. M. Fellous, "Offline Reactivation of Experience-Dependent Neuronal Firing Patterns in the Rat Ventral Tegmental Area," *Journal of Neurophysiology* 114 (2015): 1183–95.

10. G. B. Feld et al., "Dopamine D2-Like Receptor Activation Wipes Out Preferential Consolidation of High Over Low Reward Memories during Human Sleep," *Journal of Cognitive Neuroscience* 26 (2014): 2310–20.

11. C. C. Hong et al., "fMRI Evidence for Multisensory Recruitment Associated with Rapid Eye Movements during Sleep," *Human Brain Mapping* 30 (2009): 1705–22.

12. C. W. Wu et al., "Variations in Connectivity in the Sensorimotor and Default-Mode Networks during the First Nocturnal Sleep Cycle," *Brain Connect* 2 (2012): 177–90; H. M. Chow et al., "Rhythmic Alternating Patterns of Brain Activity Distinguish Rapid Eye Movement Sleep from Other States of Consciousness," *Proceedings of the National Academy of Sciences of the USA* 110 (2012): 10300–5; K. C. Fox et al., "Dreaming as Mind Wandering: Evidence from Functional Neuroimaging and First-Person Content Reports," *Frontiers in Human Neuroscience* 30 (2013).

13. M. Solms, *The Neuropsychology of Dreams: A Clinico-Anatomical Study* (New Jersey: Lawrence Erlbaum Associates, 1997).

14. M. E. Raichle et al., "A Default Mode of Brain Function," *Proceedings of the National Academy of Sciences of the USA* 98 (2001): 676–82.

15. Wu et al., "Variations in Connectivity in the Sensorimotor and Default-Mode Networks"; J. B. Eichenlaub et al., "Resting Brain Activity Varies with Dream Recall Frequency between Subjects," *Neuropsychopharmacology* 39 (2014): 1594–602.

16. T. Koike, S. Kan, M. Misaki, and S. Miyauchi, "Connectivity Pattern Changes in Default-Mode Network with Deep Non-REM and REM Sleep," *Neuroscience Research* 69 (2011): 322–30.

17. K. C. Fox et al., "Dreaming as Mind Wandering: Evidence from Functional Neuroimaging and First-Person Content Reports," *Frontiers in Human Neuroscience* 7 (2013).

18. *The Bhagavad Gita*, trans. W. J. Johnson (Oxford: Oxford University Press, 2004), 12.

19. F. Palhano-Fontes et al., "The Psychedelic State Induced by Ayahuasca Modulates the Activity and Connectivity of the Default Mode Network," *PLOS One* 10 (2015): e0118143.

20. R. L. Carhart-Harris et al., "Neural Correlates of the Psychedelic State as Determined by fMRI Studies with Psilocybin," *Proceedings of the National Academy of Sciences of the USA* 109 (2012): 2138–43.

21. R. L. Carhart-Harris et al., "Neural Correlates of the LSD Experience Revealed by Multimodal Neuroimaging," *Proceedings of the National Academy of Sciences of the USA* 113 (2016): 4853–58.

22. J. Speth et al., "Decreased Mental Time Travel to the Past Correlates with Default-Mode Network Disintegration under Lysergic Acid Diethylamide," *Journal of Psychopharmacology* 30 (2016): 344–53.

23. J. A. Brefczynski-Lewis et al., "Neural Correlates of Attentional Expertise in Long-Term Meditation Practitioners," *Proceedings of the National Academy of Sciences of the USA* 104 (2007): 11483–88; J. A. Brewer et al., "Meditation Experience Is Associated with Differences in Default Mode Network Activity and Connectivity," *Proceedings of the National Academy of Sciences of the USA* 108 (2011): 20254–59; A. Sood and D. T. Jones, "On Mind Wandering, Attention, Brain Networks, and Meditation," *Explore (ny)* 9 (2013): 136–41.

24. W. James, *The Varieties of Religious Experience: A Study in Human Nature* (Scotts Valley, CA: CreateSpace Independent Publishing Platform, 2009); A. Watts, "Psychedelics and Religious Experience," *California Law Review* 56 (1968): 74–85; J. Riba et al., "Increased Frontal and Paralimbic Activation Following Ayahuasca, the Pan-Amazonian Inebriant," *Psychopharmacology (Berlin)* 186 (2006): 93–98.

25. Henry M. Vyner, *The Healthy Mind Interviews: The Dalai Lama, Lopon Tenzin Namdak, Lopon Thekchoke*, vol. 4 (Kathmandu, Nepal: Vajra Publications), 66.

26. Brewer et al., "Meditation Experience Is Associated with Differences."

27. Carhart-Harris, "Neural Correlates of the Psychedelic State."

28. W. Hasenkamp, C. D. Wilson-Mendenhall, E. Duncan, and L. W. Barsalou, "Mind Wandering and Attention during Focused Meditation: A Fine-Grained Temporal Analysis of Fluctuating Cognitive States," *Neuroimage* 59 (2012): 750–60.

29. C. Colace, "Drug Dreams in Cocaine Addiction," *Drug and Alcohol Review* 25 (2006): 177; C. Colace, "Are the Wish-Fulfillment Dreams of Children the Royal Road for Looking at the Functions of Dreams?" *Neuropsychoanalysis* 15 (2013): 161–75.

30. E. Tulving, "Memory and Consciousness," *Canadian Psychology / Psychologie canadienne* 26 (1985): 1–12.

31. "Na Janela" online festival, May 24, 2020, https://www.youtube.com/watch?v=95tOtpk4Bnw.

32. E. J. Wamsley et al., "Dreaming of a Learning Task Is Associated with Enhanced Sleep-Dependent Memory Consolidation," *Current Biology* 20 (2010): 850–55.

33. B. M. A. Pritzker, *Native American Encyclopedia: History, Culture, and Peoples* (Oxford: Oxford University Press, 2000).

34. J. G. Neihardt, *Black Elk Speaks* (Lincoln: University of Nebraska Press, 2014), 53; J. G. Neihardt, *The Sixth Grandfather: Black Elk's Teachings Given to John G. Neihardt* (Lincoln: University of Nebraska Press, 1985), 53.

35. Plutarch, *Lives from Plutarch,* trans. J. W. McFarland (New York: Random House, 1967).

14. Desires, Emotions, and Nightmares (pages 267-288)

1. J. K. Boehnlein, J. D. Kinzie, R. Ben, and J. Fleck, "One-Year Follow-Up Study of Posttraumatic Stress Disorder among Survivors of Cambodian Concentration Camps," *American Journal of Psychiatry* 142 (1985): 956–59; A. Aron, "The Collective Nightmare of Central American Refugees," in *Trauma and Dreams,* ed. D. Barrett (Cambridge: Harvard University Press, 1996): 140–47; E. M. Menke and J. D. Wagner, "The Experience of Homeless Female-Headed Families," *Issues in Mental Health Nursing* 18 (1997): 315–30; T. C. Neylan et al., "Sleep Disturbances in the Vietnam Generation: Findings from a Nationally Representative Sample of Male Vietnam Veterans," *American Journal of Psychiatry* 155 (1998): 929–33; K. Esposito, A. Benitez, L. Barza, and T. Mellman, "Evaluation of Dream Content in Combat-Related PTSD," *Journal of Traumatic Stress* 12 (1999): 681–87; L. Wittmann, M. Schredl, and M. Kramer, "Dreaming in Posttraumatic Stress Disorder: A Critical Review of Phenomenology, Psychophysiology and Treatment," *Psychotherapy and Psychosomatics* 76 (2007): 25–39; J. Davis-Berman, "Older Women in the Homeless Shelter: Personal Per-

spectives and Practice Ideas," *Journal of Women & Aging* 23 (2011): 360–74; K. E. Miller, J. A. Brownlow, S. Woodward, and P. R. Gehrman, "Sleep and Dreaming in Posttraumatic Stress Disorder," *Current Psychiatry Reports* 19 (2017): 71.

2. R. Maor, T. Dayan, H. Ferguson-Gow, and K. E. Jones, "Temporal Niche Expansion in Mammals from a Nocturnal Ancestor after Dinosaur Extinction," *Nature Ecology and Evolution* 1 (2017): 1889–95.

3. A. Revonsuo, "The Reinterpretation of Dreams: An Evolutionary Hypothesis of the Function of Dreaming," *Behavioral and Brain Sciences* 23 (2000): 877–901; K. Valli et al., "The Threat Simulation Theory of the Evolutionary Function of Dreaming: Evidence from Dreams of Traumatized Children," *Consciousness and Cognition* 14 (2005): 188–218.

4. C. R. Marmar et al., "Course of Posttraumatic Stress Disorder 40 Years After the Vietnam War: Findings from the National Vietnam Veterans Longitudinal Study," *JAMA Psychiatry* 72 (2015): 875–81.

5. R. J. Ross et al., "Rapid Eye Movement Sleep Disturbance in Posttraumatic Stress Disorder," *Biological Psychiatry* 35 (1994): 195–202; R. J. Ross et al., "Rapid Eye Movement Sleep Changes during the Adaptation Night in Combat Veterans with Posttraumatic Stress Disorder," *Biological Psychiatry* 45 (1999): 938–41.

6. R. E. Brown et al., "Control of Sleep and Wakefulness," *Physiological Reviews* 92 (2012): 1087–187.

7. J. Froissart, *Chronicles,* trans. Geoffrey Brereton (London: Penguin Classics, 1978), 275.

8. Neylan et al., "Sleep Disturbances in the Vietnam Generation"; Esposito et al., "Evaluation of Dream Content"; B. J. Schreuder, M. van Egmond, W. C. Kleijn, and A. T. Visser, "Daily Reports of Posttraumatic Nightmares and Anxiety Dreams in Dutch War Victims," *Journal of Anxiety Disorders* 12 (1998): 511–24.

9. J. A. Meerloo, "Persecution Trauma and the Reconditioning of Emotional Life: A Brief Survey," *American Journal of Psychiatry* 125 (1969): 1187–91; R. F. Mollica, G. Wyshak, and J. Lavelle, "The Psychosocial Impact of War Trauma and Torture on Southeast Asian Refugees," *American Journal of Psychiatry* 144 (1987): 1567–72; U. H. Peters, "Psychological Sequelae of Persecution: The Survivor Syndrome," *Fortschritte der Neurologie-Psychiatrie* 57 (1989): 169–91; U. H. Peters, "The Stasi Persecution Syndrome," *Fortschritte der Neurologie-Psychiatrie* 59 (1991): 251–65; T. A. Roesler, D. Savin, and C. Grosz, "Family Therapy of Extrafamilial Sexual Abuse," *Journal of the American Academy of Child and Adolescent Psychiatry* 32 (1993): 967–70; I. M. Steine et al., "Cumulative Childhood Maltreatment and Its Dose-Response Relation with Adult Symptomatology: Findings in a Sample of Adult Survivors of Sexual Abuse," *Child Abuse & Neglect* 65 (2017): 99–111.

10. Anon., "The Dream of Dumuzid," in *The Electronic Text Corpus of Sumerian Literature,* vol. 1.4.3 (Oxford: Oxford University Press).

11. D. Kopenawa and B. Albert, *The Falling Sky: Words of a Yanomami Shaman,* trans. N. Elliott and A. Dundy (Cambridge: Belknap, Harvard University Press, 2013), 37.

12. M. Desseilles, T. T. Dang-Vu, V. Sterpenich, and S. Schwartz, "Cognitive and Emotional Processes during Dreaming: A Neuroimaging View," *Consciousness and Cognition* 20 (2011): 998–1008.

13. D. Brown, *Bury My Heart at Wounded Knee: An Indian History of the American West* (New York: Fall River Press, 2014).

14. B. Drury and T. Clavin, *The Heart of Everything That Is: The Untold Story of Red Cloud, An American Legend* (New York: Simon & Schuster, 2013).

15. D. Brown, *The Fetterman Massacre: Formerly Fort Phil Kearny, an American Saga* (Lincoln: University of Nebraska Press, 1984).

16. S. D. Smith, *Give Me Eighty Men: Women and the Myth of the Fetterman Fight* (Lincoln: University of Nebraska Press, 2010), xix.

17. F. C. Carrington, *My Army Life and the Fort Phil. Kearney Massacre: With an Account of the Celebration of "Wyoming Opened"* (Books for Libraries, 1971), 86.

18. Drury and Clavin, *The Heart of Everything That Is.*

19. G. E. Hyde, *Life of George Bent: Written from His Letters* (Norman: University of Oklahoma Press, 1968); M. Kenny, "Roman Nose, Cheyenne: A Brief Biography," *Wičazo Ša Review* 5 (1989): 9–30.

20. *Folha de São Paulo,* June, 22, 2009, https://www1.folha.uol.com.br/fsp/brasil/fc2206200911.htm.

21. G. J. Vermeij, "Unsuccessful Predation and Evolution," *The American Naturalist* 120 (1982): 701–20; G. B. Schaller, *The Deer and the Tiger: A Study of Wildlife in India* (Chicago: The University of Chicago Press, 1984); W. Hayward et al., "Prey Preferences of the Leopard (*Panthera pardus*)," *Journal of Zoology* 270 (2006): 298–313.

22. Aron, "The Collective Nightmare of Central American Refugees."

23. S. Pinker, *The Better Angels of Our Nature: A History of Violence and Inhumanity* (London: Allen Lane, Penguin Books, 2011).

24. O. Flanagan, "Deconstructing Dreams: The Spandrels of Sleep," *Journal of Philosophy* 92 (1995): 5–27.

25. P. Gay, *Freud: A Life for Our Time* (London: J. M. Dent & Sons, 1988).

15. The Probabilistic Oracle (pages 289-319)

1. D. Brown, *Bury My Heart at Wounded Knee: An Indian History of the American West* (New York: Fall River, 2014), 289.

2. Ibid.

3. R. J. DeMallie, " 'These Have No Ears': Narrative and the Ethnohistorical Method," *Ethnohistory* 40, no. 4 (1993), 515–38.

4. Brown, *Bury My Heart at Wounded Knee*, 289.

5. R. M. Utley, *The Last Days of the Sioux Nation*, The Lamar Series in Western History (New Haven: Yale University Press, 2004).

6. W. K. Morehead, "The Death of Sitting Bull, and a Tragedy at Wounded Knee," *The American Indian in the United States Period: 1850–1914* (New York: Andover, 1914), 123–32.

7. Polybius, *The Histories*, trans. W. R. Paton., Loeb Classical Library 4 (Cambridge: Harvard University Press), 105; http://penelope.uchicago.edu/Thayer/E/Roman/Texts/Polybius/10*.html.

8. Plutarch, *Lives from Plutarch*, trans. J. W. McFarland (New York: Random House, 1967).

9. Suetonius, *The Twelve Caesars*, eds. R. Graves and M. Grant (London: Penguin, 2003).

10. N. C. Rattenborg, S. L. Lima, and C. J. Amlaner, "Facultative Control of Avian Unihemispheric Sleep under the Risk of Predation," *Behavioral and Brain Research* 105 (1999): 163–72; N. C. Rattenborg, S. L. Lima, and C. J. Amlaner, "Half-Awake to the Risk of Predation," *Nature* 397 (1999): 397–98.

11. K. Semendeferi et al., "Prefrontal Cortex in Humans and Apes: A Comparative Study of Area 10," *American Journal of Physical Anthropology* 114 (2001): 224–41.

12. E. Koechlin and A. Hyafil, "Anterior Prefrontal Function and the Limits of Human Decision-Making," *Science* 318 (2007): 594–98.

13. L. W. Swanson, J. D. Hahn, and O. Sporns, "Organizing Principles for the Cerebral Cortex Network of Commissural and Association Connections," *Proceedings of the National Academy of Sciences of the USA* 114 (2017): E9692–701.

14. G. Edelman, *Neural Darwinism: The Theory of Neuronal Group Selection* (New York: Basic Books, 1987).

15. G. Edelman, *Bright Air, Brilliant Fire: On the Matter of the Mind* (New York: Basic Books, 1992).

16. S. Dehaene et al., "Cerebral Mechanisms of Word Masking and Unconscious Repetition Priming," *Nature Neuroscience* 4 (2001): 752–58; C. Sergent, S. Baillet, and S. Dehaene, "Timing of the Brain Events Underlying Access to Consciousness during the Attentional Blink," *Nature Neuroscience* 8 (2005): 1391–400.

17. A. Del Cul, S. Dehaene, and M. Leboyer, "Preserved Subliminal Processing and Impaired Conscious Access in Schizophrenia," *Archives of General Psychiatry* 63 (2006): 1313–23.

18. B. J. Baars, "How Does a Serial, Integrated and Very Limited Stream of

Consciousness Emerge from a Nervous System That Is Mostly Unconscious, Distributed, Parallel and of Enormous Capacity?" *Ciba Foundation Symposium* 174 (1993): 282–90.

19. S. Dehaene, C. Sergent, and J. P. Changeux, "A Neuronal Network Model Linking Subjective Reports and Objective Physiological Data during Conscious Perception," *Proceedings of the National Academy of Sciences of the USA* 100 (2003): 8520–25.

20. S. Dehaene et al., "A Neuronal Network Model"; S. Dehaene, M. Kerszberg, and J. P. Changeux, "A Neuronal Model of a Global Workspace in Effortful Cognitive Tasks," *Proceedings of the National Academy of Sciences of the USA* 95 (1998): 14529–34; S. Dehaene and J. P. Changeux, "Ongoing Spontaneous Activity Controls Access to Consciousness: A Neuronal Model for Inattentional Blindness," *PLOS Biology* 3 (2005): e141; S. Dehaene et al., "Conscious, Preconscious, and Subliminal Processing: A Testable Taxonomy," *Trends in Cognitive Sciences* 10 (2006): 204–11.

21. C. C. Hong et al., "fMRI Evidence for Multisensory Recruitment Associated with Rapid Eye Movements during Sleep," *Human Brain Mapping* 30 (2009): 1705–22.

22. S. Freud, *The Interpretation of Dreams*; "Formulations on Two Principles of Mental Functioning," "On the History of the Psychoanalytic Movement," and "Mourning and Melancholia," *The Ego and the Id*, in *The Standard Edition of the Complete Psychological Works of Sigmund Freud*, ed. J. Strachey et al., vols. 4, 5, 12, 14, 19 (London: Hogarth Press, 1953).

23. C. Darwin, *The Origin of Species* (Oxford: Oxford University Press, 1996).

24. C. Darwin, *The Expression of the Emotions in Man and Animals* (London: Penguin Classics, 2009).

25. C. Hobaiter, R. W. Byrne, and K. Zuberbühler, "Wild Chimpanzees' Use of Single and Combined Vocal and Gestural Signals," *Behavioral Ecology and Sociobiology* 71 (2017).

26. S. Savage-Rumbaugh et al., "Spontaneous Symbol Acquisition and Communicative Use by Pygmy Chimpanzees *(Pan paniscus)*," *Journal of Experimental Psychology: General* 115 (1986): 211–35; K. Gillespie-Lynch, P. M. Greenfield, H. Lyn, and S. Savage-Rumbaugh, "Gestural and Symbolic Development among Apes and Humans: Support for a Multimodal Theory of Language Evolution," *Frontiers in Psychology* 5 (2014).

27. M. S. Seidenberg and L. A. Petitto, "Communication, Symbolic Communication, and Language in Child and Chimpanzee: Comment on Savage-Rumbaugh, McDonald, Sevcik, Hopkins, and Rupert (1986)," *Journal of Experimental Psychology: General* 116 (1987): 279–87.

28. C. S. Peirce, *The Essential Peirce: Selected Philosophical Writings*, two vols. (Bloomington: Indiana University Press, 1992 & 1998).

29. R. M. Seyfarth, D. L. Cheney, and P. Marler, "Monkey Responses to Three

Different Alarm Calls: Evidence of Predator Classification and Semantic Communication," *Science* 210 (1980): 801–3.

30. K. Zuberbühler, "Local Variation in Semantic Knowledge in Wild Diana Monkey Groups," *Animal Behavior* 59 (2000): 917–27; K. Zuberbühler, "Predator-Specific Alarm Calls in Campbell's Monkeys, *Cercopithecus campbelli*," *Behavioral Ecology and Sociobiology* 50 (2001): 414–22; A. M. Schel et al., "Chimpanzee Alarm Call Production Meets Key Criteria for Intentionality," *PLOS One* 8 (2013): e76674; P. Beynon and O. A. E. Rasa, "Do Dwarf Mongooses Have a Language? Warning Vocalisations Transmit Complex Information," *Suid-Afrikaanse Tydskr vir Wet* 85 (1989): 447–50; C. N. Slobodchikoff, J. Kiriazis, C. Fischer, and E. Creef, "Semantic Information Distinguishing Individual Predators in the Alarm Calls of Gunnison's Prairie Dogs," *Animal Behavior* 42 (1991): 713–19; E. Greene and T. Meagher, "Red Squirrels, *Tamiasciurus hudsonicus*, Produce Predator-Class Specific Alarm Calls," *Animal Behavior* 55 (1998): 511–18; C. Evans and L. Evans, "Chicken Food Calls Are Functionally Referential," *Animal Behavior* 58 (1999): 307–19; M. B. Manser, "The Acoustic Structure of Suricates' Alarm Calls Varies with Predator Type and the Level of Response Urgency," *Proceedings of the Royal Society B: Biological Sciences* 268 (2001): 2315–24; L. M. Herman et al., "The Bottlenosed Dolphin's (*Tursiops truncatus*) Understanding of Gestures as Symbolic Representations of Body Parts," *Animal Learning & Behavior* 29 (2001): 250–64.

31. J. Queiroz and S. Ribeiro, *The Biological Substrate of Icons, Indexes and Symbols in Animal Communication*, The Peirce Seminar Papers (New York: Berghahn Books, 2002), 69–78; S. Ribeiro et al., "Symbols Are Not Uniquely Human," *Biosystems* 90 (2007): 263–72.

32. Peirce, *The Essential Peirce.*

33. S. Engesser, A. R. Ridley, and S. W. Townsend, "Meaningful Call Combinations and Compositional Processing in the Southern Pied Babbler," *Proceedings of the National Academy of Sciences of the USA* 113 (2016): 5976–81; K. Arnold and K. Zuberbühler, "Language Evolution: Semantic Combinations in Primate Calls," *Nature* 441 (2006): 303; C. Coye, K. Ouattara, K. Zuberbühler, and A. Lemasson, "Suffixation Influences Receivers' Behaviour in Non-Human Primates," *Proceedings of the Royal Society B: Biological Sciences* 282 (2015): 20150265; K. Ouattara, A. Lemasson, and K. Zuberbühler, "Campbell's Monkeys Concatenate Vocalizations into Context-Specific Call Sequences," *Proceedings of the National Academy of Sciences of the USA* 106 (2009): 22026–31; K. Ouattara, A. Lemasson, and K. Zuberbühler, "Campbell's Monkeys Use Affixation to Alter Call Meaning," *PLOS One* 4 (2009): e7808; P. Fedurek, K. Zuberbühler, and C. D. Dahl, "Sequential Information in a Great Ape Utterance," *Scientific Reports* 6 (2016): 38226.

34. D. J. Povinelli and T. M. Preuss, "Theory of Mind: Evolutionary History of a Cognitive Specialization," *Trends in Neuroscience* 18 (1995): 418–24; J. Koster-Hale and R. Saxe, "Theory of Mind: A Neural Prediction Problem," *Neuron* 79 (2013): 836–48; H. Meunier, "Do Monkeys Have a Theory of Mind? How to Answer the Question?" *Neuroscience & Biobehavioral Review* 82 (2017): 110–23.

35. S. J. Waller, "Sound and Rock Art," *Nature* 363 (1993): 501; S. J. Waller, "The Divine Echo Twin Depicted at Echoing Rock Art Sites: Acoustic Testing to Substantiate Interpretations," in *American Indian Rock Art*, eds. A. Quinlan and A. McConnell, vol. 32 (2006): 63–74.

36. R. Q. Quiroga et al., "Invariant Visual Representation by Single Neurons in the Human Brain," *Nature* 435 (2005): 1102–7.

37. Ibid.; R. Q. Quiroga, "Concept Cells: The Building Blocks of Declarative Memory Functions," *Nature Reviews Neuroscience* 13 (2012): 587–97.

38. P. Ariès, *Western Attitudes Toward Death from the Middle Ages to the Present* (Baltimore: Johns Hopkins University Press, 1974); P. Metcalf and R. Huntington, *Celebrations of Death: The Anthropology of Mortuary Ritual* (Cambridge: Cambridge University Press, 1991); M. P. Pearson, *The Archaeology of Death and Burial* (College Station: Texas A&M University Press, 2000); B. A. Conklin, *Consuming Grief: Compassionate Cannibalism in an Amazonian Society* (Austin: University of Texas Press, 2001); A. C. G. M. Robben, *Death, Mourning, and Burial: A Cross-Cultural Reader* (London: Wiley-Blackwell, 2005); V. Brown, *The Reaper's Garden: Death and Power in the World of Atlantic Slavery* (Cambridge: Harvard University Press, 2010).

39. B. J. King, *How Animals Grieve* (Chicago: The University of Chicago Press, 2013).

40. J. R. Anderson, A. Gillies, and L. C. Lock, "Pan Thanatology," *Current Biology* 20 (2010): R349–51.

41. P. J. Fashing et al., "Death among Geladas (*Theropithecus gelada*): A Broader Perspective on Mummified Infants and Primate Thanatology," *American Journal of Primatology* 73 (2011): 405–9.

42. E. Viveiros de Castro, "A floresta de cristal: notas sobre a ontologia dos espíritos amazônicos," *Cadernos de Campo* 14/15 (2006): 319–38.

43. E. Durkheim, *The Elementary Forms of Religious Life,* trans. C. Cosman (Oxford: Oxford University Press, 2001); L. Costa and C. Fausto, in *The International Encyclopedia of Anthropology,* ed. H. Callan (New York: John Wiley & Sons, 2018).

44. L. M. Rival, *Trekking Through History: The Huaorani of Amazonian Ecuador* (New York: Columbia University Press, 2002).

45. P. Descola and J. Lloyd, *Beyond Nature and Culture* (Chicago: The University of Chicago Press, 2013); L. Costa and C. Fausto, "The Return of the

Animists: Recent Studies of Amazonian Ontologies," *Religion and Society: Advances in Research* 1 (2010): 89–109.

46. E. B. Tylor, *Primitive Culture* (London: John Murray, 1871).

47. E. Viveiros de Castro, "Cosmological Deixis and Amerindian Perspectivism," *Journal of the Royal Anthropological Institute* 4 (1998): 469–88.

48. C. Lévi-Strauss, *The Savage Mind* (Oxford: Oxford University Press, 1994).

49. E. B. Viveiros de Castro, "Perspectivism and Multinaturalism in Indigenous America," in *The Land Within: Indigenous Territory and the Perception of Environment*, eds. A. Surrallés and P. García Hierro (Copenhagen: IWGIA, 2005).

50. Viveiros de Castro, "Cosmological Deixis and Amerindian Perspectivism," 469–88.

51. D. Kopenawa and B. Albert, *The Falling Sky: Words of a Yanomami Shaman*, trans. N. Elliott and A. Dundy (Cambridge: Belknap, Harvard University Press, 2013), 140–41.

52. T. S. Lima, "Two and Its Many: Reflections on Perspectivism in a Tupi Cosmology," *Ethnos* 64 (1999): 107–31.

53. T. S. Lima, *Um peixe olhou para mim. O povo Yudjá e a perspectiva* (São Paulo: Unesp, 2005).

54. F. Boas, *Contributions to the Ethnology of the Kwakiutl*, vol. 3 (New York: Columbia University Contributions to Anthropology, 1925); C. F. Feest, "Dream of One of Twins: On Kwakiutl Dream Culture," *Studien zur Kulturkunde* 119 (2001): 138–53.

55. M. A. Gonçalves, *O mundo inacabado: Ação e criação em uma cosmologia amazônica* (Rio de Janeiro: UFRJ, 2001), 277.

16. Missing the Dead, and the Inner World of Culture (pages 320-331)

1. P. McNamara, *The Neuroscience of Religious Experience* (Cambridge: Cambridge University Press, 2009).

2. A. K. Petersen et al., *Evolution, Cognition, and the History of Religion: A New Synthesis*, Supplements to Method & Theory in the Study of Religion, vol. 13 (Leiden: Brill, 2018).

3. R. Bouckaert et al., "Mapping the Origins and Expansion of the Indo-European Language Family," *Science* 337 (2012): 957–60.

4. I. Mota, "Jogo do bicho é ilegal, mas mobiliza a paixão do povo," *O Liberal*, Belém, August 6, 2017.

5. C. G. Jung, *The Red Book*, trans. S. Shamdasani (New York: W. W. Norton & Co., 2009).

6. C. G. Jung, in *Psychology Audiobooks* (Kino, 1990).

7. *The World Within: C. G. Jung in His Own Words*, directed by Suzanne Wagner (Bosustow Video Productions, 1990).

8. C. Riches, "Man Strangled His Wife After Nightmare," *Express*, London, July 30, 2010.
9. C. K. Morewedge and M. I. Norton, "When Dreaming Is Believing: The (Motivated) Interpretation of Dreams," *Journal of Personality and Social Psychology* 96 (2009): 249–64.
10. Antonio Guerreiro, in personal interview with author, September 27, 2018.
11. M. Perrin, ed., *Antropologia y experiencias del sueño* (Quito: MLAL/Abya-Yala, 1990).
12. E. Hartmann, "Making Connections in a Safe Place: Is Dreaming Psychotherapy?" *Dreaming* 5 (1995): 213–28.
13. B. O. Rothbaum, E. A. Meadows, P. Resick, and D. W. Foy, "Cognitive-behavioral Therapy," in *Effective Treatments for PTSD: Practice Guidelines from the International Society for Traumatic Stress Studies*, eds. T. M. Keane, E. B. Foa., and M. J. Friedman (New York: Guilford, 2000), 320–25.
14. J. M. Kane et al., "Comprehensive Versus Usual Community Care for First-Episode Psychosis: 2-Year Outcomes from the NIHM RAISE Early Treatment Program," *American Journal of Psychiatry* 173 (2016): 362–72.
15. J. Fuentes et al., "Enhanced Therapeutic Alliance Modulates Pain Intensity and Muscle Pain Sensitivity in Patients with Chronic Low Back Pain: An Experimental Controlled Study," *Physical Therapy* 94 (2014): 477–89.
16. K. Nader, G. E. Schafe, and J. E. Le Doux, "Fear Memories Require Protein Synthesis in the Amygdala for Reconsolidation after Retrieval," *Nature* 406 (2000): 722–26.
17. S. J. Sara, "Retrieval and Reconsolidation: Toward a Neurobiology of Remembering," *Learning & Memory* 7 (2000): 73–84; J. Graff et al., "Epigenetic Priming of Memory Updating during Reconsolidation to Attenuate Remote Fear Memories," *Cell* 156 (2014): 261–76.
18. M. Solms, "Reconsolidation: Turning Consciousness into Memory," *Behavioral and Brain Sciences* 38 (2015): e24.

17. Does Dreaming Have a Future? (pages 332-347)

1. A. Maury, *Le Sommeil et les rêves* (Paris: Didier, 1865).
2. N. Malcolm, "Dreaming and Skepticism," *The Philosophical Review* 65 (1956): 14–37.
3. D. C. Dennett, "Are Dreams Experiences?" *The Philosophical Review* 85 (1976): 151–71.
4. T. M. Mitchell et al., "Predicting Human Brain Activity Associated with the Meanings of Nouns," *Science* 320 (2008): 1191–95; K. N. Kay, T. Naselaris, R. J. Prenger, and J. L. Gallant, "Identifying Natural Images from Human Brain Activity," *Nature* 452 (2008): 352–55; T. Naselaris et al., "Bayesian Reconstruction of Natural Images from Human Brain Activity," *Neuron*

63 (2009): 902–15; A. G. Huth et al., "Natural Speech Reveals the Semantic Maps that Tile Human Cerebral Cortex," *Nature* 532 (2016): 453–58.

5. T. Çukur, S. Nishimoto, A. G. Huth, and J. L. Gallant, "Attention during Natural Vision Warps Semantic Representation across the Human Brain," *Nature Neuroscience* 16 (2016): 763–70.

6. T. Horikawa, M. Tamaki, Y. Miyawaki, and Y. Kamitani, "Neural Decoding of Visual Imagery during Sleep," *Science* 340 (2013).

7. F. Siclari et al., "The Neural Correlates of Dreaming," *Nature Neuroscience* 20 (2017): 872–78.

8. E. Tagliazucchi et al., "Increased Global Functional Connectivity Correlates with LSD-Induced Ego Dissolution," *Current Biology* 26 (2016): 1043–50; R. Kraehenmann, "Dreams and Psychedelics: Neurophenomenological Comparison and Therapeutic Implications," *Current Neuropharmacology* 15 (2017): 1032–42; R. Kraehenmann et al., "Dreamlike Effects of LSD on Waking Imagery in Humans Depend on Serotonin 2A Receptor Activation," *Psychopharmacology (Berlin)* 234 (2017): 2031–46; C. Sanz et al., "The Experience Elicited by Hallucinogens Presents the Highest Similarity to Dreaming within a Large Database of Psychoactive Substance Reports," *Frontiers in Neuroscience* 12 (2018): 7.

9. R. Kraehenmann et al., "LSD Increases Primary Process Thinking via Serotonin 2A Receptor Activation," *Frontiers in Pharmacology* 8 (2017): 814; K. H. Preller et al., "Changes in Global and Thalamic Brain Connectivity in LSD-Induced Altered States of Consciousness Are Attributable to the 5-HT2A Receptor," *Elife* 7 (2018).

10. A. Cipriani et al., "Comparative Efficacy and Acceptability of 21 Antidepressant Drugs for the Acute Treatment of Adults with Major Depressive Disorder: A Systematic Review and Network Meta-Analysis," *The Lancet* 391 (2018): 1357–66.

11. R. S. El-Mallakh, Y. Gao, and R. Jeannie Roberts, "Tardive Dysphoria: The Role of Long Term Antidepressant Use in Inducing Chronic Depression," *Medical Hypotheses* 76 (2011): 769–73; R. S. El-Mallakh, Y. Gao, B. T. Briscoe, and R. J. Roberts, "Antidepressant-Induced Tardive Dysphoria," *Psychotherapy and Psychosomatics* 80 (2011): 57–59.

12. I. Kirsch, *The Emperor's New Drugs: Exploding the Antidepressant Myth* (New York: Basic Books, 2010).

13. R. R. Griffiths et al., "Psilocybin Produces Substantial and Sustained Decreases in Depression and Anxiety in Patients with Life-Threatening Cancer: A Randomized Double-Blind Trial," *Journal of Psychopharmacology* 30 (2016): 1181–97; R. L. Carhart-Harris et al., "Psilocybin with Psychological Support for Treatment-Resistant Depression: An Open-Label Feasibility Study," *The Lancet Psychiatry* 3 (2016): 619–27; S. Ross et al., "Rapid and Sus-

tained Symptom Reduction Following Psilocybin Treatment for Anxiety and Depression in Patients with Life-Threatening Cancer: A Randomized Controlled Trial," *Journal of Psychopharmacology* 30 (2016): 1165–80; R. L. Carhart-Harris et al., "Psilocybin with Psychological Support for Treatment-Resistant Depression: Six-Month Follow-Up," *Psychopharmacology (Berlin)* 235 (2018): 399–408.

14. T. Lyons and R. L. Carhart-Harris, "Increased Nature Relatedness and Decreased Authoritarian Political Views after Psilocybin for Treatment-Resistant Depression," *Journal of Psychopharmacology* 32 (2018): 811–19.

15. L. Roseman et al., "Increased Amygdala Responses to Emotional Faces after Psilocybin for Treatment-Resistant Depression," *Neuropharmacology* 142 (2017): 263–69; J. B. Stroud et al., "Psilocybin with Psychological Support Improves Emotional Face Recognition in Treatment-Resistant Depression," *Psychopharmacology (Berlin)* 235 (2018): 459–66.

16. L. Roseman, D. J. Nutt, and R. L. Carhart-Harris, "Quality of Acute Psychedelic Experience Predicts Therapeutic Efficacy of Psilocybin for Treatment-Resistant Depression," *Frontiers in Pharmacology* 8 (2017): 974.

17. J. C. Bouso et al., "MDMA-Assisted Psychotherapy Using Low Doses in a Small Sample of Women with Chronic Posttraumatic Stress Disorder," *Journal of Psychoactive Drugs* 40 (2008): 225–36; M. C. Mithoefer, C. S. Grob, and T. D. Brewerton, "Novel Psychopharmacological Therapies for Psychiatric Disorders: Psilocybin and MDMA," *The Lancet Psychiatry* 3 (2016): 481–88; M. T. Wagner et al., "Therapeutic Effect of Increased Openness: Investigating Mechanism of Action in MDMA-Assisted Psychotherapy," *Journal of Psychopharmacology* 31 (2017): 967–74; M. C. Mithoefer et al., "3,4-Methylenedioxymethamphetamine (MDMA)-Assisted Psychotherapy for Post-Traumatic Stress Disorder in Military Veterans, Firefighters, and Police Officers: A Randomised, Double-Blind, Dose-Response, Phase 2 Clinical Trial," *The Lancet Psychiatry* 5 (2018): 486–97.

18. D. J. Nutt, L. A. King, L. D. Phillips, on behalf of the Independent Scientific Committee on Drugs, "Drug Harms in the UK: A Multicriteria Decision Analysis," *The Lancet* 376 (2010): 1558–65.

19. Mithoefer et al., "3,4-Methylenedioxymethamphetamine (MDMA)-Assisted Psychotherapy."

20. L. Osório et al., "Antidepressant Effects of a Single Dose of Ayahuasca in Patients with Recurrent Depression: A Preliminary Report," *Revista Brasileira de Psiquiatria* 37 (2015): 13– 20; R. F. Sanches et al., "Antidepressant Effects of a Single Dose of Ayahuasca in Patients With Recurrent Depression: A SPECT Study," *Journal of Clinical Psychopharmacology* 36 (2016): 77–81.

21. F. Palhano-Fontes et al., "Rapid Antidepressant Effects of the Psychedelic

Ayahuasca in Treatment-Resistant Depression: A Randomized Placebo-Controlled Trial," *Psychological Medicine* (2018): 1–9.

22. F. Palhano-Fontes, *Os efeitos antidepressivos da ayahuasca, suas bases neurais e relação com a experiência psicodélica* (doctoral thesis, Universidade Federal do Rio Grande do Norte, 2017).

23. V. Dakic et al., "Harmine Stimulates Proliferation of Human Neural Progenitors," *PeerJ* 4 (2016): e2727; V. Dakic et al., "Short Term Changes in the Proteome of Human Cerebral Organoids Induced by 5-Methoxy-N,N-Dimethyltryptamine," *BioRxiv*, 2017.

24. R. V. Lima da Cruz, T. C. Moulin, L. L. Petiz, and R. N. Leao, "A Single Dose of 5-MeO-DMT Stimulates Cell Proliferation, Neuronal Survivability, Morphological and Functional Changes in Adult Mice Ventral Dentate Gyrus," *Frontiers in Molecular Neuroscience* 11 (2018): 312.

25. C. Ly et al., "Psychedelics Promote Structural and Functional Neural Plasticity," *Cell Reports* 23 (2018): 3170–82.

26. E. Labigalini Jr., L. R. Rodrigues, and D. X. Da Silveira, "Therapeutic Use of Cannabis by Crack Addicts in Brazil," *Journal of Psychoactive Drugs* 31 (1999): 451–55; G. Thomas et al., "Ayahuasca-Assisted Therapy for Addiction: Results from a Preliminary Observational Study in Canada," *Current Drug Abuse Reviews* 6 (2013): 30–42; B. C. Labate and C. Cavnar, eds., *The Therapeutic Use of Ayahuasca* (New York: Springer, 2014).

27. T. R. Insel and P. Summergrad, "Plenary Panel: Future of Psychedelic Psychiatry," https://www.youtube.com/embed/_oZ_v3QFQDE?list=PL4FovNNTozFSw5gRe_zVTAvNIwjYD_AIU?ecver=2.

28. P. S. Goldman-Rakic, "The Prefrontal Landscape: Implications of Functional Architecture for Understanding Human Mentation and the Central Executive," *Philosophical Transactions of the Royal Society of London: Series B, Biological Sciences* 351 (1995): 1445–53; J. Panksepp, *Affective Neuroscience: The Foundations of Human and Animal Emotions* (Oxford: Oxford University Press, 1998); F. Barcelo, S. Suwazono, and R. T. Knight, "Prefrontal Modulation of Visual Processing in Humans," *Nature Neuroscience* 3 (2000): 399–403; B. Levine et al., "The Functional Neuroanatomy of Episodic and Semantic Autobiographical Remembering: A Prospective Functional MRI Study," *Journal of Cognitive Neuroscience* 16 (2004): 1633–46; R. Q. Quiroga, "Concept Cells: The Building Blocks of Declarative Memory Functions," *Nature Review of Neuroscience* 13 (2012): 587–97; P. Martinelli, M. Sperduti, and P. Piolino, "Neural Substrates of the Self-Memory System: New Insights from a Meta-Analysis," *Human Brain Mapping* 34 (2013): 1515–29; M. L. Andermann and B. B. Lowell, "Toward a Wiring Diagram Understanding of Appetite Control," *Neuron* 95 (2017): 757–78; W. Han et al., "A Neural Circuit for Gut-Induced Reward," *Cell* 175 (2018): 887–88.

29. M. Minsky, "Why Freud Was the First Good AI Theorist," in *The Transhumanist Reader: Classical and Contemporary Essays on the Science, Technology, and Philosophy of the Human Future,* eds. M. More and N. Vita-More (Chichester: John Wiley-Blackwell, 2013), 167–76.
30. A. A. Abbass, J. T. Hancock, J. Henderson, and S. Kisely, "Short-Term Psychodynamic Psychotherapies for Common Mental Disorders," *Cochrane Database of Systematic Reviews* 4 (2006): CD0046; J. Panksepp et al., "Affective Neuroscience Strategies for Understanding and Treating Depression: From Preclinical Models to Three Novel Therapeutics," *Clinical Psychological Science* 2 (2014): 472–94.
31. R. Stickgold et al., "Replaying the Game: Hypnagogic Images in Normals and Amnesics," *Science* 290 (2000): 350–53; E. J. Wamsley et al., "Cognitive Replay of Visuomotor Learning at Sleep Onset: Temporal Dynamics and Relationship to Task Performance," *Sleep* 33 (2010): 59–68; E. J. Wamsley et al., "Dreaming of a Learning Task Is Associated with Enhanced Sleep-Dependent Memory Consolidation," *Current Biology* 20 (2010): 850–55.
32. M. C. Anderson et al., "Neural Systems Underlying the Suppression of Unwanted Memories," *Science* 303 (2004): 232–35; B. E. Depue, T. Curran, and M. T. Banich, "Prefrontal Regions Orchestrate Suppression of Emotional Memories Via a Two-Phase Process," *Science* 317 (2007): 215–19.
33. M. Solms, "Dreaming and REM Sleep Are Controlled by Different Brain Mechanisms," *Behavioral and Brain Science* 23 (2000): 843–50, discussion, 904–1121; L. Perogamvros and S. Schwartz, "The Roles of the Reward System in Sleep and Dreaming," *Neuroscience Biobehavioral Review* 36 (2012): 1934–51.
34. N. B. Mota et al., "Speech Graphs Provide a Quantitative Measure of Thought Disorder in Psychosis," *PLOS One* 7 (2012): e34928; N. B. Mota et al., "Graph Analysis of Dream Reports Is Especially Informative about Psychosis," *Science Reports* 4 (2014): 3691; N. B. Mota, M. Copelli, and S. Ribeiro, "Thought Disorder Measured as Random Speech Structure Classifies Negative Symptoms and Schizophrenia Diagnosis 6 Months in Advance," *npj Schizophrenia* 3 (2017): 1–10.
35. J. Reinisch, *The Kinsey Institute New Report on Sex: What You Must Know to Be Sexually Literate* (New York: St. Martin's, 1991); G. Ryan, "Childhood Sexuality: A Decade of Study. Part I: Research and Curriculum Development," *Child Abuse & Neglect* 24 (2000): 33–48; W. N. Friedrich et al., "Child Sexual Behavior Inventory: Normative, Psychiatric, and Sexual Abuse Comparisons," *Child Maltreatment* 6 (2001): 37–49.
36. P. O. McGowan et al., "Epigenetic Regulation of the Glucocorticoid Receptor in Human Brain Associates with Childhood Abuse," *Nature Neuroscience* 12 (2009): 342–48; T. Zhang et al., "Epigenetic Mechanisms for the Early Environmental Regulation of Hippocampal Glucocorticoid

Receptor Gene Expression in Rodents and Humans," *Neuropsychopharmacology* 38 (2013): 111–23; C. J. Pena et al., "Early Life Stress Confers Lifelong Stress Susceptibility in Mice Via Ventral Tegmental Area OTX2," *Science* 356 (2017): 1185–88.

37. A. Huxley, *The Doors of Perception and Heaven and Hell* (London: Vintage Classics, 2004), 53–54.

38. C. G. Jung with A. Jaffé, *Memories, Dreams, Reflections* (London: William Collins, 1967), 183.

39. B. Drury and T. Clavin, *The Heart of Everything That Is: The Untold Story of Red Cloud, An American Legend* (New York: Simon & Schuster, 2013).

18. Dreaming and Destiny (pages 348-378)

1. J. L. Borges, "The Dream," in *Poems of the Night,* trans. A. Reid (New York: Penguin Books, 2010), 109.

2. C. G. Jung, "General Aspects of Dream Psychology," in *Collected Works of C. G. Jung: The Structure and Dynamics of the Psyche* (Princeton: Princeton University Press, 1916), 493.

3. F. Pessoa, *The Book of Disquiet: The Complete Edition,* trans. M. J. Costa (London: Serpent's Tail, 2018), 78.

4. H. Staden, *Primeiros registros escritos e ilustrados sobre o Brasil e seus habitantes* (São Paulo: Terceiro Nome, 1999).

5. A. F. C. Wallace and A. D'Agostino, "Dreams and the Wishes of the Soul: A Type of Psychoanalytic Theory among the Seventeenth Century Iroquois," *American Anthropologist* 60 (1958): 234–48.

6. P. Descola, *In the Society of Nature: A Native Ecology in Amazonia* (Cambridge: Cambridge University Press, 1994); P. Descola, *As lanças do crepúsculo: Relações jívaro na Alta Amazônia* (São Paulo: Cosac Naify, 2006).

7. M. Brown, "Ropes of Sand: Order and Imagery in Aguaruna Dreams," in *Dreaming: Anthropological and Psychological Interpretations,* ed. B. Tedlock (Santa Fé: School of American Research Press, 1992), 154–70.

8. M. A. Gonçalves, *O mundo inacabado: Ação e criação em uma cosmologia amazônica* (Rio de Janeiro: UFRJ, 2001).

9. Ibid, 289.

10. A. Barcelos Neto, *A arte dos sonhos: Uma iconografia ameríndia* (Lisboa: Assírio & Alvim, 2002); A. Barcelos Neto, *Apapaatai: Rituais de máscaras do Alto Xingu* (São Paulo: Edusp/Fapesp, 2008).

11. W. Kracke, "He Who Dreams. The Nocturnal Source of Transforming Power in Kagwahiv Shamanism," in *Portals of Power: Shamanism in South America,* eds. E. Jean Mattison Langdon and Gerhard Baer (Albuquerque: University of New Mexico Press, 1992), 127–48.

12. E. B. Basso, *The Kalapalo Indians of Central Brazil* (New York: Holt, 1973); E. B. Basso, *A Musical View of the Universe: Kalapalo Myth and Ritual Per-*

formances (Philadelphia: University of Pennsylvania Press, 1985); E. Basso, "The Implications of a Progressive Theory of Dreaming," in *Dreaming: Anthropological and Psychological Interpretation,* ed. Barbara Tedlock (Cambridge: Cambridge University Press, 1987), 86–104.

13. T. Gregor,"'Far, Far Away My Shadow Wandered . . .': The Dream Symbolism and Dream Theories of the Mehinaku Indians of Brazil," *American Ethnologist* 8 (1981): 709–20; T. Gregor, *O Branco dos meus Sonhos,* Anuário Antropológico, vol. 82 (Rio de Janeiro: Tempo Brasileiro, 1984).

14. Siasi/Sesai, *Quadro geral dos povos indígenas no Brasil,* 2014; https://pib.socioambiental.org/pt/Quadro_Geral_dos_Povos.

15. J. N. Xavante, B. Giaccaria, and A. Heide, *Jerônimo Xavante sonha: Contos e sonhos* (Campo Grande: Casa da Cultura, 1975); *Etenhiririapá: Cantos da tradição Xavante.* CD. (Warner Music Brasil: Quilombo Music, Rio de Janeiro, 1994); A. S. F. Eid, *A'uwê anda pelo sonho: A espiritualidade indígena e os perigos da modernidade* (São Paulo: Instituto de Estudos Superiores do Dharma, 1998); L. R. Graham, *Performing Dreams: Discourses of Immortality among the Xavante of Central Brazil* (Tucson: Fenestra Books, 2003).

16. Eid, *A'uwê anda pelo sonho,* 13.

17. B. Giaccaria and A. Heide, *Xavante: Auwê Uptabi: Povo autêntico* (São Paulo: Dom Bosco, 1972), 271.

18. K. W. Jecupé, *A terra dos mil povos: História indígena brasileira contada por um índio* (São Paulo: Peirópolis, 1998), 68.

19. J. V. Neel et al., "Studies on the Xavante Indians of the Brazilian Mato Grosso," *American Journal of Human Genetics* 16 (1964): 52–140; A. L. Silva, "Dois séculos e meio de história Xavante," in *História dos Índios no Brasil,* ed. M. Carneiro da Cunha (São Paulo: Companhia das Letras, 1992), 357–78; J. M. Monteiro, *Tupis, tapuias e historiadores: Estudos de história indígena e do indigenismo,* Tese de livre docência (Campinas: Unicamp, 2001).

20. J. R. Welch, R. V. Santos, N. M. Flowers, and C. E. A. Coimbra Jr., *Na primeira margem do rio: território e ecologia do povo xavante de Wedezé* (Brasilia: FUNAI, 2013).

21. D. Tserewahoú, *Wai'á rini: O poder do sonho,* in Indigenous Video Makers (Video nas Aldeias, 2001), 48 mins; https://www.youtube.com/watch?v=t44ZPqoYyCU.

22. C. Aldunate, "Mapuche: Gente de la Tierra," in *Culturas de Chile etnografía: Sociedades indígenas contemporáneas y su ideología,* eds. V. Schiappacasse et al. (Santiago: Andrés Bello, 1996), 111–34.

23. S. Montecino, *Palabra dicha: estudios sobre género, identidades, mestizaje* (Santiago: Universidad de Chile, Facultad de Ciencias Sociales, 1997).

24. J. Bengoa, *Historia del pueblo mapuche (siglos XIX y XX)* (Neuquén: Sur, 1987); R. Foerster and S. Montecino, *Organizaciones, líderes y contiendas mapuches: 1900–1970* (Santiago: Centro de Estudios de la Mujer, 1988).

25. R. Foerster, *Martín Painemal Huenchual: Vida de un dirigente mapuche* (Santiago: Academia de Humanismo Cristiano, 1983).

26. K. G. Shiratori, *O acontecimento onírico ameríndio: O tempo desarticulado e as veredas dos possíveis* (master's dissertation, Museu Nacional, Universidade Federal do Rio de Janeiro, 2013).

27. O. Villas Bôas and C. Villas-Bôas, *Xingu: Indians and Their Myths* (New York: Farrar, Straus and Giroux, 1973).

28. A. Assunção, "500 anos de desencontros," *IstoÉ*, São Paulo, n. 1555, July 21, 1999: 7–11.

29. H. Brody, *Maps and Dreams* (Madeira Park: Douglas & McIntyre, 1981), 267.

30. C. Dean, *The Australian Aboriginal 'Dreamtime': Its History, Cosmogenesis, Cosmology and Ontology* (Victoria: Gamahucher, 1996).

31. A. P. Elkin, "Elements of Australian Aboriginal Philosophy," *Oceania* 9 (1969): 85–98; W. E. H. Stanner, "Religion, Totemism and Symbolism," in *Religion in Aboriginal Australia*, ed. M. Charlesworth (Queensland: University of Queensland Press, 1989).

32. T. T. Rinpoche, "Ancient Tibetan Dream Wisdom," Tarab Institute International, 2013, http://www.tarab-institute.org/articles/ancient-tibetan-dream-wisdom.

33. B. Lee, *Bruce Lee Striking Thoughts: Bruce Lee's Wisdom for Daily Living* (Clarendon: Tuttle, 2015), 177.

34. H. Benson et al., "Body Temperature Changes during the Practice of G Tum-Mo Yoga," *Nature* 295 (1982): 234–36; J. Daubenmier et al., "Follow Your Breath: Respiratory Interoceptive Accuracy in Experienced Meditators," *Psychophysiology* 50 (2013): 777–89; B. Bornemann and T. Singer, "Taking Time to Feel Our Body: Steady Increases in Heartbeat Perception Accuracy and Decreases in Alexithymia over 9 Months of Contemplative Mental Training," *Psychophysiology* 54 (2017): 469–82.

35. G. S. Sparrow, *Lucid Dreaming: Dawning of the Clear Light* (Virginia Beach: A.R.E. Press, 1982); P. Garfield, *Pathway to Ecstasy: The Way of the Dream Mandala* (New Jersey: Prentice Hall, 1990).

36. L. D'Hervey de Saint-Denys, *Dreams and How to Guide Them* (London: Duckworth, 1982).

37. F. Van Eeden, "A Study of Dreams," *Proceedings of the Society for Psychical Research* 26 (1913): 431–61.

38. K. M. T. Hearne, *Lucid Dreams: An Electro-Physiological and Psychological Study* (doctoral thesis, University of Liverpool, 1978).

39. S. LaBerge, *Lucid Dreaming: An Exploratory Study of Consciousness during Sleep* (doctoral thesis, Stanford University, 1980).

40. S. P. LaBerge, L. E. Nagel, W. C. Dement, and V. P. J. Zarcone, "Lucid Dreaming Verified by Volitional Communication during REM Sleep," *Per-*

ceptual and Motor Skills 52 (1981): 727–32; S. LaBerge, J. Owens, L. Nagel, and W. C. Dement, "This Is Dream: Induction of Lucid Dreams by Verbal Suggestion during REM Sleep," *Journal of Sleep Research* 10 (1981); S. LaBerge, "Lucid Dreaming as a Learnable Skill: A Case Study," *Perceptual and Motor Skills* 51 (1980): 1039–42; S. LaBerge, L. Levitan, R. Rich, and W. C. Dement, "Induction of Lucid Dreaming by Light Stimulation during REM Sleep," *Journal of Sleep Research* 17 (1988): 104; S. LaBerge and W. C. Dement, "Voluntary Control of Respiration during REM Sleep," *Journal of Sleep Research* (1982): 11; S. LaBerge, L. Levitan, and W. C. Dement, "Lucid Dreaming: Physiological Correlates of Consciousness during REM Sleep," *Journal of Mind and Behavior* 7 (1986): 251–58; A. Brylowski, L. Levitan, and S. LaBerge, "H-Reflex Suppression and Autonomic Activation during Lucid REM Sleep: A Case Study," *Sleep* 12 (1898): 374–78.

41. R. Stepansky et al., "Austrian Dream Behavior: Results of a Representative Population Survey," *Dreaming* 8 (1998): 23–30; M. Schredl and D. Erlacher, "Lucid Dreaming Frequency and Personality," *Personality and Individual Differences* 37 (2004); C. K. C. Yu, "Dream Intensity Inventory and Chinese People's Dream Experience Frequencies," *Dreaming* 18 (2008): 94–111; M. Schredl and D. Erlacher, "Frequency of Lucid Dreaming in a Representative German Sample," *Perceptual and Motor Skills* 112 (2011): 104–8; S. A. Mota-Rolim et al., "Dream Characteristics in a Brazilian Sample: An Online Survey Focusing on Lucid Dreaming," *Frontiers in Human Neuroscience* 7 (2013): 836.

42. S. LaBerge, K. LaMarca, B. Baird, "Pre-Sleep Treatment with Galantamine Stimulates Lucid Dreaming: A Double-Blind, Placebo-Controlled, Crossover Study," *PLOS One* 13 (2018): e0201246; M. Dresler et al., "Volitional Components of Consciousness Vary across Wakefulness, Dreaming and Lucid Dreaming," *Frontiers in Psychology* 4 (2014): 987.

43. U. Voss, R. Holzmann, I. Tuin, and J. A. Hobson, "Lucid Dreaming: A State of Consciousness with Features of Both Waking and Non-Lucid Dreaming," *Sleep* 32 (2009): 1191–200.

44. M. Dresler et al., "Neural Correlates of Dream Lucidity Obtained from Contrasting Lucid *versus* Non-Lucid REM Sleep: A Combined EEG/fMRI Case Study," *Sleep* 35 (2012): 1017–20.

45. M. Dresler et al., "Dreamed Movement Elicits Activation in the Sensorimotor Cortex," *Current Biology* 21 (2011): 1833–37.

46. T. Stumbrys, D. Erlacher, and M. Schredl, "Testing the Involvement of the Prefrontal Cortex in Lucid Dreaming: A TDCS Study," *Consciousness and Cognition* 22 (2013): 1214–22.

47. U. Voss et al., "Induction of Self Awareness in Dreams through Frontal Low Current Stimulation of Gamma Activity," *Nature Neuroscience* 17 (2014): 810–12.

48. D. Brown, *Bury My Heart at Wounded Knee: An Indian History of the American West* (New York: Fall River Press, 2014).

49. E. Erlacher, T. Stumbrys, and M. Schredl, "Frequency of Lucid Dreams and Lucid Dream Practice in German Athletes," *Imagination, Cognition and Personality* 31 (2012): 237–46; D. Erlacher and M. Schredl, "Practicing a Motor Task in a Lucid Dream Enhances Subsequent Performance: A Pilot Study," *The Sport Psychologist* 24 (2010): 157–67; M. Schädlich, D. Erlacher, and M. Schredl, "Improvement of Darts Performance Following Lucid Dream Practice Depends on the Number of Distractions while Rehearsing within the Dream: A Sleep Laboratory Pilot Study," *Journal of Sports Sciences* 35 (2017): 2365–72.

50. D. Erlacher, M. Schädlich, T. Stumbrys, and M. Schredl, "Time for Actions in Lucid Dreams: Effects of Task Modality, Length, and Complexity," *Frontiers in Psychology* 4 (2013): 1013; S. LaBerge, B. Baird, and P. G. Zimbardo, "Smooth Tracking of Visual Targets Distinguishes Lucid REM Sleep Dreaming and Waking Perception from Imagination," *Nature Communications* 9 (2018): 3298.

51. P. Tholey, "Consciousness and Abilities of Dream Characters Observed during Lucid Dreaming," *Perceptual and Motor Skills* 68 (2018): 567–78; T. Stumbrys and M. Daniels, "An Exploratory Study of Creative Problem Solving in Lucid Dreams: Preliminary Findings and Methodological Considerations," *International Journal of Dream Research* 3 (2010): 121–29; T. Stumbrys, D. Erlacher, and S. Schmidt, "Lucid Dream Mathematics: An Explorative Online Study of Arithmetic Abilities of Dream Characters," *International Journal of Dream Research* 4 (2011): 35–40.

52. S. LaBerge, "Lucid Dreaming and the Yoga of the Dream State: A Psychophysiological Perspective," in *Buddhism and Science: Breaking New Ground*, ed. B. A. Wallace (New York: Columbia University Press, 2003), 233–58.

53. A. L. Zadra and R. O. Pihl, "Lucid Dreaming as a Treatment for Recurrent Nightmares," *Psychotherapy and Psychosomatics* 66 (1997): 50–55; M. Zappaterra, L. Jim, and S. Pangarkar, "Chronic Pain Resolution after a Lucid Dream: A Case for Neural Plasticity?" *Medical Hypotheses* 82 (2014): 286–90; N. B. Mota et al., "Psychosis and the Control of Lucid Dreaming," *Frontiers in Psychology* 7 (2016): 294.

54. M. Sigman, *The Secret Life of the Mind: How Your Brain Thinks, Feels, and Decides* (Boston: Little, Brown and Company, 2017), 288.

55. IPCC, "Global Warming of 1.5°C," IPCC Report, 2018, https://www.ipcc.ch/sr15/; R. S. Nerem et al., "Climate-Change-Driven Accelerated Sea-Level Rise Detected in the Altimeter Era," *Proceedings of the National Academy of Sciences of the USA* 115 (2018): 2022–25.

56. D. Kopenawa and B. Albert, *The Falling Sky: Words of a Yanomami Shaman,*

trans. N. Elliott and A. Dundy (Cambridge: Belknap, Harvard University Press, 2013), 406–407.

57. Ibid., 275.
58. N. Obradovich, R. Migliorini, S. C. Mednick, and J. H. Fowler, "Nighttime Temperature and Human Sleep Loss in a Changing Climate," *Science Advances* 3 (2017): e1601555.

Index

Page numbers in *italics* refer to illustrations.

nightmares, recurring, 3–5, 7, 10–11, 27,
    112, 250, 326, 379
  lucid dreams and, 372–73
  PTSD and, 135, 140, 268–73, 284–85
night terrors, 135
Nimrod, King of Assyria, 14
Nindub, 53
Nineteen Eighty-Four (Orwell), 223–24
Ninurta, 53
Noah, 14, 374
non-REM sleep, 131–33, *130*, 185, 193–94.
    See also slow-wave sleep
noradrenaline, 133–34, 141, 164, 195–96,
    210, 260, 338
noradrenaline receptors, 150
Norse sagas, 48–50, *49*
Northwestern University, 165, 171
Nottebohm, Fernando, 203–5
Nourishment for Young Painters (Dürer),
    218–19
Nubia, 43
nucleus accumbens, 151, *252*
Nutt, David, 143, 256
Nyx, 6
Nzinga, Queen of Angola, 159, 379

Obama, Barack, 13
occipital cortex, 368
octopuses, 304
Odyssey, The (Homer), 15, 62, *64*, 158,
    222, 236, 322–23
Oedipus complex, 24
Ojibwe, 17, 32
O'Keefe, John, 183, 191
old age, 35, 98–99, 112, 127, 196–98
Oldowan technology, 310
Olson, David, 340
omens, 36, 148
Oneiroi, 6
Oneirokritika (Artemidorus), 15
opium, 222
oracles, 14, 17–19, 294–301, 312, 324–26
  probabilistic, 326
oral tradition, 60, 158

orexin, 135
Organization of Behavior, The (Hebb),
    178
Orishas, 244
Orwell, George, 223–25
Osiris, 47
ostrich, 121
Ottoman Caliphate, 68
Ouroboros, 225
oxygenation of blood, 192

paganism, 69
Paha Sapa (Black Hills), 275–76,
    279–80, 291, 296, 297
pain, 252, 372–73
Painemal, Martín, 358
Pakistan, 68
Paleolithic period, 33–35, 39–41, 48, 59,
    97, 154, 259, 286, 308, 309–10, 316,
    321, 323
Palestine, 41
Palhano, Fernanda, 256, 257, 340
Paller, Ken, 171
Palo Duro Canyon attack, 295
Pan, 33
Panguilef, Manuel Aburto, 358
Papua New Guinea, 320
paranoid schizophrenia, 145–46, 155
parents, 35, 97, 146, 153, 155. *See also*
    fathers; mothers
parietal cortex, 368
Parintintín, 351
Paris, of Troy, 51
Parkinson's disease, 136
Parnassus, Mount, 15
Passavanti, Jacopo, 72
Pastinha, Mestre, 96
Patroclus, 322
Paul, Apostle, 68, 73, 206
Pavlides, Constantine "Gus," 182–83,
    185, 188, 191–92, 205–8, 243
Pawnee, 262
Payne, Jessica, 274
"Pebble from Aruanda, A" (song), 96

Illustration Credits

Pictures, graphs, and infographics were adapted by Luiz Iria.

Page
32 *(top)* Published in *The Cave Artists,* by Ann Sieveking. London: Thames & Hudson, 1979; *(bottom)* Granger, NYC/Alamy/Fotoarena.
49 *Queen Ragnhild's Dream* (1899), illustration by Erik Werenskiold.
64 Adapted from Diuk, C. G., et al. "A Quantitative Philology of Introspection" in *Frontiers in Integrative Neuroscience* 6, p. 80, 2012.
101, 116, and 130 Adapted from Bear, M. F., et al. *Neuroscience: Exploring the Brain.* Philadelphia: Lippincott Williams & Wilkins, 2007.
156 Adapted from Hartmann, E. *The Biology of Dreaming.* Boston State Hospital monograph series. Boston: C. C. Thomas, 1967.
181 Adapted from Mota, N. B., et al. "Graph Analysis of Dream Reports Is Especially Informative About Psychosis" in *Scientific Reports* 4, p. 3,691, 2014. Woodcut by Vera Tollendal Ribeiro.
184 *(top)* Adapted from Winson, J. "The Meaning of Dreams" in *Scientific American* 263, pp. 86–96, 1990. By permission of Patricia J. Wynne. *(bottom)* Adapted from Pavlides, C., and Winson, J. "Influences of Hippocampal Place Cell Firing in the Awake State on the Activity of These Cells During Subsequent Sleep Episodes" in *Journal of Neuroscience* 9, pp. 2,907–18, 1989.
189 Adapted from Wilson, M. A., and McNaughton, B. L. "Reactivation of Hippocampal Ensemble Memories During Sleep" in *Science* 265, pp. 676–69, 1994.
252 Adapted from Hyman, J. M., et al. "Stimulation in Hippocampal Region CA1 in Behaving Rats Yields Long-Term Potentiation When Delivered to the Peak of Theta and Long-Term Depression When Delivered to the Trough" in *Journal of Neuroscience* 23, pp. 11,725–31, 2003.

336 Reproduced from Horikawa, T., et al. "Neural Decoding of Visual Imagery during Sleep" in *Science* 340, 2013.

Insert

1 Travel Pix / Alamy / Fotoarena.
2 *(top)* Alamy / Fotoarena; *(bottom)* mdsharma / Shutterstock.
3 TheBiblePeople / Alamy / Fotoarena.
4 Adapted from Ribeiro, S., Goyal, V., Mello, C.V., and Pavlides, C. "Brain Gene Expression During REM Sleep Depends on Prior Waking Experience" in *Learning & Memory* 6, pp. 500–508, 1999.
5 *(top)* Albrecht Dürer, *Dream Vision*, June 1525, watercolor, 300 x 425 mm. Vienna, Graphische Sammlung Albertina. Album / akg-images / Fotoarena; *(bottom)* Marc Chagall, *Le Songe de Jacob*, 1960–6, oil on canvas, 195 x 278 cm. Nice, Musée National Marc Chagall. © Chagall, Marc / AUTVIS, Brazil, 2019. Album / akg-images / Fotoarena.
6 Salvador Dalí, *Sueño causado por el vuelo de una abeja alrededor de una granada un segundo antes del despertar*, 1944, oil on canvas, 51 x 41 cm. Madrid, Museo Nacional Thyssen-Bornemisza. © Salvador Dalí, Fundación Gala-Salvador Dalí / AUTVIS, Brazil, 2019. Album / Joseph Martin / Fotoarena.
7 Granger / Fotoarena.
8 Reproduced from Quiroga, R.Q. "Concept Cells: The Building Blocks of Declarative Memory Functions" in *Nature Reviews Neuroscience* 13, pp. 587–97, 2012.

A Note About the Author

Sidarta Ribeiro is the founder and vice director of the Brain Institute at Universidade Federal do Rio Grande do Norte in Brazil, where he currently is professor of neuroscience. He received a PhD in Animal Behavior from Rockefeller University. His research topics encompass memory, sleep and dreams, neuroplasticity, symbolic competence in non-human animals, computational psychiatry, and psychedelics.

A Note on the Type

This book was set in Monotype Dante, a typeface designed by Giovanni Mardersteig (1892–1977). Originally cut for hand composition by Charles Malin between 1946 and 1952, its first use was in an edition of Boccaccio's *Trattatello in laude di Dante* that appeared in 1954. The Monotype Corporation's version of Dante followed in 1957. Although modeled on the Aldine type used for Pietro Cardinal Bembo's treatise *De Aetna* in 1495, Dante is a thoroughly modern interpretation of the venerable face.

Composed by North Market Street Graphics,
Lancaster, Pennsylvania

Printed and bound by Berryville Graphics,
Berryville, Virginia

Designed by Betty Lew